Labor, Markets, and Agricultural Production

Labor, Markets, and Agricultural Production

Jan Douwe van der Ploeg

Translated by Ann Long

Routledge
Taylor & Francis Group
LONDON AND NEW YORK

First published 1990 by Westview Press

Published 2018 by Routledge
52 Vanderbilt Avenue, New York, NY 10017
2 Park Square, Milton Park, Abingdon, Oxon OX14 4RN

Routledge is an imprint of the Taylor & Francis Group, an informa business

Library of Congress Cataloging-in-Publication Data
Ploeg, Jan Douwe van der, 1950–
Labor, markets, and agricultural production / Jan Douwe van der Ploeg; translated by Ann Long.
p. cm.—(Westview special studies in agriculture science and policy)
Includes bibliographical references and index.
ISBN 0-8133-7827-3
1. Agricultural productivity—Case studies. 2. Produce trade Case studies. 3. Agricultural laborers—Case studies. I. Title. II. Series.
HD1433.P57 1990
338.1′6—dc20 90-12404
CIP

ISBN 13: 978-0-367-01336-3 (hbk)

Contents

Foreword, *Norman Long* vii
Acknowledgments xiii

1 Heterogeneity and Styles of Farming 1

Reproduction, Incorporation, and Differential Commoditization, 12
The Labor Process in Agriculture, 26
The Coordination of Domains, 28
Notes, 33

2 Dairy Farming in Emilia Romagna, Italy 37

Goals in Farming: The I- and E-Options, 40
The I-Calculus, 54
The E-Calculus, 67
Thin and Fat Cows, 73
The E-Calculus and Its Link to a High Degree of Incorporation, 76
Craftsmanship as a Specific Structuration of the Farm Labor Process, 78
Craftsmanship and the Creation of a Frontier Function, 92
Entrepreneurship as a Specific Structuration of the Farm Labor Process, 95
On TATE: The Technological-Administrative Task Environment, 106
Social Relations of Production and the Farm Labor Process, 116
The I-Calculus and Its Relation to Incorporation and Institutionalization, 120
Mountain Farmers and Intensification, 124
Styles of Farming as Social Constructions, 126
Back to the Mountains, 131

Concerning Scale and Social Relations of Production: A Supplementary Argument, 134
Scale: Capitalist and Cooperative Farms, 138
Notes, 144

3 **Potato Production in the Peruvian Highlands** 151

The Farmers and Their Enterprises, 154
Labor, 159
Oxen, Livestock, and Dung, 163
Capital, 163
Farm Labor: The Production of Soil Fertility, 174
Farm Labor: The Reproduction of Seed Potatoes, 183
Farm Labor: The Process of Potato Production, 193
Harvest Estimates and Per Hectare Yields, 196
Towards an Explanation of Heterogeneity in Farm Practices, 197
Notes, 202

4 **Peasant Struggles, Unions, and Cooperatives** 205

Luchadores del Dos de Enero, 205
Profit and Loss: A Historical Analysis, 225
Trabajemos y Luchemos ("Let Us Work and Fight"): Towards a Progressive Intensification of Production, 228
The Luchadores Plan, 246
Social Struggle and Autonomy as Prerequisites for Intensification: A Comparative Analysis, 252
Notes, 255

5 **Commoditization and the Social Relations of Production** 259

Value and Commoditization as a Differential Process, 262
The Impact of Commoditization, 268
Similarities and Differences, 272
Commoditization and the Social Relations of Production, 277
Commoditization and the Reproduction of the Agrarian Question, 281
Notes, 283

Bibliography 285
About the Book and Author 299
Index 301

Foreword

It is a pleasure to introduce this thought-provoking and innovative book that challenges present models and analytical perspectives on agricultural development, particularly as it comes from the Wageningen stable of rural and development sociology, where for the past few years a group of us have been struggling with some of the same theoretical and empirical problems. Its pedigree, however, goes back to the founding fathers of Wageningen sociology: to Rudie van Lier, who—though deeply skeptical of man's ability to steer the course of social change—thought that a major task of development sociology was to develop a critical and reflexive theory of intervention processes, and to E. W. Hofstee, whose contribution to the understanding of differential farming and cultural patterns in the Netherlands represents an important foundation for contemporary rural studies. The book also builds upon the stimulating work of Bruno Benvenuti (until recently also based at Wageningen), who developed an analytical approach to the study of institutional incorporation in agricultural contexts in Europe.

Yet, at the same time, the work is very much a product of the author's own distinctive background, field experience, and personal involvement with farmers and farmers' organizations in the Netherlands, Peru, and Italy. Unlike many practicing rural sociologists, Jan Douwe van der Ploeg is a countryman born and bred who early in life came to know the meaning of agrarian struggle. He comes from Friesland, from a family with strong farming and farmers' union connections, and although he now spends much of his time in mental labor, with pen rather than pitchfork in hand, he is never happier (as this book clearly reveals) than when discussing the concept of the "good cow" or the social significance of manure or varieties of potatoes. He also delights in being where the action is, whether this be with members of a Peruvian cooperative defending their right to work and to treat land as a collective asset and not simply as a commodity or with Italian dairy farmers seeking to combat the negative effects of market incorporation by devising strategies aimed at retaining control over the organization of their own labor processes.

Woven throughout the analysis are three crucial issues central to developing a critique of existing theoretical approaches to agricultural development: the significance of heterogeneity in farming practice and farm organization; the analysis and importance of commoditization and institutional incorporation for shaping farm labor processes and farmer strategies; and the importance of farmer organization and political struggle for the outcomes of incorporative processes.

While previous research has recognized diversity in farming, it has frequently been assumed that this diversity is especially characteristic of "backward" or "low-output" forms of agriculture that will eventually give way to "modern," "high-output," and technologically more standardized types. Diversity is reduced, therefore, to what van der Ploeg calls "a secondary characteristic of agricultural systems," representing either the survival of previous (and now less appropriate) forms of production that one day will wither away altogether or minority patterns that deviate from the central tendencies. Either way, of course, heterogeneity is side-tracked theoretically, leaving us with theories (whether couched in modernization or political economy terms) that concentrate upon explaining general statistical tendencies and assume normatively that empirical diversity is an expression of varying degrees of entrepreneurial success or of the uneven pattern of capitalist development. Van der Ploeg counters this rather warped type of reasoning by documenting the significance of variability in both so-called peripheral and central economies. He argues that heterogeneity in farming practice and farm organization is a structural feature of agricultural production everywhere and that it requires analysis and theorization.

The way forward, he suggests, is through recognizing that farmers themselves play a critical role in the construction and reorganization of farm practice. That is, they are, as Giddens succinctly put it, "knowledgeable and capable" social actors who set about resolving their own problematic situations through mobilizing resources and relationships and through attempting to impose their own normative definitions on the physical and social world around them. Differences in farm practice, then, are the result of differences in farmer strategy, rationality, and access to internal and external resources. A crucial point in his argument, however, is that there are likely to be several equally viable and profitable solutions that entail different degrees of market and institutional integration, different time perspectives, and different cultural commitments. Hence, we find among dairy farmers in Italy two contrasting types of farming logic that underpin the difference between strategies based upon "intensification" and those based upon scale enlargement and relative "extensification." Similar contrasts are evident among potato producers in highland Peru. Somewhat paradoxically, however, the technological and credit package introduced by government agencies designed to intensify production has led to strategies that are characterized by short-term planning, high costs and risks, dependence upon limited input and output markets, and generally lower benefits; in short, the strategies are characterized by several negative features associated with extensification. Improved technology and credit have also resulted in the destruction of various relatively "cheap" forms of non-wage labor.

These findings prompt van der Ploeg, at various points in his analysis, to embark upon a theoretical reappraisal of commoditization, giving attention to the ways in which "the means and objects of labor increasingly enter the process of production as commodities." Building upon an earlier critique of certain commoditization models (see Long et al., 1986; and Long and van der Ploeg, 1988), he argues that commoditization is a highly variable historical process that takes many forms and is inherently contradictory in its outcomes. The chapters devoted to the Italian and Peruvian cases provide a detailed picture of the many mechanisms by which commoditization penetrates the farm production process. But they also advance his central theoretical position that despite the pressures of the market and the state, farmers nevertheless possess the capability of containing or resisting commodity relations. They can, that is, adjust their farming strategies in order to benefit from non-commodity forms and from local agricultural knowledge and practice, thus creating some degree of autonomy vis-à-vis external institutions. This process is illustrated by van der Ploeg's finely honed comparison of "I-" and "E"-type Italian farmers who adopt different strategies with regard to cattle breeding, fodder provision, and mechanization. It is also shown in his account of how some Peruvian peasant producers draw upon their extensive knowledge (*art de la localité*) of different phenotypical conditions and genotypical varieties to select and exchange seed potatoes rather than purchasing "improved" varieties developed and promoted by scientific research establishments. Certain other types of farmers internalize the normative standards and follow the procedures of "modern" scientific farming in order to actively seek what they see as the benefits of commoditization.

Yet, whichever strategy predominates, agricultural decision making and the organization of the social relations of production in agriculture rest firmly in the hands of the farmer himself, even if he chooses to submit himself to the vicissitudes of the market and of price mechanisms. Therefore, farmers are active strategists who attempt to come to grips, both cognitively and organizationally, with the problems they face, rather than merely cogs in the wheels of change.

This line of argument rejects linear and externalist interpretations of agrarian development by according human agency a central role in the drama of agrarian development, but it also carries the implication that commoditization and institutionalization become real in their consequences only when introduced and translated by specific actors. The latter, of course, include not only farmers but also others—such as government bureaucrats and technicians, traders, agribusiness personnel, and development workers—who intervene directly or indirectly in farming practice and farmer decision making. Van der Ploeg's analysis also lays the essential groundwork for a detailed examination of the strategic action and ideologies of intervening parties who seek to establish economic, political, or ideological hegemony over the rural producer.

Van der Ploeg considers in depth the struggle "from below." He explores this topic through a fascinating and gripping portrayal of the long and militant struggle against the state by a Peruvian farming cooperative that was founded as part of the Velasco land reform of the early 1970s. Here he aims to characterize the nature, benefits, and internal dynamics of peasant-managed agricultural development and to document the continuous struggle that takes place in an effort to protect jobs and to secure local control over land utilization and the management of the production process. Van der Ploeg concludes that the general success of the cooperative in fending off the advances of the state, as well as in resisting the more subtle encroachment of commodity relations, rested principally upon two factors. First, when faced by government and bank pressure to reduce and rationalize labor power, the cooperative was able to develop a broad-based pattern of popular support (which at critical confrontations stretched even beyond the formal membership of the cooperative) for a strategy of maintaining all individuals in employment. Second, the type of intensified agricultural regime that it implemented facilitated a degree of autonomy from markets and external institutions.

This final element in van der Ploeg's argument reveals the "populist" strand in his thinking. Sustainable agricultural development, he believes, depends fundamentally upon the ability of local producer groups to maintain or secure control over the organization of their own labor processes and over other critical resources. Only by empowering themselves in this way through the development of their organizational capacities can they, in the face of the increasing threats of commoditization and scientification of agriculture, effectively protect their own life-worlds and interests, shape their future life chances, and guarantee the continuation of adaptive variety in the art and craft of farming.

While this message may seem to belong to the optimism of the late 1970s when "participation" and "participatory research" were the catchwords, van der Ploeg's deep and genuine concerns for the future of small-scale producers and for the survival of diversified forms of agriculture are founded upon meticulous field research and a thorough theoretical grasp of the nature and analysis of agrarian change. His point of view is not based upon empty slogans but is grounded in a systematic understanding of the contradictory tendencies of intervention and farmer organization. Furthermore, he is sensitive to the need to analyze closely the room for maneuver or the space for change that pertains to particular political contexts. He does not therefore rule out the strategic importance of piecemeal change and relatively small gains. In the end, his theoretical emphasis on actor strategy and rationality joins forces with a political standpoint that recognizes the crucial role of "everyday forms of resistance" and the many ways in which the struggles between different agrarian actors "make history."

This book, then, takes us into the battleground of agrarian problems where theory and practice confront each other. It is of great credit to Jan Douwe van der Ploeg that he is in the forefront of present debates and that he is so articulate analytically in presenting his point of view. This work, I believe, merits serious attention not only from researchers and students of agrarian development but also from development practitioners, planners, and politicians, whose actions shape, though do not determine, the course of agricultural development.

Norman Long
The Agricultural University
Wageningen

Acknowledgments

This book is based on research that was initially financed by the Netherlands Foundation for the Advancement of Tropical Research (WOTRO) and later supported by the Agricultural University of Wageningen and Leiden University in the Netherlands. Most of the research was carried out with Eppo Bolhuis, who—apart from being a good and stubborn agricultural economist—proved to be a fine companion. Throughout the research period (1980–1986), several other friends and colleagues, whom I would like to thank, made important contributions: Jaap Nieuwenhuizen, Bennedetto Franceschi, Ruffo Carcamo, Gerard Grimsbergen, Ferdinand Zoutendijk, Wies Gijbels, Jan de Rooy, Dirk Roep, Cees de Roest, Enrico Bussi, Giorgi Manenti, Francesco Badini, Corrado Pignagnoli, and Giorgi Pilone.

During the field research and writing of the first results (in Dutch and Italian), I could always count upon the critical comments and support of Benno Galjart, Wouter Tims, Henk Thomas, and Bruno Benvenuti. Indispensable institutional assistance, especially for the construction of long-term sets of data, was given by AIPA (Associazione per l'Istruzione Agricola Professionale) in Bologna; CRPA (Centro di Ricerche Produzzione Animale) in Reggio Emilia; ERSA (Ente Regionale di Svilluppo Agricolo) in Bologna; the ministries of agriculture in Rome and Lima; the universities of Parma, Bologna, and Portici near Naples; and, finally, the PRODERM project in Cuzco. Another type of institutional help, as indispensable as the aforementioned, was provided by the communities of Catacaos and Chacán in Peru; the Sindicato Unico de Trabajadores de la CAP Luchadores in Piura, Peru; the Confederación Campesina del Peru (CCP); and the Coldiretti and Confcoltivatori in Italy. Additional funding for translation was provided by Landbouw Export Bureau (LEB) from Wageningen.

I wish to thank my friends Norman Long and Bruno Benvenuti for the stimulating discussions we had on previous drafts of this book. Finally I would like to thank Ann Long, who translated the manuscript and helped me edit and clarify many ambiguities. Any shortcomings and other inadequacies are, of course, my own responsibility.

Jan Douwe van der Ploeg

1

Heterogeneity and Styles of Farming

There are a great many ways to farm, greater even, if that's possible, than the number of erudite models that have been devised for understanding, managing, and possibly neutralizing such diversity. Such models include Grigg's "agricultural systems" (1974), Dumont's "*types d'agriculture*" (1970), the "ecosystems" of Geertz (1963), the "*bedrijfstypen*" of the Dutch Agro-Economic Institute, the "*aziende tipiche*" of Italian research from Medici (1934) to de Benedictis and Cosentino (1979) and Brusco (1979), the "land-labor institutions" of South America described by Pearse (1976), and the "styles of farming" identified by Hofstee (1985). And the list could be greatly extended.

The intricacies and implications of the different classifications are equally diverse. Large geographical units, ranging from zones, provinces and states, to countries and even subcontinents, are generally the starting point for comparison and further elaboration. Some research traditions depict the differences *between* various systems by assuming a certain homogeneity *within* given geographical areas. Others highlight the heterogeneity found within different production zones. Table 1.1 presents a tentative overview of this heterogeneity. The table is based on one of the most obvious forms that diversity can take within what are otherwise relatively homogeneous agricultural areas, i.e., on the highly differing production results per object of labor (where object of labor can refer to a unit of land, herd of cattle, etc.). Under similar conditions (of an ecological, economic and technical kind), such different physical levels of productivity imply a varying input of production factors and non-factor inputs per object of labor and highly different levels of technical efficiency.[1] A greater insertion of production factors and inputs per labor object is often associated with higher technical efficiency. This is an argument which I shall take up later. In such a case, we speak of an *intensive* style of agricultural practice. When a

Table 1.1. Hectare Yields for Different Styles of Agriculture

Agricultural System	Ha. Yields with Extensive Style of Agriculture	Ha. Yield with Intensive Style of Agriculture
Rice in S. of Guinea-Bissau	820 kg/ha	1,410 kg/ha
Groundnuts in Guinea-Bissau	1,500 kg/ha unpeeled	2,200 kg/ha unpeeled
Cotton around Bagoue Ivory Coast	878 kg/ha	1,364 kg/ha
Food prod. Senoufozone Ivory Coast	28 mil F/ha	31 mil F/ha
Food prod. in thinly pop. Ignamezone, Ivory Coast	2,850 Kcal/ha	3,240 Kcal/ha
Cocoa, Nigeria	300 lbs/acre	550 lbs/acre
Potatoes, Anta Pampa, Peru	1,820 kg/topo	3,180 kg/topo
Cotton, coop, & communal, Bajo Piura, Peru		
- on good land	10.7 cargas/ha	12.1 cargas/ha
- on poor land	7.1 cargas/ha	8.4 cargas/ha
<u>Minifundia</u> agriculture, Antioquia, Colombia	100 (index)	153 (index)
Mixed agriculture, Campania, Italy	3.08 milj.l/ha	5.46 milj.l/ha
Dairy farming, Po plain, Italy	4.16 milj.l/ha	6.43 milj.l/ha

<u>Source</u>: Based on Cabral, 1956; van der Ploeg, 1981; Peltre-Wurtz and Steck, 1979; SEDES, 1965; Leroy, 1979; Galletti et al., 1956; van de Ploeg, 1977; CEC, 1976 and 1977; Bolhuis and van der Ploeg, 1985; this study Chapters 2 and 3. Average input of production factors per hectare as well as average technical efficiency were calculated for each agricultural system. When both were high then the farm was classified as belonging to the intensive style, and when both were low, as belonging to the extensive style. In all sets of data at least 74% of the farms could be classified as belonging to one or the other style.

relatively low input of production factors is combined with a relatively low level of technical efficiency, we speak of an *extensive* style of agricultural practice.[2]

It is customary in comparative research to present an agricultural system's level of development in terms of the average production of grain equivalents per hectare. At present, in most West African countries, this figure amounts to 1.5 ton gr.eq./ha. In The Netherlands, it is 10 ton gr.eq./ha. This last figure is even more impressive when compared to the 4.5 ton gr.eq/ha. obtained eighty years ago. In Italy, where higher temperatures slow down the conversion of energy and nutrients in biomass, a lower level is achieved, namely, an average production of 7 ton gr.eq/ha. While not disputing the significance of average differences, the data summarized in Table 1.1 demonstrate unequivocally that at each level of development a degree of diversity can be identified. The key question, however, is whether such diversity is structurally meaningful, or whether the distribution is simply random. This question is not new: indeed, one might even postulate that agricultural science advances through the repeated reconceptualization of such differences. Some theories maintain that they are essential as a starting point for analysis, while others, theoretically at least, see them as somewhat irrelevant.

Until the 1950s, diversity between, and especially within, agricultural areas was classified and understood in terms of intensive and extensive styles of agriculture practice. Thus the concept of "intensification" meant the ongoing development of intensive styles of farming; it referred to the progressive raising of intensity levels. The term "extensification" referred to the opposite tendency.

But these terms were anything but neutral. Intensive agriculture stood for better agriculture. It was not only desirable; it equaled progress. "Good farming," wrote Graham Brade-Birks (1950:XVI), "means farming so carried out as to produce the maximum economic output from the land." He described this "good farming" as "intensive farming," directed to "those practices designed to produce a very high output." Technical efficiency and economic results followed logically from each other. In contrast, extensive farming, "the practice of using the minimum amounts of labor, cultivation and manure," was referred to as "a low standard of farming." An authoritative Dutch author of the time, Minderhoud, wrote "intensification is rooted in the rule of raising the net yield, while extensification lies in saving labor and capital: thus one has to take a reduction of yield in kind for love" (1948:45).

Also interesting, in retrospect, is the debate which was then taking place over the farm economy ratio behind both processes of development. Minderhoud criticized those who "calculate at which intensity total production costs per kg are the lowest and give the impression that the farmer must strive for that." Contained in this whole issue is the question of which of the concepts corresponds most accurately to actual relations in agriculture and can thus be used as normative, as goals at the enterprise level. In short, what we now know as an established theory, namely, neoclassical agricultural economics, was then still subject to fundamental differences of opinion. Thus, according to Minderhoud, the proposition that the farmer must strive for the lowest price per unit of end product, "the problem is incorrectly posed." Situations vary and thus "the manner in which the soil can be rationally exploited" also varies. His comment that "many American writers, . . . take the circumstance of sufficient ground, but insufficient labor and capital as a starting point for consideration, with the consequence that West European readers find them difficult to follow," is a telling one (1948:52).

Minderhoud and Brade-Birks are exponents of a tradition which is difficult to reconstruct in retrospect and therefore may provoke surprise. There existed a broad consensus in which the bipolar dimensions just noted were taken as obvious, for the most part as an undisputed parameter for the ordering of diversity and as an analytical starting point for developing an understanding of the differential processes which such diversity brought about. Contained in such ideas was the unquestioned norm about the nature and direction of further agrarian development—namely, progressive intensification.

The 1950s marked a gradual but definitive turning point: a new paradigm became dominant. The core of the new tendency was neoclassical agricultural economics, a model for examining, describing and, if necessary, reorganizing the adjustment of agrarian enterprise behavior to market and price relations. In essence, this model comes down to projecting current price relations on the farm enterprise in order to specify precisely the "optimum" solution. The model implies a situation that is both thoroughly *atomized* and completely *static*. Development at farm enterprise level (a shifting "optimum") is only possible within this model by the grace of changing market relations and technical progress, mostly understood as the taking up of externally-developed innovations. It is striking that, with the emergence of this neoclassical model, concern for a dynamic that could be produced within the farm itself disappears from the literature. Agriculture appears increasingly as "the text-book paradigm of neoclassical perfect competition" (Lipton, 1968), and so the concept of intensification logically loses its meaning. Intensity level becomes a derivative of the enterprise as an economic organization orientated to the market. Indeed, particular agrarian subsectors can then be defined as "too intensive" (Galletti et al., 1956:308). Finally, categories such as intensive and extensive completely disappear from applied agricultural science. New parameters rule.

The agro-economic definition of the optimum—given a series of assumptions, such as the same economic and institutional conditions for all enterprises, the same technological level, and a striving for profit maximization by each and every producer—means that "only one point would be observable on the production surface" (Yotopoulos, 1974:265). Empirical diversity, from being the first tenet for theoretical construction, now becomes a residual factor, or worse still, an "anomaly."

In the recent literature, insofar as diversity and its implications get any attention, one or more of the following factors are usually referred to:

1. Diversity in hectare yields between otherwise "identical" enterprises would primarily point to variation in micro-ecological conditions: the vagaries of climate and soil are to be understood, as one Peruvian author expressed it, as "Satan and Messiah"; as prosperity for some, adversity for others.
2. A more elaborated view (which often rests on tautological argumentation) reduces differences in hectare yields to differences in price and cost levels.

One encounters a clear description of both viewpoints in Mellor (1968:259): "Studies that demonstrate peasant farmers to be, on the average, in good economic adjustment with their environment normally include considerable variability around that average. It is usually not clear to what extent such variability arises because some peasant farmers are not in optimal economic adjustment and to what extent the environment itself differs significantly from one farmer to another. Certainly the latter is true in part. Soils and other physical features differ widely even within small areas. . . . Perhaps even more important, costs of labor and capital differ substantially from one farm to another." With a different relation between available land and family size, farmers feel either more or less pressure, thus goes the reasoning, "to squeeze the last bit out of the farm (so that hectare yields rise rather than fall), *and thus they in effect act as though labor were more (or less) expensive to them than to other farmers*" (italics added).

3. In the most recent literature, besides assumptions over the subjective evaluation of factor costs, a *risk factor* is also introduced (in principle differing from farmer to farmer) in order to reconcile the assumption of an optimum to the actual distribution around that optimum. Later I shall examine this reasoning more closely.
4. A fourth, more instrumental explanation, links differences in hectare yield to different rhythms for adopting external innovations (more productive varieties, fertilizer, etc.)—a problem localized primarily in the psyche of the individual farmer.
5. Differences are also related to imperfections in the institutional environment—to inadequate commercialization structures, to a

lack of transport facilities, and to extension and credit mechanisms which reach only a part of the potential number of clients. These factors would explain the poor performance of some farmers and thus some of the differences between them.

6. Finally, there is a simplistic but often used argument in the NW European research tradition that "good" and "bad" hectare yields can be reduced to differences in craftsmanship. "Satan and Messiah" speak here not through soil and climate but through the (randomly bestowed) distribution of individual talents among farmers.

Without going into each of these arguments separately, we can observe that they have one characteristic in common. They are, in essence, all factors which are exogenous to the model; i.e., they are *residual factors*. Rather than seeing such factors as a possible falsification of the model itself, they are redefined, at least theoretically, as a question of extraneous conditions, such as those listed above.

The pioneering work of Hayami and Ruttan (1971) represents, in certain respects, a break with this agro-economic paradigm. They take diversity, so clearly evident in international comparative research, to be the starting point for their analysis. Their model also makes diversity—as being the *product* of different agrarian development patterns—acceptable. The parameters within which this diversity is investigated and exposed are reduced, in principle, to two: production per unit of land and the input of labor per unit of land. A combination of both terms specifies production per labor force—an approximation of income.

Production per labor unit can be raised within their model in different ways. First, by *intensification*, i.e., by raising production per unit of land. Such intensification, particularly if it forms a systematic pattern over time, is referred to in the literature as the "Japanese model." A second, an opposite strategy, is by *scale enlargement*, i.e., raising the land/man ratio or lowering the labor input per unit of land, referred to as the "American model." All kinds of in-between forms are, of course, possible.

Taking individual countries as the units of research, Hayami and Ruttan made a number of international comparisons, the assumption being that the pattern constructed for a country as a whole is a meaningful average. The explanation for the occurrence of one or another of the above patterns is, as Hayami and Ruttan assumed, to be sought in relative factor prices: i.e., the price of labor, capital, and land in relation to each other. If land is cheap and labor relatively expensive then agricultural development will follow the path of scale enlargement, especially if labor-saving technology is available. If land is scarce and labor overabundant, then intensification is to be expected. This explanation, theoretically speaking, is a repetition of the neoclassical model, in this case applied to countries, each country representing a specific constellation of relative factor prices. Beyond that, the model implicitly assumes that relative factor prices are "always and everywhere" translated in the same unilinear way in terms of the direction and rhythm of agrarian development.

The meaning and scope of these assumptions can be examined by applying Ruttan and Hayami's analytical model to a homogeneous agricultural area and seeing to what extent divergent development patterns can be identified within such an area. Such an experiment demands first a more precise definition of the term "homogeneity."

Homogeneity in economic terms means that the relative factor prices, to which Ruttan and Hayami attached such importance, will be virtually the same for all units of research, which implies that the research units will have the same internal structure and dimensions. Homogeneity in the institutional and technological sense means that all enterprises have the same access to credit and marketing facilities as well as to whatever technology is available. Homogeneity in the ecological sense speaks for itself. It implies that differences in agricultural style and development will not be due to differences of soil type, micro climate, etc. "Satan and Messiah" would thus be the same for everyone, certainly in the longer term.

If homogeneity in the above sense is satisfied but differences in enterprise patterns are still, nevertheless, observable (hypothetically and schematically represented in Figure 1.1), then several things may be concluded:

1. Diversity (symbolized by the cross-sectional analyses, Figure 1.1 at moment t=5 and t=10) will not be so much a chance phenomenon but a product of different development patterns that can be logically explained.
2. Other relations and processes besides relative factor prices play an important role in the emergence of a particular development pattern. Perhaps there are structural patterns which account for factor prices weighing heavily in some enterprises and being of less or no importance in others.
3. Finally, "induced technological change," the strategy for speeding up agrarian growth proposed by Ruttan (1973; 1977), would be incomplete. Besides the availability and diffusion of innovations as such, the specific demand for and selection of particular innovations would also emerge as essential factors.

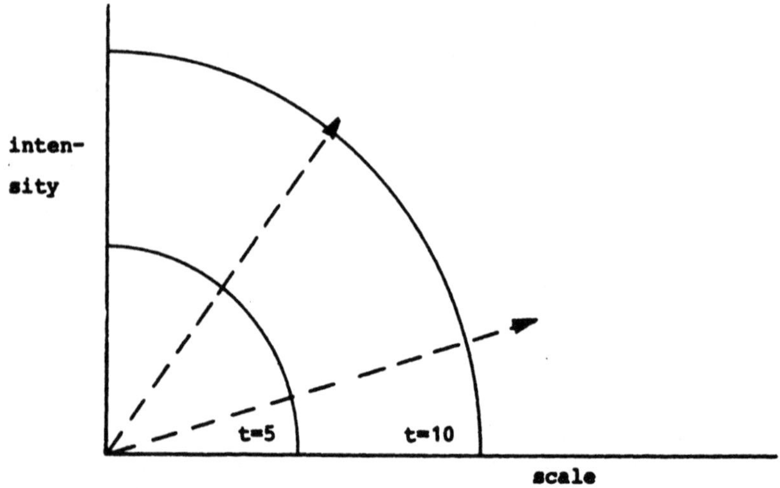

Figure 1.1 Hypothetical sketch of different enterprise development patterns in a homogeneous setting

These hypothetical conclusions suggest that there are differences in the dynamics and rationalities within enterprises, which lead one to question what kind of relations might lead to such differentiation.

In the several districts where we carried out our research, differences in development patterns do indeed appear. In this book three different agricultural systems are discussed: dairy farming in Emilia Romagna, Italy; potato cultivation in the southern highlands of Peru; and cooperatively organized agriculture on the northern coast of Peru, where the cultivation of export crops is prominent. In each of these three areas, a great heterogeneity can be observed, both in the technical production structure of the farms as well as in hectare yields. Some production cooperatives, for example, achieve a gross value of production per hectare (GVP/ha) 30% or 40% higher than the "average" cooperative under what are otherwise similar circumstances. Such differences are not coincidental. They are, as we shall see later, the result of the way in which such cooperatives are linked into their economic environment and of the social struggle that some cooperatives pursue: a struggle for the right to work, which in present day agriculture is becoming increasingly marginalized. Likewise, substantial heterogeneity is to be found in potato production, as practiced on the smallholdings in southern Peru. Some farmers harvest 5 tons per hectare, while others achieve a production of more than 15 tons. This variance is significant. For the farmers in the highlands, this is not due to a blind Satan or Messiah or to the unpredictability of the rains, frosts or diseases. As they themselves say, "It is the rich farmers who can bring home a fine

Table 1.2. Diversity of Dairy Farming in Emilia Romagna. ERSA data. plain and mountain (n=75, n=59 resp.)

	Plain		Mountains	
	Mean	SD as % of mean	Mean	SD as % of mean
Labor input/ha (man/ha)	0.23	48	0.16	50
Depreciation/ha (in 1,000 lire)	319.52	71	141.00	70
Variable costs/ha (ditto)	2584.79	39	891.54	64
(lab+cap+inputs)/ha (ditto)	5244.54	34	2589.64	45
GVP/ha (ditto)	5464.04	42	1611.13	48
Herd value/ha (ditto)	2577.89	51	983.89	44
Machine value/ha (ditto)	1305.45	113	648.48	85
Cow value per AA (ditto)	958.55	16	953.47	17
Fodder per AA (ditto)	775.81	32	706.25	61
GVP/per AA (ditto)	1369.15	25	1055.33	32
Yield/cow (in ql)	44.10	23	31.71	27
Food prod./ha food crop (in 1,000 lire)	1046.53	64	520.48	81
Animal prod./ha food crop (in 1,000 lire)	5134.17	87	1230.10	52

AA = Adult Animal, a statistical category that allows for a classification of different kinds of animals to the same quantitative unit.

harvest. They are the ones who can work the land as it should be worked."

If one should turn one's thoughts from the micro to the macro level, then again particular trends can be observed which show the strategic meaning of yields. Sugarcane and cotton, typical crops for the large production cooperatives in Peru, showed an absolute drop in hectare yields in the 1970s, while the opposite trend, i.e., a sustained increase, can be seen for a crop such as beans, produced mostly by the small-scale farming sector.

Marked differences in hectare yields do not only manifest themselves in the Third World. One can see them just as well in the so-called "modern" agricultural areas of the European Community. One such region, Emilia Romagna, discussed in more detail later in this study, is composed of the provinces of Parma and Reggio Emilia, Italy, a dairying area noted for its highly productive form of agriculture. For the specialized family farms located on the plain there, all operating under the same ecological conditions, one finds considerable variance over a whole range of relevant farm characteristics.

Thus, one finds for the average labor input per hectare (see Table 1.2) a standard deviation of 48%; for depreciation per hectare (an indication of the capital input) we find a standard deviation as high as 71%. Inputs (measured as variable costs per hectare) also vary noticeably, with a standard deviation of 39%. It should therefore be no surprise that gross production value per hectare should also strongly vary (i.e., 42%). Such variance does not disappear when the research area is narrowed (for example, to the level of a single province, where one might expect ecological levels to be less varied), but instead becomes even greater. Also, within groups of farms of the same size, variance appears to be large. In other words, between what are almost identical enterprises, enormous differences in intensity are visible.

Analysis of constant samples relating to these dairy farms in northern Italy shows that, during the period researched (1970–1980), diversity actually increased. There are two clearly diverging development patterns to be seen. There is a group of farms which intensified substantially over time—i.e., production per unit of land rose substantially—while "scale" (the relation between land worked and available labor) remained relatively stable, and a second group which developed to a large extent in terms of scale—and hectare yields rose to a lesser extent than those of the first group. For this reason, I refer to the latter as *scale-enlargement combined with relative extensification.*

These development patterns, which may be seen as opposite poles between which a whole gamut of development patterns might fit, will be illustrated through an analysis of twenty four dairy farms located in the Pianura Parmense (an outstanding example of a homogeneous agricultural area). In 1970, at the beginning of the research period, these farms were relatively homogeneous. They were all roughly the same size, about 25 hectares, and they could all be considered as dynamic, well-functioning family farms.

In the following ten-year period many changes took place, changes which as usual can be excellently described in terms of *average trends.* The average size of the farm grew, average production per farm rose, average input of labor fell, and mechanization increased. In other words, the Parma data are *pars pro toto* par excellence of what appears to be the universal image of western agriculture.[3] During this same period, six farms closed their dairies, a fact which is also congruent with the general development of Italian agriculture. However, for the remaining eighteen farms, there was certainly no uniform pattern of development. The average rise in gross value of production per hectare during the research period amounted to 2.53 million lire (current rates). Some farms, however, achieved a noticeably higher increase while others remained clearly behind. Table 1.3 charts these differences. The farms are divided into three subgroups according to their GVP/ha: those with a less than average rise, those with an average rise, and those with a much higher than average rise in GVP/ha. A clear difference in development patterns is observable. The differences in GVP/ha, which

Table 1.3. Divergent Patterns of Development, Parma Data

	Intensity			Scale	
	delta GVP/ha	GVP/ha 70/71	GVP/ha 79/80	ha/man 70/71	ha/man 79/80
Intensifying subgroup (n=6)	>3.00 mil.lire	0.90 (0.26)	4.76 (0.27)	4.92	6.02
Middle group (n=6)	2.15 - 3.00	0.79 (0.20)	3.34 (0.42)	6.84	7.8
Relatively extensifying group (n=6)	<2.15	0.67 (0.10)	2.47 (0.29)	7.66	10.59

were minor in 1970/71, were substantial in 1978/79. The range observable in 1979/80 is the result of diverging patterns of farm development.[4]

The third subgroup is intriguing. Already relatively extensive, these farms achieved a rise in GVP/ha which, judged by the intensive group, remained far behind what the real potential could be. This relatively extensive group, however, raised scale the most (taken here to be the relation between area and labor force: ha/man): from 7.66 ha/man to 10.59!

Compared to the middle group, intensity in the third subgroup dropped from 85% in 1970/71 to 74% in 1979/80. In contrast, scale rose (as a percentage for the group) from 112% to 135%! Among the intensive group intensity rose from 114% to 143%, while scale remained relatively constant, 72% in 1970/71, and to 77% in 1979/80. If one presented these different patterns of development in graph form, the picture hypothesized in Figure 1.1 would emerge.

Each pattern is economically viable in the sense that it provides a reasonable income per labor force: 18.39 million lire in 1979/80 for relatively extensive, large scale enterprises, and 16.28 million lire for the intensive group.[5] The middle group earned the least, 15.10 million lire. Historical trend analysis shows that the middle group also achieved the lowest increase in income.

An important part of the heterogeneity identified in various agricultural systems can be analyzed in terms of different styles of farming. These styles, as the previous analysis shows, are to a large extent the result of different patterns of farm development reproduced through time. Style of farming (or soil use), can be defined as a valid structure of relations between producers, objects of labor, and means. "Valid" means that at least those directly concerned consider the structure as an adequate means for making a living. Characteristic of each farming style are its productive results per object of labor as realized through the specific interrelations between the direct producer, his objects of labor, and means. We shall conceptualize the physical production levels achieved as both characteristic of, and the structural outcome of, the way in which farming is organized (subject to a series of conditions to be specified later, among which is a degree of continuity through time).[6] Thus we speak of an intensive style of agriculture if the production level per object of labor is high, and of a relatively extensive style when the production level per object of labor is relatively low.[7]

How many styles of farming there are, the degree to which they explain variance, and the extent to which they are linked to structural differences in agricultural development patterns, are questions which a priori are not easily answered. They are themes for further research.

The interaction between direct producer, his objects of labor, and means can also be defined in an immediate sense as *farm labor*. This might appear confusing, but it is not. Farm labor and styles of farming, though not to be separated, are nonetheless distinguishable. One can consider farm labor as the continuous reproduction (and sometimes the gradually produced change) of a style of farming, and a style of farming as the material result of farm labor—a result of the antecedent practices and decisions laid down in the setting up of and the operating and developing of any particular farm. In summary, *a particular style of farming is the product of a specific structuring of farm labor.* A style of farming can rightly be defined as a "social construction," at least if its construction (the "construing moment") is located within the farm labor process. Farm labor as a conscious activity assumes the formation and use of goals towards which organization and development of production will be oriented; it assumes also the development of the means or qualities to effectively pursue such goals. Ends and means will together form a rational pattern.

Following Weber, Mannheim and others, different forms of rationality can be distinguished, with substantial and functional rationalities being the two classical opposite poles. Different ends can thus be combined, through different ratios, with different means. Applied to the notion of farm labor, a whole gamut of theoretically possible structurations and thus styles of farming, can arise, each noticeably different, with its own characteristic development pattern.

Reproduction, Incorporation, and Differential Commoditization

In this book I will develop the thesis that different development patterns each assume a specific pattern of reproduction. I shall discuss two such patterns, that of relatively autonomous, historically guaranteed reproduction, and that of market-dependent reproduction.

Farm labor cannot be reduced to the production of certain end products such as milk, meat, potatoes, and grain, for it is also interwoven, in a variety of different ways, with reproduction. To produce milk, the necessary labor power, objects of labor, and means all have to be reproduced. One can even go a step further and say that the social relations of production themselves, i.e., the relations under which production takes place and which give the production process its concrete form,[8] must also be reproduced in the production process.

The unity of production and reproduction in farming can be symbolized in all kinds of ways. A cow must be pregnant to produce milk. Only then can there be any talk of (potentially high) milk yields. And in reverse, the calf that is to be born will take care of future milk production provided that she in her turn is pregnant. The same goes for a more mundane affair—potato cultivation. A potato variety can only yield a "fine production" on fields that are well cared for. And in reverse, only through a "fine production" can a better seed potato be selected.

Relatively Autonomous, Historically Guaranteed Reproduction

In the apparently most simple reproduction pattern, that of relatively autonomous historically guaranteed reproduction, the labor force, objects of labor, and the means necessary for each production cycle are the material result of the preceding cycle. Each cycle assumes the availability of land, labor, capital and all kinds of inputs produced in the preceding cycle. Production then depends on reproduction in the previous cycle, just as production in the present cycle lays the basis for future cycles. The reproduction process (and thus the production cycle) is thus said to be historically guaranteed.

This process is not difficult to imagine. Take, for example, a dairy farm. The hay barn is full of the hay that was collected from the fields in the last cycle. In the present cycle, the fields will be sufficiently fertilized to harvest enough feed and fodder for the coming cycle. The family houses enough labor—farmer, wife, children and maybe others. Their knowledge and skills have been picked up in the experience of other cycles, and in the coming cycle their labor power both materially and qualitatively will again be reproduced. The means for taking care of the replacements that are sometimes necessary are found on the farm. Finally the patriarch keeps somewhere the proverbial sock with savings to take care of the unexpected emergency. The coming harvest can likewise reproduce or add to such reserve funds. In summary, at the beginning of a production cycle, the farmer has at his disposal all

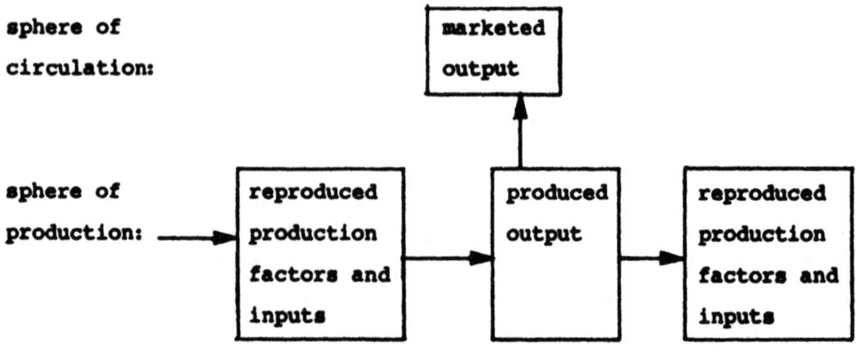

Figure 1.2 Pattern of autonomous, historically guaranteed reproduction

the necessary prerequisites to produce. Production and reproduction are thus historically guaranteed.

This can be schematically projected as in Figure 1.2. The figure illustrates that the production cycle depends on the production factors and non-factor inputs reproduced in the previous cycle. Production is realized with the help of these production factors and inputs. Part of this production is sold in the various markets, and part serves to safeguard the following cycle. In principle, the ratio between the part marketed and the part that provides for the following cycle is variable. Short- and long-term perspectives and interests must here be balanced against each other. Once such a decision is made, it follows that each cycle will begin with a given amount of production factors and inputs. With this the farmer must work in such a way that a maximum output is reached, for in this way both short-term interests and long-term needs (the reproduction of production factors and inputs for the benefit of future cycles) are optimized. The key for achieving this goal, of course, is to raise technical efficiency.[9]

Production of a marketable surplus plays a crucial role in this pattern. The production process is geared to the creation of commodities, and at the same time, to the guaranteeing of future cycles. Production and reproduction form a specific unity in this respect. The production process is geared to the creation of commodities, and thus to the market. Reproduction, however, goes on outside of the market, as it does not depend on buying the necessary labor, objects of labor, nor the means. The means of production are relatively autonomously produced. They do not appear as commodities in the production process but as use values, their specific value being that they assure production (i.e., one's own future and that of the farm). Hence, production depends *not* on the market but on a relatively autonomous and historically guaranteed reproduction.

The reproduction process may very well interact with markets in some respects. However, this is not a question of market dependency. With the money gained through the marketing of some commodities, those elements which cannot be reproduced in the labor process itself are bought. Iron is the *cause-célèbre* of earlier agrarian historical debates of this kind. Iron was practically always needed, but impossible to produce oneself. Thus, from ancient times onwards, each agricultural system had a *culture d'or*, as Marc Bloch called it, a crop cultivated for exchange—exchange for gold in order to buy iron.

Orientation towards commodity production certainly cannot be considered a secondary characteristic of this pattern. Market developments were frequently decisive for both progress and misery. A telling example is the so-called *Intensitätsinsel* of early Europe: the proximity to city consumption markets gave impulse to a considerable intensifying of agriculture around the cities, so creating these "islands of intensive agriculture."[10] Even today, one hears in the daily speech of farmers references to this relationship. Farmers in northern Italy speak of a "*mercato che tira*," of a market which draws farming onward. It is an expression which fits perfectly with the pattern of historically guaranteed, relatively autonomous reproduction.

However, it must be stressed that this pattern should not simply be identified with the past. An important part of contemporary farming can, as I will presently show, be analyzed and understood through this pattern. Relatively autonomous historically guaranteed reproduction is often to be better understood as the outcome of the far-reaching and long-lasting emancipation processes of the farming people themselves than as a leftover of earlier relations.

Take Friesland, for example. There was in Friesland, in what was later to become the birthplace of cattle breeding, a period in which hundreds of ships left the harbor laden with manure. The soil fertility of local fields was not improved because the manure was sold as a commodity to Holland. The same happened with cattle. Animals which appeared resistant to disease or were especially good stock breeding animals were not used for the improvement of local herds. They were sold. Hay was also sold and exported. The pattern of historically guaranteed relatively autonomous reproduction became established only later, between 1860 and 1890. A certain decommoditization then occurred when farmers tried to gradually improve the production process and to ensure continuity. Then, hay, manure and good cows were no longer sold but kept on the farm to produce more and better milk and butter.

This change came only after intervening social and economic change. From the 1830s on there was a profound struggle under way in Friesian farming organizations for complete power: farmers even took the upper hand in organizations which were primarily controlled by the city bourgeoisie and rural aristocracy.[11] Radical changes were also occurring in a cultural sense. As Hofstee (1985) demonstrated for arable farming in neighboring Groningen, long-term perspectives began to prevail over medium-term interests. The idea that farmers could make progress over the longer term through their own means was essential for improving stock, for raising soil fertility, and for creating the infrastructure needed for irrigation and drainage. It also seemed to be typical that the means seized upon for the gradual improvement of the production process (for the raising of technical efficiency) fitted closely within the pattern of relatively autonomous historically guaranteed reproduction. Van Zanden, who investigated the enormous increase in productivity in "peasant agriculture" on the sandy soils of the eastern Netherlands, showed clearly that this important development, which occurred between 1850 and 1900, did not proceed via a growing division of labor between farm and external institutions nor via exogenously induced technical progress. Quite the opposite, "The production packet (of the farms) was earlier much more mixed than specialized. . . . There was no strong specialization of farm activities for which they had a relative advantage. On the contrary, on the output side we see a de-specialization" (1985:183). Growth, in essence, is autonomously generated: "the greatest part of the growth was brought by greater production of inputs (manure and fodder) *within* the farm." "Improvement in the *quality* of production factors also played an essential role"; an improvement that "was in the end made possible by a sharp *increase in labor input*" (van Zanden, 1985:184) (italics added).

The pattern of historically guaranteed relatively autonomous reproduction can be complex. A growing part of the marketed surplus will be used for the purchase of those technical means of production that cannot be produced on the farm itself. In itself, this implies no essential breach in the system (take the example of iron already mentioned). The same can be said for tractors, chemical fertilizers, etc., for the schema does not refer to an imaginary autarky but to a specific structuring of the reproduction process. Almost all purchases of the various technical means of production can fit within this structure if they follow the pattern of historically guaranteed reproduction. Buying a tractor, building a new cow shed, etc., will thus be based on savings and not, for example, on loans. In this way the tractor will appear in the production process as a "means with which to lighten or improve labor" and *not* as a commodity. A primary example of this phenomenon is formed by the proverbial cycle of *fat and thin years*. It is after a number of "fat" years, years in which farmers are able to save, that extensive investment

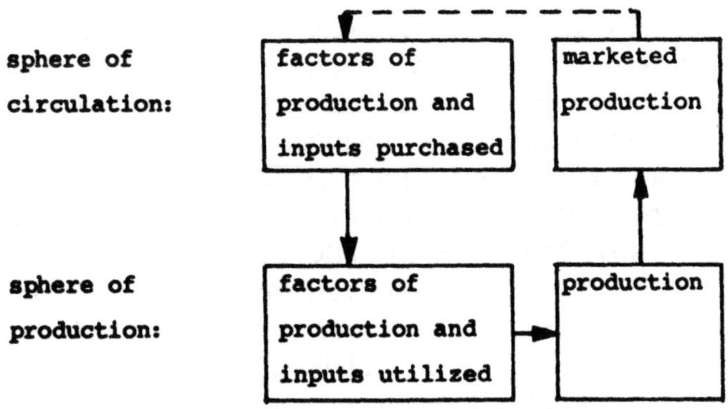

Figure 1.3 The pattern of market-dependent reproduction

is carried out. Such investments depend on savings, and they thus remain within the logic of historically guaranteed relatively autonomous reproduction. If such a wave of investment leads to a boom in production and a subsequent fall in prices, then "thin" years dawn.

Market-Dependent Reproduction

A contrasting pattern or schema to that of historically guaranteed reproduction is that of market dependent reproduction. A simplified sketch of such a schema is presented in Figure 1.3. First, it is characteristic of this pattern that the necessary production factors and inputs are not the result of preceding cycles but are mobilized wholly, or to a large extent, through the relevant markets (capital, labor, food, cattle markets, etc.). The production factors and inputs used thus appear in the production process, in a direct sense, as commodities. The level of inputs is by nature variable: it will depend on economic considerations in which cost/benefit relations are crucial. Second, the total output is considered to be marketable. Farm conduct is no longer so directly oriented towards materially assuring the following cycle. The total production is sold, and the costs incurred (production factors and inputs mobilized on the market) must in the first instance be valorized. Reproduction is thus not only market dependent but equally future dependent. Harvest prices at the end of the cycle will determine the extent to which the production process can be reproduced. Only if the costs can be valorized[12] can a following cycle be organized. Reproduction thus indeed becomes in essence, market-dependent. The latter implies that in the actual process of production, economic efficiency must be optimized: the production process must be organized to comply with prevailing market relations and in such a way that the difference between costs and benefits is optimal.[13]

In historically guaranteed relatively autonomous reproduction, the market is indeed an *outlet*. The market is not a structuring principle that determines farming in a thorough way as it is in market-dependent reproduction, where commodity relations penetrate to the heart of the production process and exercise a strongly directive effect.

Market-dependent reproduction can arise for many reasons. After several consecutive bad harvests or after extreme exploitation by others, the reproduction of a farmer's own resources (of the necessary production factors and inputs) may become inadequate, and the required resources will then have to be mobilized via the respective markets. Other processes can also, however, play a decisive role. I shall discuss some of these processes on the basis of a more general exposition of the process of externalization in agriculture.

The Process of Externalization in Agriculture. Farm labor entails an extremely wide range of tasks. Let us take again the dairy farm as an example, beginning with the production of the green fodder, silage, hay, and concentrates needed. For this, all the fields must be fertilized. Thus, besides fodder production, reproduction of soil fertility emerges as an important task in the labor process. Combined with this is the storage and conservation of at least part of the fodder. Then, of course, there is the feeding itself. Another series of tasks is related to caring for the herd. Cows have to be milked and young cattle reared. The milk will be processed into butter and cheese, and the whey, mixed with meal from the grain fields, is taken to the pigs. Pigs after slaughter are preserved as meat for the family or they can be sold on the consumption market. This sketch of tasks can be endlessly lengthened and detailed. In grain cultivation alone, 400 decisions are identifiable. Each decision has sometimes small, sometimes far-reaching consequences for the level of yields and costs incurred. And each decision relates to a specific task. In ploughing, for example, decisions must be made concerning the depth of the furrow, the direction and manner of ploughing, and so on.[14]

The first point to be made here is that all these tasks must be closely coordinated in farm labor. Choice of seed, of the way to work the land (including ploughing), the timing of fertilizing, and the composition of manure must all be carefully interrelated. If they are not, excessively high costs or wastage can result, as well as disappointing harvests. One can go a step further: to an important degree the basis for endogenous progress lies in this continuous interrelating and coordinating of tasks. The farmer can, through his own labor process, and through continual observation, interpretation, evaluation and manipulation, identify positive and negative consequences and translate the insights gained from this process into possible improvements in the coordination of tasks.

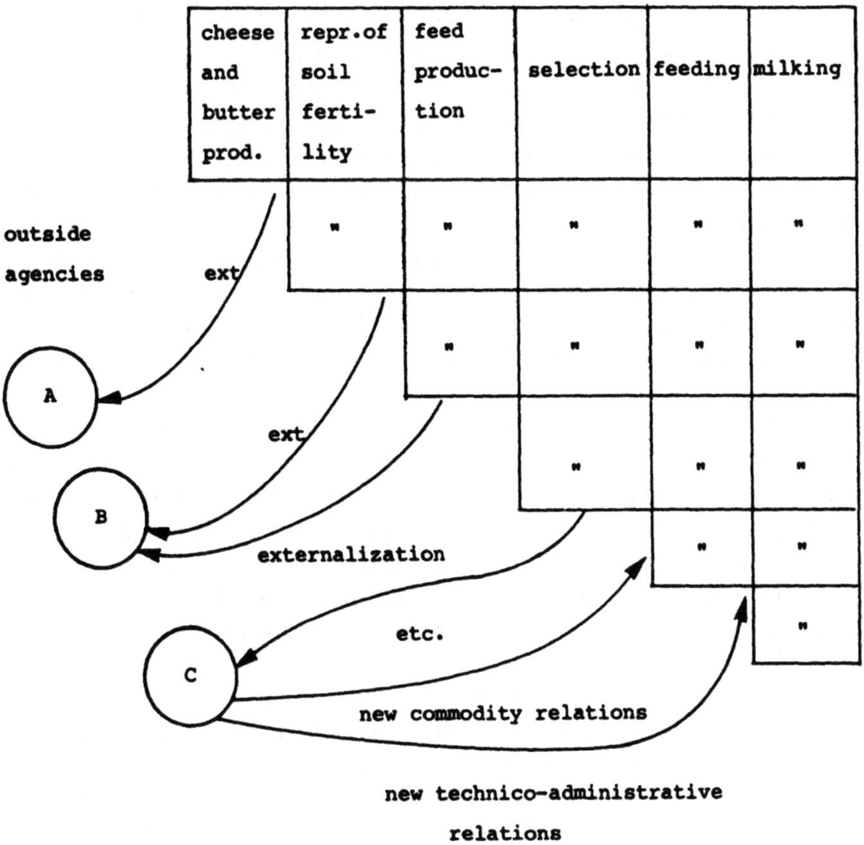

Figure 1.4 The process of externalization

The need for a close coordination of tasks holds a fortiori if production and reproduction are systematically combined in the labor process. The farmer can then, through a gradual improvement of his objects of labor (land, cattle, plant material, etc.), lay the groundwork for higher production levels in cycles to come. The same holds for the reproduction of labor and the means of production. The quantity, and above all, the quality of labor can be raised in the reproduction process just as the means can be improved.

The process of externalization means a gradual, or sometimes an abrupt, shift of particular farm tasks to external institutions, as schematized in Figure 1.4. The reproduction of soil fertility, for example, can be almost wholly delegated to such institutions. Then the means will no longer be produced on the farm (in the form of matured and mixed animal dung, green manure, specific intercropping schemes, etc.) but will have to be bought in the form of fertilizer. The knowledge needed to fertilize adequately can also be externalized. Institutions that specialize in this task, will take soil samples and subsequently translate these into fertilizing instructions. The task of fertilizing is thus reduced to closely following the externally delivered prescriptions. Even the task of fertilizing itself can be externalized by employing a contract worker. A similar exposition is possible in connection with most other tasks. Externalization can be carried to such a degree that, in the end, few tasks remain on the farm.

New relations arise in and through this externalization process. What originally was organizationally tied to the labor process on the farm becomes divorced from it in the externalization process. As a result a new unity must be created via commodity and technical administrative relations.

On 40% of Dutch farms, milk cows will be reared on the farm itself. On another 30%, a majority of the cows will be bought. On such farms the task of rearing good milk cows is externalized, either to other farms which specialize in the breeding and rearing of heifers for calving or to specialized institutions which produce high value genetic material. Thus on farms where breeding is externalized, animals for milking will have to be purchased and will therefore appear in the cow shed as commodities. The farm must enter systematically into commodity relations with external institutions. But that is not all. Because bought cows are no longer the result of the farmer's own labor, he will lack the necessary knowledge about these animals. If tasks are to be coordinated, then feeding, milking, and other tasks must be in tune with the genetic potential of the animal. That is, coordination requires insights concerning the parentage and ancestry of the animal, how earlier generations reacted to particular methods of feeding, etc. Where farm labor encompasses a large and coherent scale of tasks, insights and procedures flow together in one knowledge system. With the process of externalization a break in this system occurs. Insights, experiences and methods must now be communicated to the farmer by external institutions as "directions for use," which specify exactly how the purchased objects of labor and means must be used. It is in this way that technical administrative relations arise.[15]

It goes without saying that with the advance of the process of externalization, market-dependent reproduction (illustrated in Figure 1.3) becomes dominant.

The same process also has radical consequences for the farm labor process; some qualities will become superfluous or will be subjected to a certain dequalification. At the same time, new qualities will become necessary (Lacroix, 1981). In historically guaranteed relatively autonomous reproduction, the raising of technical efficiency is crucial; in such a context, craftsmanship develops, which is fundamental to progress.

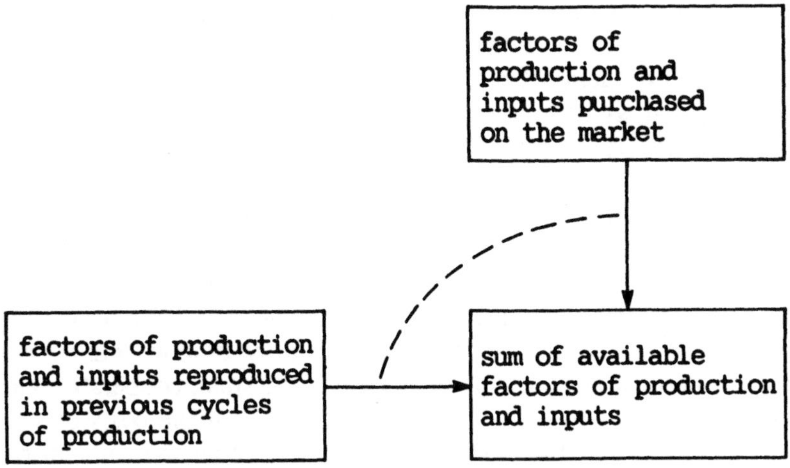

Figure 1.5 The mobilization of production factors and non-factor inputs

In market-dependent reproduction, however, economic efficiency becomes strategic; i.e., what is important is the degree to which farm conduct corresponds to prevailing market relations and results in a certain profit. And for this, entrepreneurship is needed.

Towards an Operationalization. These two patterns of reproduction are of course ideal typical models. They represent two opposite poles of the farming continuum. There is no question, however, of a unilinear movement. The degree to which commodity relations penetrate the labor process, and the degree to which markets are thus a coercive principle in farm practice, are variable, in terms of both time and space.

In order to operationalize this variability in the relation between agriculture and markets, the schemas are combined in Figure 1.5. The figure shows that some of the necessary production factors and inputs are reproduced on the farm, while others are mobilized through the market. The production is thus partly autonomous and partly market-dependent, and the ratio between them can vary widely. In this study the ratio of resources mobilized via the market against the total of resources committed in an enterprise is defined and examined as the degree of incorporation.

Degree of incorporation represents the extent to which a farm is incorporated into markets on the supply side. A high level of incorporation refers to a high degree of market dependency, and a low level implies a significant degree of autonomy from such markets. The notion of incorporation level can be applied to the totality of resources employed or to separate production factors and inputs, in which case we speak of incorporation into the labor market, for example, or incorporation into the market for fertilizers, genetic material, etc. Degree of incorporation is, at the same time, an expression of the extent to which the various production factors and inputs (labor, objects of labor, and means) appear as commodities in the production process. The higher the level of incorporation, the higher the actual commoditization of the elements which provide the basis for the labor and production process. In order to illustrate this argument, I will once again return to Emilia Romagna, one of the regions discussed in detail in the next chapter.

Dairy farms in the region deal with several supply markets and, naturally, with the outlet markets for milk and meat. The farms in the provinces of Parma and Reggio Emilia mainly produce milk for processing into the famous Parmesan cheese in the small cooperative cheese factories. The returns from this cheese (set against the costs of storage while maturing and the delivered milk), determine the milk price on the market. Prices for meat are set by the cooperative slaughterhouses to which superfluous young cattle and old cows for the market, are usually delivered. They indirectly follow the movements of the European meat market.

On the input side of dairy farms, a large number of supply markets can be identified, although, unlike outlet markets, the degree to which supply markets are linked to enterprises varies considerably. There are markets for production factors, for land (where a number of different mechanisms can be identified, such as buying and selling, tenancy, and renting), for labor, and for capital. The capital market, as the various "actors" know, can be divided into markets for short-, medium- and long-term loans. The commercial houses and consortia operate mainly in the short-term loan markets. They supply concentrates, fertilizer, diesel oil and veterinary medicines, often on credit to be repaid at the end of the agricultural year. Farmers mostly use the market for medium-term loans for buying machinery, farm implements, vehicles, and sometimes cattle. This form of credit is usually provided by the banks. Such loans are frequently given an interest subsidy by the regional government service responsible for the administration of EC funds for agricultural development. Long-term loans, often used for the purchase of land and the construction of new cow sheds and installations, are likewise obtained through the formal bank circuit and almost always qualify for an interest subsidy. The conditions which farmers must satisfy for such loans, however, are many and far-reaching.

Besides the above markets there are still several others. The market for machine services and contract workers is one of them, supplying services such as harvesting, the spreading of fertilizer, deep ploughing and others. The extent to which a farmer must call upon this market depends on whether he has at his disposal the necessary machines,

Table 1.4. Degrees of Incorporation in Dairy Farming, Emilia Romagna, Italy (Averages and Standard Deviations)

	Plain		Mountains	
Incorporation in the market for:	M%	(s)	M%	(s)
Labor	9.1	(22.8)	0.1	(0.4)
Contract work	30.7	(28.5)	10.0	(12.5)
Credit, short-term	4.6	(16.3)	1.9	(10.4)
Credit, medium-term	11.1	(50.5)	3.4	(10.8)
Credit, long-term	2.4	(3.4)	2.4	(7.6)
Land	28.7	(37.8)	20.2	(30.2)
Fodder and feeds	43.8	(18.2)	37.8	(16.7)
Cattle	7.2	(9.0)	7.6	(11.1)
Overall degree of commoditization	26.0	(15.0)	15.1	(8.3)

farm implements and labor. Historically speaking the labor market has been replaced to some extent by the market for machine services.

A particularly important market is for fodder and concentrates. The transport of fodder to Reggio and Parma from the far south of Italy and France has become a common sight, as it has elsewhere in Europe. An extensive system of labor division has developed around specialized dairy farms, with the surrounding zones devoting themselves to supplying them with fodder. The market for concentrates is dominated by huge agri-business groups, that maintain a trading machine of global proportions for their supplies.The extent to which a farm is dependent on the market for roughage and concentrates can, to an important degree, be regulated from the farm itself and will vary according to the number of cattle, availability of grassland and the farms own intensity of food production.

A last important market is that for genetic material. As in the Netherlands, some farms in Emilia Romagna breed their own animals while other farms buy replacement stock.

The wide variation to be found in practice between farms and the markets described here is not coincidental, as we shall show later, but is to some extent the result of the conscious behavior of the different producers.

An analysis similar to that illustrated in Figure 1.5 was applied to relations between farms and markets in Emilia Romagna. For each of the 134 farms, the degree of incorporation into the respective markets was calculated. Table 1.4 summarizes these data and presents the average incorporation level and standard deviation for each market.

The data are specified according to whether the farm is in the mountains (n=59) or on the plain (n=75).

The range around each average is considerable, suggesting that the relation between agriculture and the different supply markets is highly differentiated. This differentiation is clearly shown when the different forms of incorporation per farm are summed up in a synthetic index. Some farms then appear as outstanding examples of incorporation, while others appear relatively autonomous. In addition, historical analysis of the available data shows that the degree of market dependency is even less of a constant. On some farms the degree of incorporation increased markedly, while on others it decreased.

The partial and synthetic degree of incorporation investigated in Emilia Romagna can be calculated in the same way for other agricultural systems. This subject is discussed in Chapters 3 and 4. Adjustments have to be made, for as agricultural practice varies so do the relevant markets. The line that is followed, however, remains the same: differential incorporation patterns are related to the degree to which reproduction is market-dependent or of the relatively autonomous historically guaranteed type. The analysis that follows is devoted to examining the influence of increasing incorporation (i.e., of increasing commoditization) on the structuring of the labor process in agriculture and thus on the enterprise development pattern achieved by the farmer through his labor.

The concept of incorporation and the notion of differential commoditization that it entails play a strategic role in the following analysis. Before proceeding, some theoretical and methodological observations should first be made, particularly as the reasoning that I have followed up to now deviates on several points from the usual discourse followed in the social and agricultural sciences.

The reasoning usually followed in relation to the notion of commoditization depends heavily on a wholly archaic idea of self-provisioning.[16] Self-provisioning is then the degree to which farming families can supply their own consumption needs from what is produced on the farm. The concept implies that an important part of production is thus not marketed, but directly consumed. Reducing the level of self-provisioning and increasing commoditization are seen as identical. The shortcomings of reasoning in such a way speak for themselves. It limits the notion of reproduction simply to that of labor power[17] and ignores the need for the farm to reproduce also itself and the necessary means, relations and conditions that go with this.

Tradition (which implies that in European agriculture, for example, commoditization is "complete," while in many parts of Africa and South America, commoditization scarcely exists) is replete with strongly deductive reasoning: if production is oriented to the production of commodities, then it follows that all the elements used in the production of these commodities must themselves be considered as commodities.[18] Some go even further to argue that insofar as agriculture takes place in "generalized commodity economies," each product that is the result of farm labor and each element that goes into producing that product must be considered a commodity, irrespective of the actual exchange processes that are involved.[19] In the final chapter of this book I go deeper into these ideas. I will confront such theoretical constructions with empirical findings taken from the analyses that follow, based on the notion of differential commoditization, in which the actual level of incorporation is shown to be one of the most important indicators.

Current incorporation theories need some refinement. Such theories (see Pearse, 1968, among others) too readily conceptualize the process of incorporation as a unilinear and an all-embracing process that is strongly deterministic and centralistic in nature (see Long, 1984, for a more detailed critique). In the following study I limit myself to those forms of market and institutional incorporation (the terms are from Pearse, 1968) which have a direct bearing on agriculture. In this I consider the incorporation process not as a unilinear pattern of development but as a process that waxes and wanes, as a process that embraces several dimensions in which the farmers concerned, and their wives, play an important, active and conscious role as decision makers.

In addition to these general theoretical points, some methodological details must be discussed. I shall do that with reference to the data summarized in Table 1.4. Behind the data lie certain hidden choices. Suppose a farmer has too few resources to accomplish a certain task satisfactorily (e.g., getting the fodder in). There are usually a number of alternatives. He could mobilize extra labor on the market, contract out the task concerned, or he could take out a loan and buy a machine which would enable him to carry out the task himself. It is clear that whatever the choice, a particular pattern of incorporation will always occur. The dependency which arises will relate to either the labor, the machine services or the capital market. Thus, a specific incorporation pattern can never be taken as a simple externally determined given: the farmer as a conscious decision maker plays an active and important role in the constitution of the specific incorporation pattern.

A good illustration of this can be taken from farming in Emilia Romagna. On the smaller farms, farmers often find themselves short of fodder. Buying fodder (i.e., fodder appearing directly as a commodity, a fixed cost in the shed), is seen by many farmers, however, as extremely undesirable, as we shall see later. Some farmers create a way around this problem by using their fields mainly for tomato cultivation. The gross production value and the gross added value achieved in this way are extremely high. With the returns, they subsequently purchase the necessary fodder. They explain the method by stating that feed thus obtained "does not enter the cow shed as a cost; it is already paid for through the work in the fields, in the tomato production." So in a situation which objectively would appear to lead to a high level of dependency on the market for fodder and concentrates (a relatively large herd but a restricted area for growing fodder), the structure of historically guaranteed relatively autonomous reproduction is again to some extent reinstated. In the stall at least, the farmer is able to neutralize to a degree the direct impact of commodity relations. The other side of the coin, of course, is a sharp rise in the level of incorporation into the labor market: tomato cultivation requires a legion of temporary workers. Again, the specific form of incorporation can and is effectively influenced by the farmers own, conscious participation. That they seize upon such a possibility is hardly surprising considering the many consequences that go with different forms of incorporation.

However, farmers are not the only actors who operate in the markets and actively try to accomplish certain forms of incorporation. Banks, industry, commerce and extension services try equally to effect particular forms of labor division (and thus particular forms of incorporation and institutionalization).

The Labor Process in Agriculture

The labor process in agriculture is always characterized by a specific and close coordination of technical, economic and organizational parameters. It would be fundamentally incorrect, however, to say that technical and economic parameters are determining. As mentioned before, a multitude of tasks can be identified which hold a degree of flexibility regarding their implementation. For example, take hay tedding, one of those apparently insignificant tasks whose purpose is to homogenize the quality of the hay and speed up its drying. If this step is neglected, then the grass that lies underneath will dry less well and will form mold or foment.

The timing of hay tedding is crucial. Also crucial is how quickly it occurs (we are assuming the technical parameters as given—i.e, the tractor and the implements). If it is done too quickly then there is a danger that some of the drying grass (the driest) will be ground to powder, which would mean a poorer harvest. The number of workings is therefore also of great importance. How much and when, and at what speed, are all important decisions to be made, while bearing in mind the particular field and the specific quantity and quality of the hay that lies drying there.

Like hay tedding, each farm task possesses some degree of flexibility and can be performed in a variety of ways. Even given the technological and economic parameters (price of hay and labor, in the above example), a series of decisions are needed to specify how the work should be finally done.

If one puts together all the stages identifiable in a particular labor process, an extremely complex matrix containing a complex whole of interlinking tasks emerges, each with its own degree of flexibility and particular procedure. An exciting complication is that procedures cannot be wholly or to any great degree specified in an a priori way. Decisions that are crucial for the end result can be made only during the labor process itself. Therein lies the craftsmanship of farm labor: the interaction between direct producer and labor object; i.e., the continual observation, interpretation and evaluation of one's own labor in order to be able to re-adapt it. This process is in marked contrast to industrial labor, where the labor process can usually be broken down, quantified, predicted and therefore planned and controlled. Interaction with living objects of labor excludes, to a large extent, such an industrialization of the agrarian labor process. The craftsmanlike nature of it and the need for a continual interaction, if not unity, of mental and manual labor, remain dominant characteristics. Thus, with hay tedding, speed can usually only be determined during the performance of the work. Unevenness of terrain or a change in the composition of the crop as it lies drying will determine whether to speed up or delay. The sight of an approaching cloud can again alter the decision. Even if a robot were available, a farmer would be unlikely to set it on a tractor to take over hay tedding from him. Too much can go wrong; the "damage risk" is too high.

Craftsmanlike organization, a continuous cycle of observation, interpretation, evaluation and reorganization remains indispensable. This can often be seen in practice in the labor process and in the division of labor that it contains. If a farmer places a high value on "good hay" in order to feed cows well, then he will almost certainly undertake the tedding himself. He will not readily give the task to inexperienced children or outside labor. Maybe he will ask his father to carry out the job, since the old man is equally likely to know all the subtleties and consequences of method, terrain and weather for this or that particular crop. A less experienced labor force could perform the task if it was the last mowing and the hay had only minimum nutritive value left, or if the farm's primary concern was beef production and the hay was to be used for fattening young bullocks. Experience, the unity of head and hand, and the craftsman's ability to use optimally the potential per labor object would then be less relevant. Technical and economic parameters are no less ambiguous, even if only for the fact that they are not equally relevant for all farms.

The number of tasks that must be coordinated, as well as the flexibility attached to each, points to the need for an organizing principle. A farmer must be able to define what is important and why and at the same time be able to translate his insights into practical procedures. In other words, goals and the capacity to translate these goals into a concrete structuring of the labor process are necessary.

The goals in question are not simply of a technical nature. Optimization of cost/benefit relations or of yields (as is nowadays increasingly assumed) is naturally never strived for as such. Like the concept of the optimum, costs and benefits are social concepts, and it is for social actors that these concepts have a specific and guiding meaning. Both the coordination of tasks and the specific definition of each separate task are always a social process. Even when it is a question of purely technical tasks, their coordination will be social.

Let me illustrate once more with hay tedding. If labor in the family is scarce and there are a number of other activities that urgently need attention (the care of calving cows, for example), then hay tedding must take place as quickly as possible. If there is a surfeit of labor on the market and the farmer is inclined to make use of it, then everything else being equal, the decision will turn out quite differently. Just as important is whether a poor hay harvest can be compensated for by buying feed produced elsewhere.

The Coordination of Domains

The social coordination of technical management assumes a number of domains. The interests and perspectives to be found in these domains may also be important in the sphere of production. In the model below, four domains (or as Vincent, 1977, calls them, "fields of activity") are presented. In addition to the domain of production and reproduction, they include the domain of family and community, and the domain of economic and institutional relations. The domain of family and community includes all the relevant social or non-commodity relations. The domain of economic and institutional relations includes the relations formed between the farm and markets or market agencies. External parameters, perspectives and interests of a political and economic nature, for example, can influence the labor process through this domain (but again, such an influence will depend on the specific relations which prevail in this domain) (see Figure 1.6).

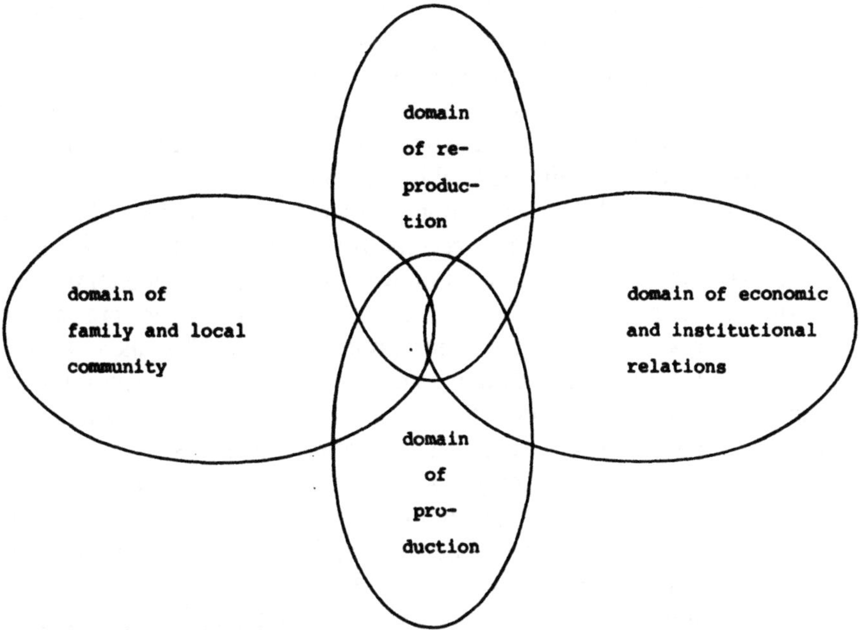

Figure 1.6 The domains of farming

The fact that the different domains of farm labor must be coordinated means that the significance of particular interests, relations and parameters holding in a particular domain will necessarily be carried over to other domains, thus becoming a precondition or guideline for the activities to be undertaken there. The coordination of domains and the inherent transference of meaning between them has always been one of the main themes of agrarian sociology. Chayanov defined the relations and processes located in the family (such as the demographic cycle and the associated labor-consumer ratios) on the one hand, and the domain of production, on the other, as an important theme for research. Chayanov argued that relations within the family determine the size, expansion and contraction of the cultivated area (noting that in Western Europe where land was scarce, the intensity of agriculture would vary). Composition of the family also affects labor input, capital formation and the level of production. The family functions, in this sense, as a social relation of production par excellence: relations within the family were determining factors for both the quantity produced and the way of producing. Characteristically enough, Chayanov noted that an analysis of farm production in terms of the dominant economic pattern (in which labor is given a clear price, namely a market price, so that profit also becomes a category that can be quantified) is not relevant, precisely because the economic relations on which the pattern is based do not penetrate (i.e., are not real) into the agrarian labor process. Though he did not make fully explicit this state of affairs, Chayanov argued eloquently that "literally before our eyes the world's agriculture, ours included, is being more and more drawn into the general circulation of the world economy, and the centers of capitalism are more and more subordinating it to their leadership" (1966:257). He pointed to the "trading links" and "credit conditions" "that convert the natural isolated family farm into one of a small commodity producer" (1966:258). Thus the domain of family and community gives way to economic and institutional relations which become the locus of the principles that direct the organization of the labor process. Or as Chayanov himself formulated it, "then the trading machine . . . begins to actively interfere in the organization of production. It lays down technical conditions, issues seed and fertilizers, determines rotation and turns its clients into technical executors of its designs and economic plan" (1966:262).

If interrelations between the family and the domain of production stand central in the analysis of Chayanov and many others who followed his footsteps, in the work of Bennett (1981) the main focus is upon relations between the domain of community and that of reproduction. The social definitions prevailing in the community (one farmer is defined as a "silver spoon," another as "someone doing a good job") reflect the activities undertaken by the farmers concerned in the specific domain of reproduction, but they also form the guidelines for the behavior of these farmers. To be seen as "a man doing a good job," implies that a farmer performs a number of work activities in a particular way. The social relations and social definitions implied in this, reproduce, as it were, the specific modes of behavior, procedures and patterns which are followed in the sphere of reproduction. Benvenuti's work (1975a and b, 1982a and b, 1985), on the other hand, focuses on the interrelations between the domain of economic and institutional relations and production. Benvenuti argues that the labor process is increasingly prescribed and sanctioned by what he calls TATE, an acronym for "technical administrative task environment." According to Benvenuti, this trend brings with it a number of demonstrable changes in the various other domains.

In the domain of production and reproduction (the domains in which agriculture in a narrow sense takes place, and in which a specific style of agriculture develops) many tasks can be identified. These must not only be coordinated with each other but also with the framework or context within which they take place, with the relevant social and economic parameters. The separate tasks, and in particular the integrated whole, derive meaning from the results they achieve in other domains. This is why, in a broader sense, farm labor can be defined as the coordination of domains in relation to each other. Activities in one domain are structured via specific goals which represent particular interests and perspectives in another domain; that is to say that one procedure for performing a task is chosen from the many and related in a specific way to other tasks. Such structuring often assumes a transfer of meaning. If in the family domain considerable value is attached to "keeping the name on the land" (Arensberg and Kimball, 1948), then such a goal has to be translated into concrete and meaningful action in the domains of production and reproduction. The same applies to the domain of economic and institutional relations. Such a goal is either made operational in such domains or it remains a meaningless dream. One should, for example, specify whether it is valid to sell land when the price on the market is high. In contrast to Ireland where Arensberg and Kimball did their work, it was usual for the Canadian farmers studied by Bennett to change farms once or twice every generation. Does such a goal mean that soil fertility must always be reproduced, and if so, to what degree and by what means? Finally, "keeping the name on the land" can even have all sorts of consequences for the way in which hay tedding is organized.

The way particular goals are translated into practice is investigated in this study in terms of patterns of farming logic, as a calculus which defines how work must be done in practice for all relevant tasks and under all conditions. A calculus enables advantages and disadvantages to be weighed against one another and enables alternatives to be thought through. A calculus, in other words, makes it possible to operationalize general goals into the daily reality and complexity of the labor process. Hofstee (1985) argued that a particular style of farming cannot be separated from the specific cultural heritage that farmers in a particular locality share which defines how farming ought to be done. A calculus, or farming logic, is here conceived of as the practical discourse that farmers follow in the organization of their labor. A certain way of working is then "logical" (as farmers themselves will not hesitate to tell you) because it appears as the concrete embodiment of what is strived for.

One of the things for which farmers strive is "progress." Although its concrete expression differs widely (from maybe an improvement of income or a reduction of labor time to a fine farm to hand over to the next generation), "progress" might well be seen as an adequate umbrella under which to summarize the diversity of immediate expressions. A strategic question of course is to what degree the potential for endogenous development is identifiable in agriculture and under which conditions farmers might make effective use of it.

A lot of ink is expended in the social sciences depicting agriculture as "intrinsically backward." Only intervention from outside can induce a certain dynamic.[20] The definitive argument here is the well-known law of diminishing returns. In the same way that Schultz argued that farmers are "efficient but poor," so, within the framework of this law, the personal goal-directed activity of farmers is also seen to be irrational. What in essence is ignored is that new "optima" can be created within the labor process itself. As already suggested, there is a potential for producing progress in an endogenous autonomous way within agricultural practice. In this connection a number of mechanisms can be indicated, of which I consider two.

To begin with, the objects of labor can be improved through the process of continual reproduction: the quality, i.e., the productive potential of the land and of animal and plant material can gradually be increased, precisely because this reproduction process is the object of goal-directed activities by the direct producer. Lacroix (1981) identified three phases in agricultural history related to the reproduction of objects of labor. First is agriculture which derives its objects of labor directly from the surrounding ecosystem. The *savoir faire paysan* in this phase commanded an extremely detailed knowledge (see also Conklin, 1955) of the variety in nature as well as a knowledge of how to optimally utilize the available natural elements. At the same time, agriculture took place within the narrow confines of given ecosystems. Only in a second phase were farmers increasingly able to succeed in shaping the objects of labor according to their own insights, thereby often transforming the given ecosystem (Bolhuis and van der Ploeg, 1985). It is a phase characterized by substantial and continuous increases of productivity. Finally, in a third stage—that of incorporated agriculture—the reproduction of labor objects is increasingly externalized, i.e., divorced from the actual labor process.

Second, lasting progress can be achieved by a close coordination of tasks in relation to each other. If the inhibiting factors can be established through careful observation and experimentation (which is an important task though one often not noticed by outsiders)[21] then a degree of progress can be achieved through appropriate interventions, i.e., by a reorganization of specific tasks.

This does not mean that endogenous growth potential in the agricultural labor process can be made absolute. It is clearly subject to a number of social influences, some with a facilitating, others with an inhibiting influence.[22] This is taken up as the central theme of the following chapters where I shall demonstrate that increasing incorporation in markets and market agencies lowers the potential to develop agriculture in an endogenous manner and de facto makes way for a growing dependence on exogenous technological development.

Notes

1. Technical efficiency is the ratio of input of production factors to output realized. According to Timmer (1970) an enterprise is technically efficient "if the firm produces on the technical production function that yields the greatest output for any given set of inputs. A failure in this regard means the firm is technically inefficient" (1970:99). And Yotopoulos states that "a firm is considered more technically efficient than another if, given the same quantities of measurable inputs, it consistently produces a larger output" (1974:270).

2. A greater or lesser number of production factors (labor and means) may be used per object of labor, as illustrated by Ishikawa (1981). Technical efficiency can vary considerably. Thus a whole gamut of combinations can emerge. Limiting ourselves to the most simple case, i.e., high or low input of production factors and high or low technical efficiency, already gives four possibilities. In this book the discussion is limited to the two most common combinations, i.e., a high input of production factors per object of labor combined with high technical efficiency (defined as the intensive style of farming) and, second, the case of low inputs and relatively low technical efficiency. Throughout the following analysis a high or low input of production factors, and high or low technical efficiency are identified and discussed only in relation to homogeneous agricultural regions. It is evident that these notions only make sense in a comparative analysis. Hence, intensive and extensive styles of farming are relative concepts, which only make sense when related to each other. As far as the "missing" combinations are concerned, one could argue that high and rising inputs combined with decreasing technical efficiency provide the analytical background of "agrarian involution" (Geertz, 1963). The opposite situation, i.e., low and decreasing inputs combined with high and rising technical efficiency, is the exceptional case to be found in some parts of modern northwest European farming (van der Ploeg, 1987), in parts of the United States (the "industrialized farm firms" identified by Gregor, 1982) and in some areas where the Green Revolution was highly successful. Chapter 3 discusses a similar intent to substitute farm labor as the driving force of continued intensification for applied science and technology.

3. This aspect is highlighted in Cantarelli and Salghetti (1983). They analyzed the same sample in terms of average trends.

4. For a detailed analysis the reader is referred to van der Ploeg and Bolhuis (1983).

5. It is striking that the agro-economic literature pays little attention to variation in yields (or GVP/ha levels) as a significant phenomenon. Although Heady and Jensen (1954) dedicated a complete chapter to "yields" in their classical handbook, they did not go further than interregional differences. Variability within a region is only related to the question of risk. Yields are not conceptualized as the structural result of a consciously planned and orga-

nized labor process. This is the more striking since in several empirical studies of the time it was concluded that "extensive cultivation gives higher average gross and net returns per hour of labour" (Galletti et al., 1956: 346 and table 138, p. 317). Schultz argued the same way: "In farming, yields are subject to much uncertainty. They cannot be controlled fully. Nor can they be foreseen accurately."

6. The problem of continuity of yield levels over time, and hence of the sustainability of farming, is amply discussed and illustrated in Bennett (1981).

7. This goes logically with the definition presented earlier: from an analytical point of view high production per object of labor is the result of both (a) a high level of production factors and non-factor inputs per labor object and (b) high technical efficiency. The same goes for the extensive style of farming where low inputs and low technical efficiency lead to relatively low production results per object of labor.

8. I am using here Poulantzas's definition of "social relations of production." This point is further elaborated in Chapter 5.

9. Detailed documentation on mechanisms for raising technical efficiency used by different groups of farmers operating under different conditions is to be found in Slicher van Bath (1960), especially as far as the seed/harvest ratio is concerned; in van Zanden (1985); in Bray (1986), who gives a beautiful description of endogenous progress in rice cultivating economies; in Hofstee, 1985; in Watson (1983); in Hosier (1951); in Fals Borda (1961); and in Barrigazzi (1980:43–84). A particularly interesting feature of this specific growth model is that increases in labor productivity followed from increases in physical productivity: hence intensification of production became the all-embracing goal in farming. See for this aspect Reynolds (1983); Ishikawa (1981); and Hayami et al. (1979). Within this line of reasoning, Yotopoulos (1977) stressed the need to structure agrarian development as "labour intensification." The main problem, however, is, as Yotopoulos indicates, that the "diseconomies of scale" might be smaller than the "financial economies of scale."

10. This particular concept is derived from the work of Riemann (1953:29). For a more general discussion, see Slicher van Bath (1960).

11. A well-documented description is to be found in the work of Spahr van der Hoek (1952, volumes I and II). Similar processes elsewhere are documented by Samaniego, in Long and Roberts (1978).

12. No matter whether the costs made in this cycle still have to be paid or the costs of the next cycle have to be pre-financed.

13. In other words: economic efficiency is to be optimized. As Messori (1984) showed in a detailed empirical study on technical and economic efficiency in dairy farming in Emilia Romagna in Italy, the two are often at odds with each other. Maximizing economic efficiency implies another strategy (and a different organization of the farm) than maximizing technical efficiency. A more general discussion of this interrelation is discussed in Yotopoulos (1974).

14. The argument is spelled out at length in van der Ploeg (1985).

15. For an ample discussion of this concept the reader is referred to the work of Benvenuti. See especially Benvenuti (1980, 1982a and b, 1985a and b); and Benvenuti and van der Ploeg (1985).

16. This is exceptionally clear in Galeski (1972). Even as a concept for the analysis of agricultural systems of the past the notion of "self-sufficiency" or "autarky" is, as demonstrated by Bloch (1939:7–16) completely inadequate.

17. In the Marxist tradition then, focus on the reproduction of labor power was narrowed to the question of whether labor was reproduced through the family (i.e., through non-wage labor) or through wage labor relations. Thus a certain "dualism" was repeatedly introduced in the analysis of agrarian formations (for an empirical critique: see the end of Chapter 2 of this book). With the gradual disappearance of wage labor in the northwest European countryside this particular analytical focus became for evident reasons powerless. Replacement of wage labor by contract work and a whole new market for machinery services largely escaped the attention of these scholars: their analysis of farming was so narrowly focused on labor only that they could not come to grips with these new empirical tendencies (see, for instance, Koning (1982); and Gorgoni (1977).

18. As argued by Chevalier (1982); Gibbon and Neocosmos (1985); and recently by Bernstein (1986).

19. Gibbon and Neocosmos in particular develop this line of reasoning.

20. See Bolhuis and van der Ploeg (1985, Chapter 2), for a detailed discussion. Such an assumption is omnipresent in today's "integrated rural development programs."

21. See Box (1982, 1984 and 1985) and also the discussion on peasant techniques for potato selection in Chapter 3 of this study.

22. An interesting discussion of the effects of social organization and the organization of time and space on the rhythm of endogenous growth is to be found in Hofstee (1985). Herrera many years ago pointed to the function of magic for regulating endogenous growth (reprinted 1984). The work of Boserup (1965) and Slicher van Bath (1960) is equally relevant.

2

Dairy Farming in Emilia Romagna, Italy

The region of Emilia Romagna, located along the river Po in northern Italy, is not only one of the most "red" but also one of the most prosperous areas of Italy. Its agriculture is commonly described as "agricoltura ricca." The term is no exaggeration. Good to very good incomes are earned on family farms there, and the dynamism of this agricultural sector is striking compared to other areas of Italy. This is not to deny that there are also problems, sometimes severe, but these occur at a level which is nonetheless one of prosperity. This situation contrasts sharply with the impoverishment and poor outlook of several other agricultural areas in the European Community, especially within Italy itself.

Agricultural diversity in Emilia Romagna is considerable. Conditions for farming change quite dramatically as one descends from the mountains in the west to the plains in the east. On the plains another change occurs as one travels from northwest to southeast: dairy farming gradually gives way to intensive fruit cultivation around Modena, and then to extensive arable farming around Ravenna and Ferrara. The distribution of cultivation systems over the region has also changed. The past ten or fifteen years have seen substantial expansion in extensive styles of farming. The interest in dairy cattle has been gradually superseded by the keeping of cattle for fattening, which in turn has given way to arable farming. Even within arable farming there has been a move towards more extensive cropping characterized by crops that require little labor.

In the heart of the region lie the provinces of Parma and Reggio nell'Emilia, where dairy farming is most concentrated. Its stability is due to the fact that the region produces a highly valued product, Parmesan cheese, known locally as "Parmigiano-Reggiano." Thanks to this product, produced in small cooperative cheese factories, milk prices in the region are substantially higher than those found elsewhere in the EC. The majority of farms in the region are family farms, followed by farms organized along capitalistic lines and by production cooperatives.

With the help of a group of agricultural scientists from different disciplines, it has been possible to construct four sets of data on the region, all relating to dairy farming. In the following text, reference will be made to these data as the Parma, the ERSA, the BOLKAP and the COOP sets of data. The four sets refer to the same time periods, some covering a four-year and others a ten-year period. This approach allowed both synchronic cross-sectional and diachronic historical analyses to be carried out, in order to verify the degree to which intensification, or scale enlargement and relative extensification, figured as the dominant pattern of development. Each set of data consists of economic and structural as well as sociological material. The latter was obtained by means of questionnaires and informal interviews and the former from account books and data obtained from university research institutes.

The first set collected, the Parma data, relates to twenty-four dairy farms, all of them located on the plain. The period researched covered the period 1970–1981, during which time six of the farmers closed their dairies in order to specialize in arable farming or market gardening. This is consistent with the general trend towards a more extensive type of farming. In the analysis that follows, repeated use is made of the remaining group of eighteen dairy farms. The detailed economic and structural data relating to these farms were collected in a systematic way by the University of Parma over a ten-year period and were checked each year with the farm head. The sociological, technical and complementary economic data were collected in 1980 and 1981. Farmers were interviewed five or six times, with visits timed to coincide with the yearly production cycle. Finally, the farm heads were all interviewed again to ensure the standardization of specific data.

The wealth of reliable sociological and economic data and its historical dimension make this set of Parma data comparatively unique, the more so as this was originally a relatively homogeneous group of farms. Towards the end of the research period, however, considerable variety was noted, suggesting differential patterns of farm development. These differential patterns also emerged from the ERSA data.[1] In that respect both sets form an ideal pars pro toto, and are therefore appropriate for research into the dynamics of different developmental patterns.[2]

The second set of data, the ERSA data, covers 134 dairy farms over a four-year period. The economic data were collected by the Ente Regionale di Svilluppo Agricola (ERSA) in Bologna and compiled by the Istituto Nazionale di Economia Agraria (INEA) in Rome. The sociological and technical data were obtained from questionnaires administered by the technicians of the regional farmers' organizations. The ERSA data refer to farms both in the mountains and on the plain and are used in the analysis mainly when there is a need to discuss the statistical significance of various interrelations. The Parma data, on the other hand, are primarily used for the setting out of qualitative arguments based on detailed ethnographic material. The questions asked in the ERSA questionnaire were based on a selection of questions developed from the interviews with farmers. Where relevant the distribution of answers from the ERSA data is given.

Both the Parma and ERSA data concern what are generally referred to as family farms, where family labor is the mainstay and wage labor is secondary or incidental. The remaining sets of data differ on precisely these points. The BOLKAP data are economic data relating to twenty-four farms organized along capitalistic lines, studied over a ten-year period by the University of Bologna. The data were supplemented by a questionnaire and interviews. The final set of data, the COOP data, concerns twenty-six production cooperatives. These last two sets, the BOLKAP and the COOP data, will be used primarily to highlight tendencies which emerge from the Parma and ERSA data and for exploring the implications. Such comparisons are particularly useful for looking at different relations of production on which patterns of farm development and organization of the labor process are based, for they allow one to see to what extent commoditization and institutionalization of family farms introduce a type of farming logic that was considered until recently to be typical only of capitalistic farming.

Heterogeneity is perhaps the best term for describing the research setting. The connections between farms, markets and market agencies vary enormously. There were also substantial differences in styles of agricultural practice. It is within such a setting that farm labor is examined as a concrete and heterogeneous phenomenon.

The structure of the following analysis is simple. Starting from the assumption that farmers are knowledgeable actors, I examine the extent to which they consciously pursue different patterns of farm development. The extensive interviews of the Parma data, which are primarily used for this purpose, lead to a description of two underlying patterns of farming logic, which I refer to as "calculi." These calculi are specific sets of interrelated "folk concepts" which play a guiding and legitimizing role in organizing the farm labor process and which, at the same time, reflect the relations between farms and the markets into which they are integrated. Two concepts central to the calculi, entrepreneurship and craftsmanship, are discussed in some depth later in the chapter.

The analysis then shifts from the labor process itself to its conditions and results. The differential degrees of incorporation and institutionalization of farm labor are analyzed as social relations of production, that is to say, as relations which effectively structure the labor process. The heterogeneity in dairy farming, and the different styles of farming practice and development patterns which such heterogeneity contains, are then discussed in relation to the differently structured processes of farm labor.

Goals in Farming: The I- and E-Options

In order to gain insight into the degree to which farmers consciously opt for intensification or for enlargement of scale and extensification, two hypothetical examples of farm management were simultaneously presented to the respondents of the Parma group:

Farm 1	*Farm 2*
20 cows	30 cows
50 ql/cow	40 ql/cow
1,000 ql	1,200 ql

The idea behind the exercise was as follows. The first farm was meant to represent an example of intensive farming. There are fewer cows but the production is higher per cow, 50 ql of milk as against 40 ql (1 ql=100 liters). In comparison, the second farm symbolizes a more extensive style of farming: yield per cow is lower, but the scale is larger as he manages more cows. Total production for the first farmer is 1,000 ql and for the second 1,200 ql. Parma farmers were told that all other factors for the two examples were the same. Respondents were then asked:

- who was the best farmer,
- which of the two would have the highest income,
- who would have the lowest costs,
- which farmer would have the best survival chances during a period of low prices,
- whether the examples were thought to be "real,"
- whether examples of both these types were to be found in their own environment,
- why some farmers did it one way and some the other, and
- which farmer was most like them.

The aim of the technique was twofold.[3] In the first place, we wished to explore whether a conscious choice for intensification or for scale enlargement and extensification existed, and if so, how such an option was distributed over the sample of respondents. Second we were especially interested in how farmers would argue for having chosen one or the other. What would be the rationale for each option? What means would be thought necessary for implementing a particular option, and in what terms would the respondents justify their choices?

One of the interesting experiences of the research was the matter of fact way in which each respondent approached this pairing of examples. It was as if the examples spoke a clear language to him, as if they referred to known and considered realities. This "self-evidence" was the more remarkable seen in relation to the reaction of the technicians who were sometimes present during the first interviews. For them, the examples represented a total absence of meaning ("you can't say anything about that, you need to know the costs, what type of cow sheds he has, has he enough fodder," etc.), but for the farmers the examples were a symbolic link to the known and meaningful in their experience, an invitation to expansive explanation.

The Parma sample, taken in 1980/81, consisted of eighteen dairy farmers, of whom eight promptly opted for the 20/50 example as illustrating the "best farmer." They described in full detail their own farms in terms of this example. I will presently return to this. Six respondents found the 30/40 farmer to be the best, the one with the highest earnings, etc., and they were also of the opinion that their own farms were like this example. Naturally, complications were not lacking. Two farmers pointed to one example as the best, but said that owing to specific circumstances, their own farms resembled more the other example. Two others hesitated, though through further discussion and analysis an assessment of these two cases was also reached. Thus two groups could be identified: first, those opting consciously for intensification as the main route to farm development, and second, those who opted for scale enlargement and relative extensification as the "right and best way." In the subsequent analysis these two groups are referred to as the I- and E-farmers respectively, and their particular view of farm development as an I- or an E-option.

The central question, of course, is what, if anything, the answers signify. Do they refer to the existence of divergent patterns, which lead farmers not only to different evaluations of the two examples but which have at the same time a guiding influence on action, on the organization and planning of labor, on production, and on farm development? I shall try to answer these questions step by step, first by validating the definitions of "best farmer" and by investigating the "course of action"

Table 2.1. Hierarchy and Relative Weighting of Various Elements in the Future Planning of Farmers Who Opt for Intensification or for Scale Enlargement (Parma, n=18)

Farmers Opting for Scale Enlargement and Relative Extensification (E-option)	Farmers Opting for Intensification (I-option)
1. Increase of farm acreage....3.56	1. Raise production per cow....3.90
2. Cost reduction..............3.31	2. Increase of farm acreage....2.90
3. Reduction of labor input....2.81	3. Cost reduction..............2.30
4. Increase of stock...........2.56	4. Increase of stock...........2.30
5. Raise production per cow....2.13	5. Reduction of labor input....1.80

which accompanies this, and subsequently, by making some links with the historical pattern of farm development and the present structure of the respondents' farms.

In order to obtain an impression of the course of action, a simple research technique was developed with which it was possible to construct a hierarchy (a relative weighting) of the various elements in the future plans farmers had for their farms. For this purpose the following list of elements was presented:

- to raise production per cow
- to reduce costs
- to increase stock
- to lower labor input
- to increase farm hectarage

Respondents were asked which of these elements they would consider valid for their own farm and which they could not implement. Subsequently they were asked to rank those elements considered valid. In analyzing the data, the element thought most important was given the highest weighting (5), and that ranked as least important was given the lowest (1). In the Table 2.1 the average course of action is represented as a function of the option chosen.

Those who opted for intensification gave the highest weighting to "raising production per cow," and those who opted for extensification ranked it last, giving the highest weighting to "increased hectarage." The picture summarized in Table 2.1 can be analyzed further. Let us begin with hectarage expansion. Those who opted for enlargement of scale, gave hectarage the highest priority, and none in this subgroup considered it invalid for their farms. Farmers who opted for intensification put hectarage expansion as second in priority and some 30%

Table 2.2. Earlier, Present and Planned Acreage, Differentiated According to I- and E-Options

	E-Option	I-Option
Farm acreage in 1970	25 ha	25 ha
Farm acreage in 1980	31 ha	27 ha
Necessary acreage as defined by farmers	54 ha	39 ha
Historical acreage expansion	6 ha	2 ha
Planned acreage expansion	23 ha	12 ha

considered it invalid. A difference of degree? Or does it mean more than that?

Each of the respondents was asked how much extra hectarage was considered necessary to achieve a reasonable farm size. It should be noted that in 1970 most of the farms in the Parma sample were roughly the same size, about 25 hectares. By 1980, those who opted for scale enlargement had increased the size of their farms to about 31 hectares. Growth was less, from 25 to 27 hectares, for those who preferred intensification. Asked about their desires concerning further increases, those who chose scale enlargement said they would like, on average, another 23 hectares, while the I-farmers thought only 12 hectares more were needed to provide what they defined as a "reasonable" farm size. In other words, those who had expanded most in the past ten years, and who now had the largest farms, were those who thought it necessary to expand most in the future to reach a reasonable farm size.[4] Table 2.2 summarizes these important differences.

One may deduce from the table that those who choose the I-option have a different perception of hectarage expansion than those who choose the E-option. One might say that the whole is more than the sum of the parts, that it is the situation thought of as ideal which gives meaning to the component parts of the course of action. A closer analysis of the interviews corroborates the view that this is true not only in a quantitative, but also in a more qualitative sense. In the core of the E-option, hectarage expansion takes on an independent meaning and is considered decisive for long-term success. In the core of the I-option, where improvement of rotation schemes or the desire to become self-sufficient in fodder might one day entail a need to increase hectarage, hectarage expansion is a relative or subordinated notion. It is considered important insofar as it is functional for other elements. Hectarage expansion is not in itself a norm. This is also true for an element such as reducing costs. Its meaning, the way it is achieved and the degree to which it is pursued all depend on the context in which the farmer places them. "Yes, of course," said one typical I-farmer, "reducing costs is often unavoidable. Then you have to do more work in order to spend less." He explained this as follows.

> "Look, if concentrates become expensive, then I grow more maize and barley myself. I then grind it and mix it with salt and such like. In this way I replace the expensive purchases from outside. Of course that involves costs in the sense that it involves much more work. . . . However, it solves a problem. The total amount of concentrates for the cows remains the same, but I spend less and thus reduce costs. . . . What else can one do in such a situation? If I didn't do it I would have to give less feed and my production would fall and also my income. . . . What help is that to anyone?"

Those who opted for scale enlargement and relative extensification see cost reduction in quite a different light:

> "To begin with you have to cost your own labor like everything else. The thing is always to keep the labor time to a minimum, so that you can do more.
>
> "Then you give less feed. Actually what does it finally matter, this fantastic production per cow? It's a question of one's pocket, what interests me is what I earn, because although it may sound strange, it is nevertheless the case, that by limiting costs, giving less feed, fertilizing less and so on, you earn money by it."

It is as if behind the I- and E-options there are two different ways of thinking—thinking which not only gives each apparently similar element a different place in farm development but also a different meaning, a different content.

There is a striking concomitance between actual farm situations and their historical development and the options and associated courses of action. This can be summarized as follows: farmers who opt for intensification are those who have intensified their farms most in the past ten years and who give the highest priority to further intensification in their plans for the future. On the other hand, farmers who opt for extensification and enlargement of scale have farms whose historical development, proposed development and present characteristics all fit with this option. These options are thematically represented in Figure 2.1.

It is not possible on the data given so far to go beyond noting the concomitance. The construction of causal relationships requires more than can be offered by the observation of a certain correspondence. Are the farmers simply "speaking the language of their farm," or is

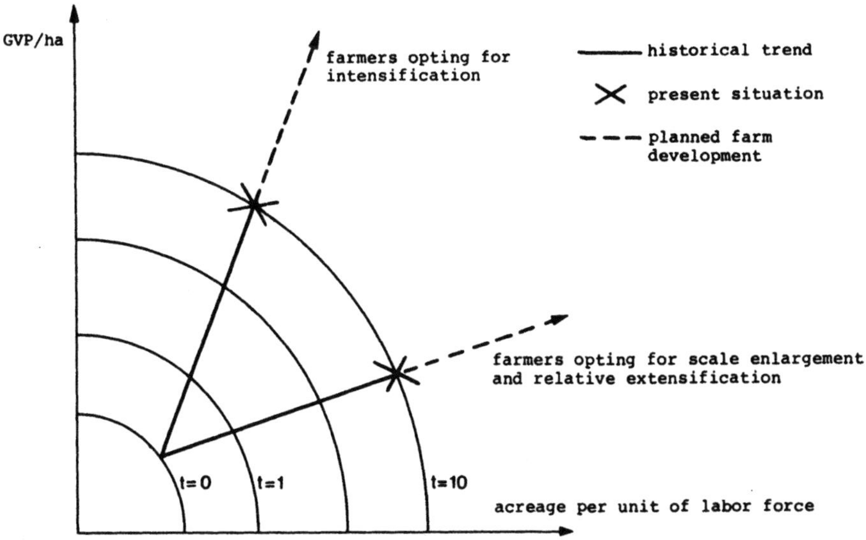

Figure 2.1 Options and farm development

the farm structured according to the options of the farmers concerned? This essential question is explored further as more elements for answering it are developed.

Some of the research techniques described above were also used in the ERSA questionnaire. The pattern of answers given by the two groups (the plain and the mountain farmers, n=75 and 59, respectively) was subsequently explored using factor analysis.

Table 2.3 summarizes the results of the first such analysis. The first and, in particular, the second factor, appear at first sight to confirm the correspondence between option and course of action. OPTIONpl is characterized by a high loading on the degree to which intensive farming is normative, labor reduction is rejected, and high priority is given to raising milk yield per cow. OPTIONp2 can be interpreted as the dimension representing scale enlargement and relative extensification. Priority is given to hectarage expansion and to an increase in stock, and none or very little attention is given to raising milk yields.

In comparison with the Parma data, which showed a striking association between the more normative and the more technical aspects (i.e., between the definition of "best farmer" and "course of action"), it might be surprising that more than one single dimension emerges from the factor-analysis on the ERSA material. There is not one simple factor representing the I-option and its course of action on the positive side and the E-option on the negative side. In such a case the I- and E-options would be mirror reflections of each other. This is definitely not the case as Table 2.3 clearly shows. Not one, but three, factors

Table 2.3. Factor Analysis Applied to the Definition of "Best Farmer" and to Elements of "Course of Action" (ERSA/Plain/n=75)

	Factor 1: Option for Intensification (OPTIONp1)	Factor 2: Option for Scale Enlargement and Relative Extensification (OPTIONp2)	Factor 3: Option for Cost Reduction (OPTIONp3)
The farmer having 20 cows and producing 50 ql per cow is the "best" and "earns the most"	.77	-.10	-.11
High priority for reduction of labor input	-.73	-.10	-.01
High priority for raising milk yield/cow	.46	-.69	.09
High priority for increase of stock	.20	.65	.46
High priority for increase of acreage	.15	.63	-.36
High priority for cost reduction	-.09	-.09	.83
Eigenvalues	1.44	1.33	1.02
Variance explained	24%	22%	17% (cum 63)

emerge. As far as the first and second are concerned, one could argue that the first factor summarizes the more normative aspects of farm development while the second expresses itself in more technical terms, or in more neutral terms, concerning the course of action. This "separation" between norm and practice which emerges from the ERSA material, although it contrasts with the Parma material, is not a complete surprise. There are some arguments which indeed validate this apparent "separation." The farmer working intensively and "the well-cared-for field" are still very much the norm in the Emilian countryside (and certainly in Parma and Reggio). In the cafes, the farmers will outbid each other to claim they have the best producing cow or the best field of luzerne. Farms are also judged by the same criteria. Riding with farmers around their home territory is a learning experience. The complexities are then also revealed. As one Parma farmer explained:

> "I take a darned good look around me. My neighbor, for example, he wants everything too beautifully done. He mows his field ten times over and he even mows the sides of the ditches, by hand. . . . Of course he gets beautiful fodder from his land. However he's wasting his time. Then there are those who muddle along. Me, I would rather work in a rough and ready way. Admittedly I do mow the ditches along the roadside where it can be seen, after all you don't want to get yourself a bad name in the neighborhood. . . . But really it makes no sense. I would be better off leasing one and a half hectares of land than stand wasting my time on one square meter."

In short, the norm is one thing and practice on a particular farm another. At most, "along the roadside," and evidently within the pub, lip service is paid to the norm, but "out of sight," there can be a sharp discrepancy between norm and practice.

Perhaps the strength of this general norm (shown up in the factor analysis as OPTIONpl) was overestimated during the Parma interviews. Presumably the longer interviews, taken over several sessions, created the trust which allowed the farmers to admit to an outsider (in this case a Dutch researcher) what was more difficult to admit to technicians of their own farmer organizations who carried out much shorter interviews—i.e., that people consciously depart from the prevailing norm that assumes that someone who farms intensively is a better farmer. This is probably how more than one factor emerged from the ERSA questionnaire. At the same time one might assume that the real options open to farmers are so complex, and sometimes even so contradictory, that they can never be represented by just one factor alone.

OPTIONp3, "the option to reduce costs," represents in some respects similar complications because again it emerges as an independent dimension (even with oblique rotation). Clearly increase in cattle stock scores relatively high on this factor, but an interpretation of the factor in terms of I- and E-options is not possible at this level. We have already seen the double-edged meaning of the term "cost reduction"; in a global strategy of intensification cost reduction will mean something different and will also be carried out differently from when it is functioning in a strategy of scale enlargement and relative extensification.

In short, although each of the three factors refers to clearly identifiable dimensions, the meaning of individual factor scores must always be interpreted within the context of the pattern that they form with scores on the remaining factors. In an analytical sense this means that the interaction and addition of OPTION factors will mean more than a factor in isolation. I will return to this point when OPTION factors are used in the explanation of other factors. Table 2.4 summarizes the results of the factor analysis applied to the mountain sample.

One sees that the structure of OPTIONm factors ("m" for mountain) differs somewhat from OPTIONp factors ("p" for plain). Thus OPTIONm1, a factor which takes up 30% of the variance, combines an explicit opting for scale enlargement and relative extensification as the norm, with a priority for reducing labor. Thus there is no reason to assume an ambiguous meaning here to the concept "cost reduction." The second factor, OPTIONm2, combines (just as OPTIONp2) "increase in cattle stock" with no priority for "raising milk yields"; in other words, scale enlargement and relative extensification. However, whereas on the plain this option was combined with a high priority for hectarage expansion, this was not the case in the mountains. This seems self-evident in view of the extremely low cattle density in the mountains where, in contrast to the plain, hectarage is seldom a problem.

The Relation Between Incorporation/Institutionalization and Goals

Having arrived at this point let us ask an important question. Is there any relationship between incorporation and institutionalization and how certain choices are made? Do commoditization and institutionalization induce such a change in the "contextual whole" or "environment" of the farm that they coincide with a shift in the goals which direct and legitimize farm labor? In other words, do the various aims refer to a completely atomized situation in which producers opt

Table 2.4. Factor Analysis Applied to the Definition of "Best Farmer" and to Elements of "Course of Action" (ERSA/Mountains/n=59)

	Factor 1: Explicit Option for Extensification (OPTIONm1)	Factor 2: Option for Increase of Stock (OPTIONm2)	Factor 3: Refusal to Expand Acreage OPTIONm3)
The farmer having 20 cows and producing 50 ql/cow is the "best" and "earns the most"	-.72	-.02	.45
High priority for reduction of labor input	.83	-.09	.22
High priority for cost reduction	.64	.09	.20
High priority for increase of stock	-.15	.82	.15
High priority for increase of milk yield/cow	-.19	-.71	.20
High priority for acreage expansion	-.21	-.01	-.91
Eigenvalues	1.77	1.35	.97
Variance explained	30%	22%	16% (68%)

for certain goals on purely individual grounds, or do the exposed goals reflect structurally anchored patterns which direct and legitimize thinking and behavior?

Incorporation was defined as the degree to which farming becomes dependent on markets for supplies. Eight such markets can be identified, at least for dairy farming in Emilia Romagna. These markets, together with the average degree and standard deviation of incorporation, were presented in the first chapter (Table 1.4).

Institutionalization of agricultural practice was defined as the degree to which tasks carried out by farmers are externally prescribed and sanctioned, i.e., the degree to which they are influenced by the technical-administrative task environment (TATE). This general concept may be subdivided threefold, i.e., into the influence which the technical-administrative task environment (TATE) has on: (1) the acquisition and processing of information, (2) the making of investment decisions, and (3) on the development of craftsmanship. Considerable variance also appears on the scales for measuring this threefold influence of TATE. One possibility for reducing the large number of variables with which incorporation and institutionalization are measured, without too much loss of information, is to use a variable cluster analysis. This method is a certain combination of "oblique component analysis" and "multiple factor analysis" with which a numeric series of variables can be ordered in clusters (Harman, 1976). Applied to the plain's data, three clusters arise. The first cluster is composed of variables Inc1, Inc3, Inc4, and Inc7 (INC1347 standing for incorporation into the markets for labor, short-term loans, medium-term loans, and fodder and concentrates). A second cluster combines Inc2, Inc5, and Inc8 (INC258, markets for machine services, long term-loans, and cows). The third cluster combined the different TATE or institutional variables.

The series of cross-tables in Figure 2.2 shows the global links between incorporation and institutionalization and goals. The first series of three tables give the average of OPTIONpl. This factor gives the degree to which the farmers on the plain opted for intensification. The overall average for factor scores is 0, with a standard deviation of 1.00. The dimensions in the cross-tables are formed from the three clusters of variables given above: INC1347 and INC258 and TATE.

With a simultaneous increase in both incorporation dimensions (INC258 and INC1347), the degree to which farmers opt for intensification falls from 0.05 to −0.38. With increasing incorporation, the option for intensification loses its normative character. Institutionalization of agricultural practice (TATE) exercises a similar influence, at least as it occurs in combination with increasing incorporation.

Degree to Which Intensification Is Opted for (average score on OPTIONp1)

INC258 \ INC 1734	$\leqslant 0$	> 0
> 0	0.07	−0.38
$\leqslant 0$	0.05	0.11

TATE \ INC258	$\leqslant 0$	> 0
> 0	0.04	−0.17
$\leqslant 0$	0.10	0.03

TATE \ INC1734	$\leqslant 0$	> 0
> 0	0.00	−0.22
$\leqslant 0$	0.13	−0.04

Degree to Which Scale Enlargement and Relative Extensification Is Opted for (average score on OPTIONp2)

INC258 \ INC1734	$\leqslant 0$	> 0
> 0	−0.28	0.14
$\leqslant 0$	−0.04	+0.44

TATE \ INC258	$\leqslant 0$	> 0
> 0	0.28	−0.29
$\leqslant 0$	−0.03	0.08

TATE \ INC1734	$\leqslant 0$	> 0
> 0	−0.17	0.33
$\leqslant 0$	−0.12	0.26

Figure 2.2 Relationship between options, incorporation, and institutionalization (ERSA/plain/n=75)

For scale enlargement and relative extensification (OPTIONp2) a similar influence of incorporation and institutionalization is present, but in the opposite direction. Thus, opting for scale enlargement and relative extensification rises with the simultaneous increase of TATE and INC1347 from −0.12 to 0.33.

Later I shall discuss the statistical significance of such relationships which also apply to the mountains. It will become clear that they are indeed significant. Increasing incorporation and institutionalization lead to a shift in the goals which direct and legitimize farm labor from intensification to an increasing preference for relative extensification and scale enlargement and the course of action that goes along with this option.

Patterns of Farming Logic: The I- and E-Calculus

When asked why they had chosen the farmers with "20 cows producing 50 ql of milk each" to be the "best farmer," farmers always referred to production level per cow and not production level per farm. For those farmers, milk production per cow and *not* total milk production per farm was the central argument in their discourse. Production per cow represents a norm, and the farmers "read" the two examples from the standpoint of this norm.

> "Naturally the farmer who milks 50 ql from his cows is the best, he has the highest production (*produzione*), and you should take good note of that. Like us, if we visit another farm then the first thing we want to know is what the produzione is. It says a great deal about a farmer and his work. . . . It's also an important yardstick for one's own business. We are proud of our produzione. It is much higher than it used to be and we have worked very hard for that. It gives a real feeling of achievement, of pride in one's work; high produzione means that things are going well. . . . Of course prices are sometimes bad but even then you are worse off if your produzione is low. And with good prices and low produzione you feel awful. For me it's simple. The first rule for a good farmer is produzione."

Produzione plays a cardinal role in this quotation. Taken literally it means the same as the English word "production," i.e., in principle it is a polyvalent concept; it can be related to the farm as a totality, to the labor force, and to the amount of land. One can speak of production per farm, production per man, etc. However, those who opted for the intensive farmer as being best handled the term in a very specific way. For them the term refers unambiguously to production per object of labor, in this case, production per cow. This is so self-evident to them that the qualifying noun is simply left out. In the second place, the quotation illustrates to what extent produzione is to be understood as a normative concept. It is norm and yardstick at the same time, "a high produzione is beautiful (*e bella*)." From this high produzione comes "pride in your work." Low produzione is "bad," it refers to "a bad farmer," a farmer who "cannot or will not work."

Given the normative meaning of produzione it is logical that it is this concept that is picked up by the typical I-farmer (and not, for example, total output) when commenting on the second example, that of 30 cows and 40 qls:

> "Forty ql . . . that's not possible, that is already a serious problem with 20 cows, but with 30 its an absolute disaster. That man has first to learn to look after his cows, before he takes more animals. I know, there are enough farmers of the kind here in this neighborhood: but in my opinion there is something missing in the logic there."

Perhaps this all seems self-evident. However, this self-evidence arises within a specific schema, within a certain logic or calculus (seen as a specific ratio linking specific goals with specific means which therefore structures the labor process). In the calculus of farmers who opt for intensification, produzione is indeed a self-evident element for judging farm situations. In their calculus it is equally self-evident *not* to consider scale enlargement as a compensation for the lower production per cow. For them, *that* is illogical ("something missing in the logic").

The degree to which this self-evidence is tied to a calculus (explored more fully later) becomes more obvious when we let the farmers who chose the 30/40 example speak for themselves. A father said, "naturally the farmer with 30 cows is better." The son added,

> "Yes, we are like the farmer in that example, more cows and less milk per beast. And we still earn, and well too. If we farmed like the man in the 20/50 example, well reckon it for yourself, we wouldn't earn a cent any more, for it's with greater numbers that the attractive margin comes. You have to look for your earnings in numbers. . . . Look, for it's like this: you need fifteen cows to keep a family, to live. So every beast beyond that is pure profit. We have gone from 30 to 60 beasts but our profit has more than doubled, come and look at our books. . . .
>
> "That second farmer there (30/40 example) has more business in his cow sheds. He can negotiate and commercialize much more. By the end of the year he will have brought more in than goes out, that's as sure as houses, his outgoings will be proportionately much lower than with the other more intensive farmer."

These are some of the more economically tinged arguments used by E-farmers for debating and "explaining" the examples. The arguments for produzione that we came across from farmers who opted for the 20/50 farmer are notably absent here. Next to their more economically colored arguments, these E-farmers also use an array of more technical arguments—arguments which interestingly enough stand at odds with the notion of produzione as a norm because they cast doubt on the assumptions underlying the concept.

> "And then I would like to see how long these cows will go along with that. In my opinion he's milking them to exhaustion, to illness or even to death. It might seem fine, 50 ql, but if after a couple of years they are milked dry, what will he do then? I would also like to know how often such a cow calves."
>
> "I would like to see the veterinary bills of that first farmer. And you know, it can of course be alright, I have also tried it, but it takes a hell of a lot of work. And then to get such a high milk yield you need to feed them on a lot of concentrates. I reckon such cows would soon get mastitis."
>
> "That isn't a good farmer, he's a tail-washer."

One can see that for this group of respondents, who consider an extensive style of farming to be the "best," produzione is no longer a norm in itself, on the contrary a high produzione is suspect and ridiculed. Suspect, because they see it as associated with a number of technical problems: mastitis, infertility, and too high a degree of replacement. Ridiculed, because it is implicitly assumed that a "farmer" wouldn't behave like that; only a caricature of a farmer, a "tail-washer" would try to get such a milk yield from a cow.

In summary, presenting the two examples brought out two entirely opposite patterns of reaction. What is normative for one is suspect for the other. For some, produzione is a yardstick separating the "good" from "bad" and a beacon for their own farm development, while for others it has no such function.

In these opposite reactions two specific patterns of farming logic lie hidden. These will now be more closely examined.

The I-Calculus

Let us first return to those farmers who opted for intensification, that is, those who defined the 20/50 example as the "best farmer." Why do they think a high produzione is so important? One said,

"If your production is higher, your income is also higher. That is logical. If it wasn't the case why would we raise produzione?"

Figure 2.3

In general, of course, this statement is not sufficient, because income can and does depend on many more factors than produzione alone. However, the relationship as sketched works better for some farms than for others. That is, some enterprises are structured in such a way that income is basically dependent on production per object of labor, while in others this factor matters much less. If we apply this idea to the Parma sample it appears that income in the group of intensive farms is indeed dependent on the Gross Value of Production per hectare. The correlation coefficient between GVP per hectare and income per unit of labor force was $r=0.41$. The same positive correlation was found between milk yield per cow and income, $r=0.32$.

If we look, on the other hand, at the extensive group of farms, then this relationship, de facto, breaks down ($r=-0.21$ and $r=0.07$, respectively). In this group, income depends heavily on scale: $r=0.90$. Agricultural practice, indeed, can be structured in several ways. One of the consequences of this is that within identifiable farm realities different relations emerge. What applies in one group (for example, the relation between produzione and income) is absent or even absurd in another. This corresponds with the differences in calculi; what is logical in one is to some extent inconceivable in the other.

Again this appears very clearly in the question of farm income. When the two examples were presented, the question "Which of the two would have the highest income and why?" was always asked. The pattern of answers is telling.

"That second farmer might as well close his dairy. He must be an old farmer who has no interest anymore. From a production of 40 ql you would have nothing over, there is no income in that."

"Only with a secondary source of income can a farmer drop to 40 ql."

"What he lives on that second farmer I don't know. The man lives from day to day, does nothing more than the minimum necessary and that's it, goodbye."

In short, those who see a solid link between produzione and income cannot imagine that one can earn anything from a lower production.

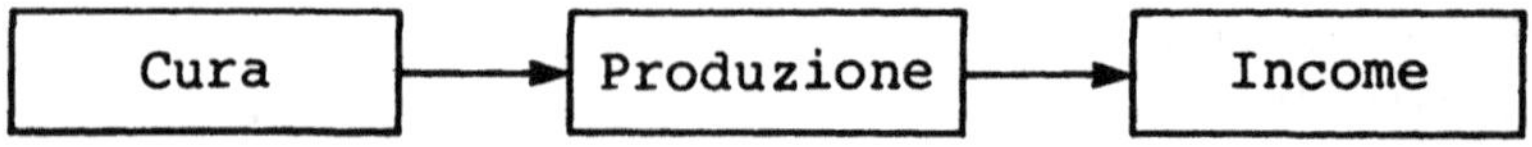

Figure 2.4

A secondary source of income is the suggestion most frequently offered to account for this. However, those who opt for the other example, and who, as we shall see later, primarily relate income to cost/benefit ratios and scale, are unable, in turn, to understand how a living can be made from 20 cows and a milk yield of 50 ql. They also suggested secondary income, but to explain precisely the opposite situation!

> "Those farmers who chase after such high production can't be thinking of the economics of it, it's a sort of luxury. They can only carry on in business because they have a secondary income."

How is a high produzione and a high income then attainable for those opting for intensification? The key to it is *cura*. Taken literally, "cura" means "care." It refers to a specific relation between the objects of labor (cows, crops, etc.) and the direct producer. Cura stands for craftsmanship. Cura refers to working in such a way that produzione is optimal. With this new element arises the chain shown in Figure 2.4.

For the farmers concerned the relation of cura to produzione is clear and direct. As one older farmer commented, "If your production is low then you haven't worked with sufficient care." What does this care consist of? The same farmer commented,

> "Everything must be kept well under control. As a farmer you must have everything in your head—the age and history of your cows, their weight, the fat content of their milk, the feed and fodder rations, fertility, everything. Books won't solve anything for you. There are dozens of factors which have to do with each other, which affect each other . . . and as a farmer you must take heed of all that and be properly prepared."

And a younger colleague added,

> "Good care is fundamental for your yields: only by immersing yourself thoroughly and by feeling and seeing everything can you recognize and prevent mistakes. Only by working continually yourself can you tell what grass is best, what must be improved, how cows react to different kinds of feed, how you can supplement fodder, in what way a certain animal needs special treatment."

Asking about the meaning of cura is a starting point for very detailed descriptions of particular fields, crops and animals. Production technique, its development, the decisions contained in it, and the experiences on which such decisions are based, are all themes which farmers who opt for intensification will happily expand upon. People who are proud of their work will gladly talk about it. I will later go further into a discussion of craftsmanship, but what is essential here is that the concept of cura refers to a specific structuring of one's own labor. The relations between the direct producer and his objects of labor and the permanent interaction of intellectual and manual labor are crucial to this specific labor process. This interaction is geared towards obtaining high produzione: a high production per labor object. "Good cura is fundamental for your yields."

What now are the conditions for cura? Why do some farmers work with care and others not? Let us look first at the conditions mentioned by farmers who opt for intensification and who define their own labor as cura. I will quote a young farmer who six years ago finished his training at the agricultural university:

> "With many farmers, let's say you have a long way to find any professionalism, any formal professional knowledge. They are mostly older farmers who are no longer so interested in picking up information. That brings a problem with it; if they need specific technical advice they have to go to the *Consorzi*, and you can guess the advice they get there. And unfortunately they don't experiment enough themselves these old farmers. . . . But then, the cura is there with some of these old farmers, and as the care is there, they succeed in obtaining the same yields as myself even without all the technical help that I have.
>
> "Their hectarage is mostly somewhat smaller, but they are always out among the cows and their plants; they manage, they try. . . . The sad thing is that you see people with good technical training from college, or like me with the faculty of agriculture behind them, who nevertheless work badly, who are very 'extensive.' They have the technical know-how but the cura is missing. . . . Generally speaking, you need two things as an intensive farmer. You need scientific information, but interpreted in a critical manner. You must certainly not swallow hook, line and sinker what the experts tell you, and if you don't have technical information at your disposal then you should at least have some experience and the will to gain experience and put it to use. And secondly your heart must be in this sort of work, without a certain *passione* you won't make it."

Passione and knowledge—where knowledge is preferably the combination of personal experience and critically approached scientific information—are two important conditions for cura. An equally important condition put forward in pretty well all the interviews is *impegno*. It is again a concept that, as well as having a normative content, also comprises a rather exact definition of an essential economic relation. "Impegno" means that you must put in not only hard work, but soul ("*chi cura, tiene che impegnarse*"). And why do farmers do the extra? Why sometimes work late into the night? Why is it necessary to spend day and night in the cowshed? One farmer replied rhetorically, "That's very simple. The incentive is to raise production."

Clearly impegno is a norm. From that comes the "logic" of defining those who get low production from their dairy as "men who would rather sit drinking in the bar, who will not or cannot work." But impegno is more: the term is also a declaration that the relation between labor input and the objects of labor is to be a stable relation. Labor input per cow is to be high and above all stable. Feeding cattle well is also seen as a stable phenomenon. Translated into economic terms, what is understood by impegno is that fixed as well as variable costs per object of labor are not to be seen as changeable entities, let alone as costs that ought to be minimized. Naturally, over the longer term, impegno is seen as being flexible. Technical progress—in terms of new sheds, mechanization of milking, transporting manure, etc.—always creates new opportunities for altering the relationship between labor force and objects of labor, without endangering cura. Certain kinds of technical progress (automatic manure removers and spreaders, for example) even make it possible to raise cura further because less time needs to be spent on secondary activities. A number of respondents indicated this:

> "Look, if you have a modern cow shed, then reasoning from your examples, you are best off keeping 30 cows and looking after them equally well so that your production remains up to the mark or is even improved."

Given the technological level, then the farmer who opts for intensification will specify the size of his herd in terms of the desired cura and in terms of the necessary impegno. That is a fundamental reason for rejecting the 30/40 example:

> "That farmer has more cows than he can cope with."

Another farmer, who has 120 cows himself, added

> "That goes beyond all logic, He is laying up trouble for himself by overstepping what's possible."

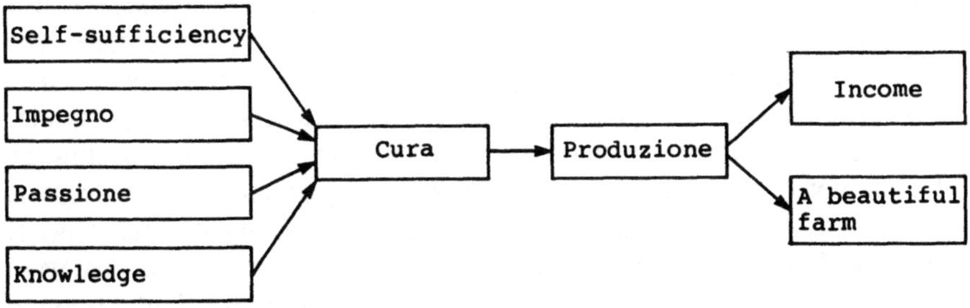

Figure 2.5 Structure of the I-calculus

This last farmer does not care single-handedly for all 120 cows. He works with two brothers and his old father. Wives are equally important in the overall labor process. When such a farmer touches upon accounting then the strategic implication of the concept of impegno becomes even clearer.

> "The first farmer can get through the work on his own, that's possible with 20 cows, but the second with 30, he would need help with the milking. The same goes with the fodder, because going from 20 to 30 cows raises the cost of feed by a half, but the yield improvement is only 20% (from 1,000 to 1,200 ql). It isn't good from any side what the second farmer does. His milking assistant costs so much per cow, the feed will bring almost similar costs with it, and the fixed costs will be about the same. He has also a bigger stall to depreciate and so on. So, whether you have 20 cows or 30 cows the costs per cow are about the same so you are better off maximizing the yields per cow. There lies the logic of the first farmer and the mistake of the second."

Thus objects of labor have a central place in their reasoning, and furthermore, labor input, input of feed, and the investment per cow are not considered manipulable as variable quantities. They are perceived as fixed relations:

> "I cannot give less attention or work to the herd just because the cheese price is low, can I? If the price goes up later then I'm left with a ruined herd."

Figure 2.5 presents a model of these relations. Apart from the terms already presented, two additional concepts are introduced. These are "self-sufficiency," perceived by the farmers as a strategic prerequisite for being able to "work with care," and "*la bell'azienda,*" the beautiful farm," which is seen as a long-term result of high produzione, just as income (*guadagno*) represents the short-term outcome. On both self-sufficiency and the beautiful farm, I will comment later. I consider this structure to be the backbone of the calculus as used by farmers who opt for intensification. It is the specific ratio binding means with ends. It is a well-integrated system of meanings through which a particular reality can be interpreted and organized. Within this system occurs the already noted "self-evidence" of the reasoning given. Each concept is locked into a network that shows where the meaning of that concept lies as well as the way in which the desired results are to be achieved.

This calculus also makes clear why certain respondents pointed without hesitation to the 20/50 example as representing the "best farmer" who "earned most," as if the reasons for this choice were self-evident:

> "Naturally the cow shed with the highest produzione is the best; if it wasn't the case, what are we talking about?"

It is because of this self-evidence and the frequent references farmers themselves make to the "logic" as they see it that I define Figure 2.5 as a calculus, or in particular as the I-calculus. All the key concepts of this calculus refer to the need and chance to intensify as a conscious farming strategy. In this sense it also means a clearly ordered scheme which reads from left to right, as a structured whole of conditions, means, intermediary goals and end goals: as a goal function. Income and reproduction of the farm enterprise is the final goal, intensification is the path, and working with cura is the particular means for achieving it. Finally, cura entails a number of clearly described prerequisites: impegno, passione, knowledge and self-sufficiency.

The congruence of ends and means in this connection is critical. Isolation of any one element makes both that element and the rest of the scheme meaningless. That is why I emphasize the calculus as being a specific ratio or logic which binds ends and means in a specific way, and although it is given here as an ideal-typical construction, such a calculus is by no means an abstraction. As will be shown later, for farmers who opt for intensification, it forms a practical guideline for many farm decisions. Not only does it structure observation and interpretation, but it also provides a beacon for guiding and legitimizing affairs.

Substantive rationality is often described in economics, sociology and anthropology as the possessing of insight into and an overview of the relevant whole. The I-calculus is a concrete form of such rationality, not because it rests on the more genial characteristics of this group of farmers, but simply because the relevant whole is so structured that it is both surveyable and insightful. The objects of labor stand central in the calculus. Both cura and produzione derive their meaning from this centrality. Produzione stands for results per labor object, and cura refers to treating objects of labor in such a way as to ensure maximal productive results. In short, farm labor itself—understood as the interaction between the direct producer and his objects of labor—is central. It forms the core of an insightful, well ordered and developable "world." It is a world which can be created and further developed on the basis of one's own experience and insight—not in any whimsical way—but normatized according to the calculus.

In the I-calculus explicit references to markets and institutions are missing. They are considered as exogenous to the domain of farm labor. A number of respondents also named self-sufficiency (*autosuficienza*) as being prerequisite for working with the I-calculus:

> "To be self-sufficient in as many respects as possible, it is essential to be able to farm well, because everything that has to be purchased is in a manner of speaking too expensive: it's a question of monopoly products; those who sell determine the price."
>
> "Your costs are commensurate with how much you have to buy, especially when things are going badly. Independent farms can always withstand adversity, they don't go broke so easily. The others are much too vulnerable."
>
> "When getting goods and services outside the farm, you are often unsure of their quality. You can counter this uncertainty by putting in more of your own labor, even make your own concentrates, keep your own bull so as not to be wholly dependent upon artificial insemination, even carry out your own maintenance and repairs."
>
> "I would rather weed myself than use herbicides. You don't know what mischief is caused by that rotten stuff."

In the eyes of these farmers, quality is more doubtful, prices are higher and induced market risks greater when market dependency is increased. That is why they prefer to be self-sufficient. Rendered in strictly micro-economic terms, one could say that in such a view, market dependency causes marginal costs to be higher and marginal profits to be lower. A certain measure of extensification would be the economic result of this double movement, and it is precisely that which is unacceptable to I-farmers. This is why they argue that one does well to be self-sufficient. But there is another reason: the micro-economically defined optimum will not only shift, but the definition of the optimum itself changes with increasing incorporation. "Farming more cheaply" or "more economically" then emerges as a definition of the optimum:

> "I personally buy and pay. I pay immediate cash for everything. I don't know if that is the best in economic terms, but that is what I do and I have no wish to do otherwise. . . . I was once in debt, but after that experience I shall never do it again. If I can't pay for something then I damned well don't buy it. I was so deeply in debt that I thought I would never get out of it. The worst thing is the enormous insecurity you feel because of the debts. One hailstorm is all it takes and who pays? And that's it precisely, if you have debts you almost no longer dare to farm well. Every cost becomes risky. You are inclined to farm more cheaply, to lower impegno, but of course that is madness. . . . No, now I make sure that everything that I put in the land is paid for; that way I have peace of mind and can farm without worries, can farm well."

"But why," I asked, "could you not produce more cheaply when you were so in debt? The risks would have been smaller."

> "No, you can't do that, you have to farm well and intensively even if you have debts. You mustn't let your produzione suffer. It doesn't work. Spending less in order to save is not a practice to be advised. The main aim of farming is produzione, you can't stray from that. Otherwise you play havoc with your income. That goes also for times of crisis. Produzione comes first, and if you have to save then you must save in the house not on the farm."

In short, increasing market dependency (the external financing of working capital in the case above) leads to a different form of optimization—to an optimization of the average cost/benefit ratio which tempts one to make a reduction in costs but which also reduces yields. It is precisely for this reason that market dependency is rejected: the core of the I-calculus would suffer and cognitive dissonance result.

An excursion into micro-economics can highlight this problem. The first graph pictured in Figure 2.6 gives a classical production function—the relationship between various input levels and the corresponding output levels.[5] If the price per unit output is known (PY1), then the relationship can also be read as "income function" (TR for total revenue). The second graph gives the marginal profit (MP), which is the extra output obtained by adding an additional input unit. Multiplied by the price per unit output, it gives the VMP (value of the marginal product). In the same way the average product (AP) can be calculated. Multiplied by the price (PY1), it gives the value of the average output (VAP). So what will a farmer do if he structures farming operations according to the I-calculus? He will strive to raise produzione, i.e., try to reach a high output level. A prerequisite for this, given his calculus, is the raising of impegno, i.e., input 1 in the graph. To what point will

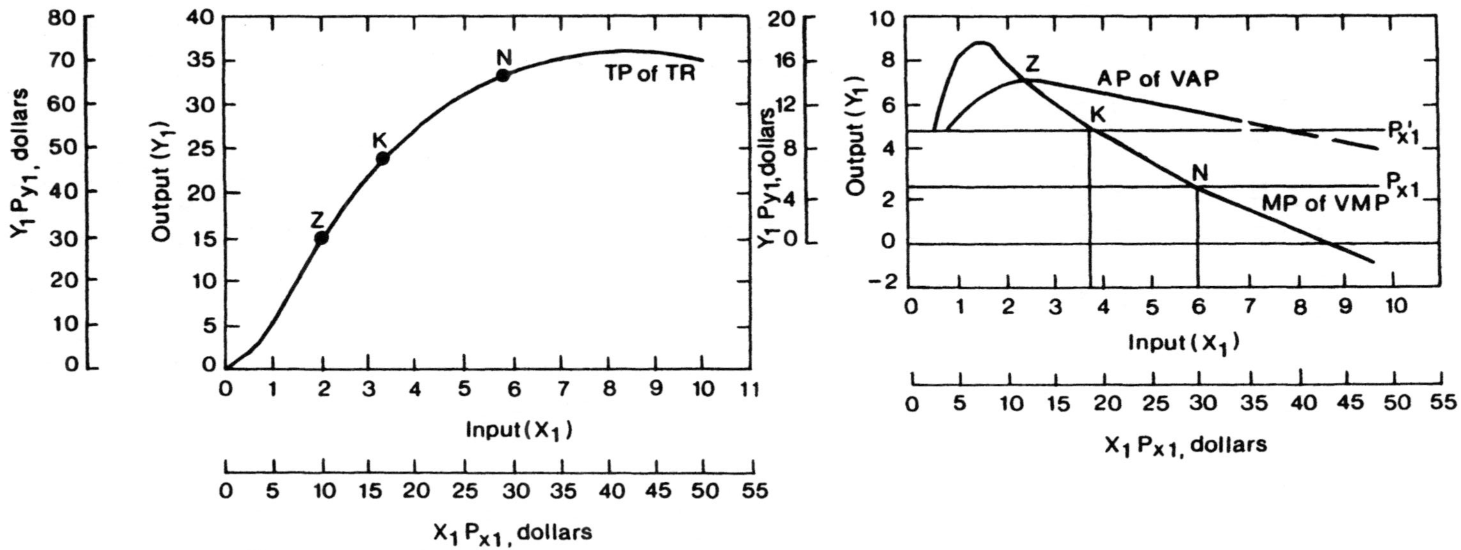

Figure 2.6 A production function and its derivations

he raise impegno and therefore produzione? Those who opt for intensification are very outspoken about this. One I-farmer said:

> ". . . to a specified point naturally. Because look, you must not allow your cost to go higher than your yields. Extra fertilizer, extra feed and so on, naturally has to give extra production, otherwise it no longer makes sense."

That is self-evident. Translated in terms of the graph, that means that you optimize to the point where the marginal costs (the Px1 line) are the same as the marginal yields (the VMP line). But note that farmers who use the I-calculus surround the exact determination of this point with a number of extra prerequisites which give a special character to their optimum. To begin with, they do not use the market prices operating at the time, which often fluctuate. They base their calculations more on the assumption of long-term trends:

> "There are those farmers who when milk and cheese prices are low, give less concentrates and hay to their animals than is proper, but I don't hold with that. When prices are low I look after my animals the same as usual. Of course I am not going to neglect my cows, because when milk prices then rise and become attractive, your cows are in poor shape and your production is low. Our cura remains always at the same level, though naturally within the limits of the possible, but cura is not determined by the cyclical movement in milk prices and even less by fluctuations in fodder prices. Last year for example I bought as much hay as always despite higher prices. That stable level of cura and impegno means indeed that earnings are now lower, but well, as soon as milk prices go up, that will compensate for it."

> "You can't base feed on calculations which follow the uncertainties of the day."

In other words the inherent risk of price fluctuation is consciously not part of their reasoning. A temporary rise in costs (in the graph presented as a development of Px1 to P'x1) is eliminated from the calculus: produzione is justifiably central. There is yet another factor related to this. It came to our notice when one of the respondents said, "Everything that you have to buy is, in a manner of speaking, too expensive" and "you must try to be as self-sufficient as possible." If this is so (and I come back to this when discussing the phenomena of thin and fat cows), then it implies, at least for the farmers who opt for intensification, that the cost of provisioning oneself is less than the Px1 line suggested by market prices. Thus the real optimum will lie beyond point N.

This rather hypothetical digression[6] is a reasonable starting point for answering, at least to a degree, the question of what happens with increasing market dependency. Why do farmers who reason according to an I-calculus think that this is such a problem? They claim that with market dependency they lose their peace of mind. Biological and economic risks would be extremely disadvantageous at the optimum for the intensive farmer (i.e., where marginal profits equal marginal risks). As Heady (1952:515) argued, "they can break him." But the risk we are talking about here is a market-induced risk, a risk which arises from market dependency. I have already shown that such a risk is consciously excluded from the I-calculus. However, with a quantitative increase in market dependency, such exclusion is no longer possible, or at least is less so. How can this risk be reduced? Not by ignoring it as in the I-calculus, but in a material sense by shifting the optimum—a shift achieved by defining the optimum differently. In theory, induced risk is at its smallest when the cost/benefit ratio is maximized, because with a maximal benefit/cost relation (VAP/Px1) there is maximum room for dealing with falling prices, rising costs and loss of production. So, through increased market dependency, one is pushed to point Z on the production function instead of being able to work towards point N. The costs per labor object (impegno), as well as yields, drop. This decrease might cause an improvement in the benefit/cost ratio, but such a concept is absent in the I-calculus. What remains then is the fall of produzione. This is precisely why an increasing market dependency is rejected in the I-calculus. It stands in the way of "good farming."

The relationship between risk factor and a less intensive style of agricultural practice has now been mentioned several times. Since the classic work of Heady on farm management economics (1954:546), we know that "subtracting a safety margin" can lead the farmer to use less fertilizer and to opt more for the "bottom curve" (see also Hazell et al. 1978:26). To these insights one might add that the relevant risk-factor can neither be traced back to psychological qualities of the farmer or entrepreneur (as in Heady, 1954, and to some extent Ortiz, 1973), nor to a general setting of market and price relations (market fluctuations, etc.). This risk factor is primarily rooted in the degree of market dependency. It emerges with commoditization.

Another particular feature of the reasoning of I-farmers is that the definition of potential benefits often goes beyond the particular reality as defined by the current situation on the markets. This is where the crucial importance of a time perspective enters the discussion. The "bell'azienda," the beautiful farm, to be built with one's own labor (cura) symbolizes the long-term perspective with which these I-farmers work. As one of them clearly expressed:

> "Yes, indeed, there is always this temptation, this endeavor to make something out of it; something you can be proud of, a beautiful farm you can hand over, God willing, to the children."

Building a fine farm often implies, as the farmer said, "going against this pressure of the market." During our conversation this respondent was sowing a bean variety in some of his fields, just to plough them under later. The explicit reason for this activity, and for not sowing a product with a high market value, was the desire to improve, over the long term and through his own means, the fertility of his soil. The same goes for so many activities in the cow shed. A lot of the work defined as typical of the "tail-washer" is nevertheless done by I-farmers. The work dedicated to "good cows" so as to secure a good offspring is an outstanding example. A particular cow might cause a lot of trouble while being milked (which implies extra work). She might well demand special feeding and privileged housing (again extra work), her milk yield might be poor, and her age preclude any thought of a good price at the slaughterhouse. Nonetheless some farmers will go on caring for such an animal, simply because her offspring might be promising. She might contribute to the "beautiful farm" he envisages. Hence, use value as defined by the farmer (a definition implying often a long time perspective) clearly dominates over the immediate exchange value as determined by the current market situation. In terms of the graphs presented in Figure 2.6, this long-term perspective as symbolized by the notion of "la bell'azienda" implies a further shift of point N to the right, since benefits are defined in a way that transcends the market and the commodity relations that go with it. Further intensification is the outcome.

In the structure of the I-calculus, terms which refer directly to the labor process as such are central. What appear to be strikingly absent are details of economic relations. Put more forcefully, the I-calculus rests, as we saw, on a strong preference for self-sufficiency. Temporary price swings are consciously excluded as relevant parameters, and benefits are defined in a way that goes beyond the market. This lack of emphasis on economic criteria becomes more striking when compared to the calculus of farmers who opt for scale enlargement and relative extensification, as described later. Terms which directly relate to market and price relations dominate in their calculus. All this does not imply, however, that the I-calculus is a-economic. What it implies is a specific interpretation of the economy, an interpretation which is consistent with a specific ordering of economic relations, i.e., the structure of historically guaranteed autonomous reproduction (see Figure 1.2 in Chapter 1). The supply of production factors is given and historically guaranteed. Variation in their input is not possible in the short term. Therefore the planning of an optimum in terms of cost/benefit relations turns out to be improbable if not superfluous. Talk of costs is here fictitious insofar as it is about monetary cost which per se must be valorized. Instead the scheme presupposes, both in the long and short term, an increase in the relation between the input of production factors and gross production. Hence the structural importance of cura and produzione. The marketable surplus and the reproduction of production factors for the following cycle can be raised by the degree to which that relationship (i.e., the technical efficiency) is high. So income (guadagno) is raised and the "bell'azienda" develops.

The I-calculus then refers to that economic reality embodied in the economic relations of autonomous historically guaranteed reproduction. That is why the absence of specific economic interpretations in the I-calculus does not mean that it is a-economic. It is completely economic in the sense that it provides a rationale for economic action in a context typified by the relative absence of market dependency on the supply side of the farm enterprise.

The E-Calculus

The E-calculus is the particular rationality used by farmers who opt for scale enlargement and relative extensification. A high level of market dependency, problematic and therefore rejected in the I-calculus, is no problem within the E-calculus; in fact, it is an advantage. The E-calculus will be reconstructed here in the same way as was the I-calculus earlier. I will therefore begin with the argument put forward by the respondents of the Parma sample, who explained why they think that the farmer with the 30 cows which give 40 ql of milk each is to be seen as the "best farmer" who "earns the most."

> "That is rather easy, such a choice. What the second farmer does is the only way to be able to continue farming. He also earns much more. Of course he earns less per cow, but that is alright. To earn less per unit isn't important, we no longer live in those times. As long as I produce more, have more cows, then that is no problem, because your income comes from numbers. From that point of view I think that the second farmer's cow shed is still not functioning well. He should have a lot more cows to milk, to reduce further the milking time per animal. I would

rather see the example with 60 or 80 milk cows, but well, compared with the 'tail-washer' one might say he is well on the way."

Another typical E-farmer expressed the same point of view:

"Look, the attractive margin emerges with the quantity."

And yet a third said,

"The second farmer (30/40) is naturally much better. He scores better results on the benefit side. He sells more milk and can sell more calves as well. What's more, he has less of a headache in his cow shed. If you drive your cows to produce 50 ql then you have a load of problems—sickness among the herd and no more rest for yourself, you must be for ever in the cow shed. Moreover you milk the cows to death forcing such a high milk yield. That first farmer by the end of the year will have only ten cows left. The rest will be finished. And you can reckon that in other respects, with feed, for example, his cost will be far too high. It is, all in all, sunshine clear, more results on the sales side and less cost is what makes a better income."

The difference from the I-calculus is immediately obvious. Production per object of labor (produzione) has absolutely no normative meaning here. It is instead suspect. It functions even less as a yardstick. This function has been replaced by total production which is compared to costs. Along with this viewpoint a new term crops up—"scale"—because "income comes from numbers." A second decisive difference with the I-calculus is that impegno, as a normative interpretation of stable relations between labor and other inputs per labor object, is entirely missing. Instead the search to reduce labor and costs (the opposite of impegno) is offered as normative:

"Look, the first farmer will lose far too much time and the management of his beasts will cost far too much."

I break this quotation here to point out an interesting detail. A farmer who opts for intensification as we saw, mostly uses the term "cura" or the verb "curare" to describe the caring for his animals. For the farmer who opts for scale enlargement this is not usually the case. That is understandable, for neither produzione nor impegno are meaningful terms in the E-calculus. The essential element that binds the whole thing together, the cura, is thus superfluous. It is also interesting that with this shift in meaningful structure the discourse also partly changes. Such farmers do not speak of cura and curare but mostly of

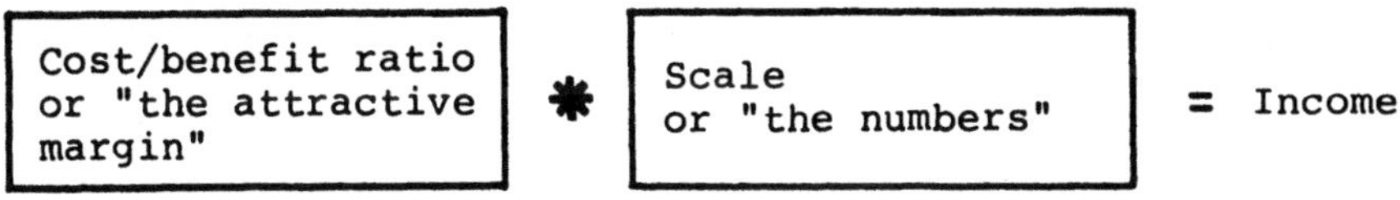

Figure 2.7 Structure of the E-calculus

controlare (to control) or of being able to *governare* (to manage) the herd. To continue with the quotation,

> "I, or better said, the second farmer, we do not need to give such a quantity of concentrates. That first farmer has to do that, his feed costs must be much higher. He also loses a lot of labor time so again he cannot keep so many beasts."

After many counter-arguments on my side, he said,

> "And even then, at the end of the year this second farmer has more income because he spends relatively less. That first farmer, that farm doesn't pay its way. They are bewitched by the cow shed and their animals and spend money by the bucketful."

Two terms are central to the E-calculus—cost/benefit ratio and scale. Together they determine income. Figure 2.7 schematically summarizes this idea. In the core of the E-calculus, improving the cost/benefit ratio means largely following the path of extensification. Several studies on the area, including that of Messori (1981), and Garoglio and Mosso (1986), confirm this tendency.[7] Some of the E-farmers are quite explicit about this in their own terms:

> "There comes a time when you learn that by lowering costs, even sometimes eliminating them, you always earn. Yes, certainly you have to reduce your costs continually whether it be concentrates, or roughage or whatever, even labor. A higher milk price disputed over in Rome or Brussels, that's crazy. The market makes the price. It is useless to go and demonstrate. As a farmer it's your job to lower your costs."

> "No, then with my rougher style (*il nostro farmale*), I may make poorer hay, and care for my beasts less well, but I've no time for all that nonsense. It costs me too much money. I have to earn, damn it! I haven't time for pottering around on one square meter; you must look for it in space. Attractive margins arise only through greater quantity. In the cow shed that is certainly the case: why should I waste my time and incur such high costs for one cow. I have to milk more cows, that's the way to farm."

Table 2.5. Benefits, Costs and Their Mutual Relations, Derived from Figure 2.6.

		$P_{y1} = \$2$			$P_{x1} = \$5$		
Inputs (X_1)	Total Product (Y_1)	Total Revenue dollars	Value of Average Product dollars	Value of Marginal Product dollars	Total Cost dollars	Net Revenue dollars	Total Revenue/ Tot. Cost
0	0	0	-	-	0	0	0
1	5	10	10	10	5	5	2
2	14	28	14	18	10	18	2.8
E 3	21	42	14	14	15	27	2.8
4	26	52	13	10	20	32	2.6
5	30	60	12	8	25	35	2.4
I→ 6	33	66	11	6	30	36	2.2
7	35	70	10	4	35	35	2.0
8	36	72	9	2	40	32	1.8
9	36	72	8	0	45	27	1.6
10	35	70	7	-2	50	20	1.4

High produzione is rejected: it is too expensive and also too "*impegnativo*"—it takes too much labor. Too much and too expensive for the returns. Too expensive also in terms of the alternative, scale enlargement. This can be illustrated in a well-structured way by means of the production function used earlier. Table 2.5 gives the figures on which Bishop and Toussaint (1958) based the previously outlined functions. Various levels of inputs (X1) and corresponding production levels (Y1) are shown. If one projects onto this prices per input and output unit (PX1 and PY1), then gross and net incomes, the value of the average product, the value of the marginal product, and the average cost/benefit ratio can be calculated. The cost/benefit ratio (given in the last column as Total Revenue/Total Cost) is maximal with an input of 2 or 3 units X1 and a corresponding production level of 14 (or 21). That figure is noticeably lower than the level of inputs for the I-calculus, where a level of inputs of 6 units and a corresponding production of 33 units was arrived at.

Assuming that other conditions remain equal, then the cost of such an operation (maximizing the cost/benefit ratio) would be a decrease in the net income per labor object unit. If one reads the table as referring to a hectare or a cow, then going from the optimum that holds when applying the I-calculus to what holds when applying the E-calculus, a halving of net income occurs: from 36 to 18 dollars. However, if we couple the second term from the E-calculus (scale enlargement) to the first term (improving the cost/benefit ratio), then quite a different picture emerges. Suppose that a farmer has 100 dollars available. Reasoning according to the I-calculus that would be sufficient to work 3.3 hectares "properly" (or to feed 3.3 cows well, etc.). The produzione is optimal then with an input of 36 dollars a unit. Therefore with 100 dollars, three units can be worked. The earnings per unit are then optimal. The net income is 3.3 × 36 = 118.80 dollars. Passing now to the E-calculus, the picture changes. The cost/benefit ratio is optimal with an input of two. That "costs" 10 dollars per labor object unit. Thus with the 100 dollars available, 10 labor object units can be worked. The falling income per unit is amply compensated for by a greater number of labor objects: income per unit of labor force rises to 10 × 18 = 180 dollars.

Of course this whole exercise is highly hypothetical and above all incomplete. It is a question of whether the technology exists to enable one unit of labor force to "work" 10 labor objects. Perhaps with a lower input, the quantity of work per labor object would also drop. But if substantially increasing objects of labor should lead to using extra labor force, then the results of the above calculations could be quite different. Even a fall in net income per unit of labor could occur. A change in fixed costs could also modify the picture, because going from 3.3 to 10 labor objects could bring with it an increase in fixed costs (depreciation, etc.). Be that as it may, the example illustrates at least the possibility that the combination of relative extensification and scale enlargement gives rise to a ratio which is absent when one of the terms is modified. It is worth noting that the mention of such a strategy is missing in most of the standard works on agrarian economy.

Take, for example, *Farm Management Economics* by Heady and Jensen (1954). Chapter 15 deals with size of farm or enterprise. The chapter gives special attention to the dairy farm. It suggests that given the fact that the cost of feed usually forms 75% of total cost, advantages of scale are difficult to achieve; "the main economics must come from building and labor" (468). Modern milking machines and automatic feeding, etc. can be installed with a larger milk herd, and building costs per cow will drop. The relevance of all that is, however, rather small, for according to Heady and Jensen, "There is little chance for economics or dis-economies in feed . . . aside from those due to good or poor management." Thus in their view there is only one optimum for feed level. Variation of feed input per cow in combination with scale enlargement is excluded: "If one is going to have 4000 rather than 2000 broilers, one will need twice as much feed . . . 200 beef cattle will require four times as much feed . . . as 50 cows. . . . Similar statements apply to sheep, beef cattle, laying flocks and dairy cows."

In short, the possibility that the advantages of scale enlargement lie in its combination with relative extensification is not considered. This omission is the more remarkable since Heady and Jensen actually indicate the occurrence of the phenomenon: "We know of some large-scale operators who would actually have greater profit if they contracted their unit and gave more attention to improved practices for their crops and livestock!"

At first sight it might be surprising that this particular phenomenon was not studied as a meaningful reality in itself. However, the characterization of all kinds of phenomena as mere deviations is unavoidable if, in a heterogeneous reality, only one schema or calculus is applied to the study and understanding of farm management. The adoption of more than one calculus allows for a breakthrough in this monolithic scheme because it implies that a great variety does not need to be reduced anymore to one optimum. On the contrary, heterogeneity can then be understood as the ever rational outcome of a variety of models of rationality. Later, when the thin and fat cows are discussed, this question will be illustrated and analyzed with an empirical example. However, before we step into the pastures, it might be useful to give some thought to the second term of the E-calculus—the term "scale," i.e., the relationship between available manpower, the size of the herd, and hectarage. Existing sociopolitical relations in Emilia preclude certain forms of scale enlargement and encourage other forms.

- First labor input can be reduced: "more than one person farms" (sometimes erroneously referred to as "more than one man farms") where part of the family (often extended) is active, are able to lower labor input to one or two people; the rest seek positions elsewhere. Part-time farming can also be considered in this light.
- A second possibility is hectarage expansion: although buying land is extremely difficult (but not impossible), the leasing and semi-legal renting of land (*sfalcio*) give considerable scope to the individual farmer interested in expansion.
- A third form, though more ambiguous to interpret, is the raising of cattle density: more cows are kept on a given hectarage, which usually entails seeking an increasing amount of fodder on the market. In fact, a claim is thus laid on agricultural land elsewhere via market mechanisms.
- A fourth form, which in present-day Emilia still exists, is the specialization of the farms: the rotation systems are simplified as far as possible. Grain, wine, tomatoes, etc., disappear from the cropping plan and luzerne and maize dominate. A certain degree of specialization is also possible in the cow shed by limiting, as far as possible, the time one spends on rearing young animals, or even by a complete externalization of this particular practice, in

order to increase the numbers of milking cows per man. Thus more remunerative objects of labor appear in the same cow shed. The scale is bigger, at least if labor input remains the same.

This list could in principle be extended and made more complex. Likewise, some farmers develop a veritable genius for accomplishing scale enlargement where it would appear impossible.

The point I would now like to move to, however, is something different—to the combination of scale enlargement and relative extensification (whereby relative extensification is the vehicle for the improvement of the cost/benefit term). To farmers who opt for it, the essence of scale enlargement lies precisely in its combination with relative extensification and the improvement of cost/benefits which accompany it. As they already stated: "It is with greater numbers that the attractive margin comes." The point is illustrated more fully in the next section.

Thin and Fat Cows

It is apparent from both our own and other research projects in the area that feed input per cow varies considerably. This phenomenon is of course also recognized by the real experts, the farmers themselves. They often relate feed input per cow to the degree to which farmers are either self-sufficient or market-dependent for food provision.

> "If fodder comes from the farm itself, then you will see that the cattle are well-fed. Yes, it's always better to produce your own feed than to buy it."

But why would you feed less if you have to buy it?

> "If you have to buy it then it costs much more and that makes you careful, you have to be, so you are less ready to feed them so they have enough."

Another dairy farmer said, in answer to the same question,

> "In my opinion, giving less fodder isn't very smart. If you sell a cow you get money for the flesh, not the bones. And your milk yield drops. No, it makes no sense to give less feed—your income also drops. Maybe you earn something but that is only superficial. You are in fact losing by it. And when I think of the farmers in this neighborhood who have to buy a lot of hay and silage, you see their cows walking around looking

decidedly thin. Of course they obviously give them less. Their reasoning is against all logic."

But the farmers who buy so much raw feed have this to say:

"Are you mad? I give enough feed to my cattle. The problem that you are talking about is actually nonsense because if the farmer who grows all his own feed gives more feed than me then he is giving his cows too much."

And after further questioning,

"You have to look at the economics of it: everything you give beyond what is necessary is money wasted. With today's fodder prices you can't afford to do silly things. If you have feed over, if you can in a manner of speaking give more feed than me, then you shouldn't waste it on the cows but sell it, or take more cows."

What such comments show is that, in practice, two different calculi are applied which find different solutions to the same problem. In the reasoning of the self-sufficient farmer, high produzione is foremost. The norm therefore applied in the cow shed is *alimentare a volonta*, "let the animals eat what they will." They can work with this norm because feed does not represent an immediate expense. Neither roughage nor hay nor self-produced concentrates are seen in this case as commodities. They are simple use values. For market-dependent farmers this is not the case: feed represents a substantial and immediate outlay. It is a commodity. You must therefore "not throw money away." If what you are striving for is a better cost/benefit ratio, then the E-calculus is applicable. The norm is *la mangiatoia deve essere pulita,* meaning literally "the manger should be clean," i.e., empty. The feed you give should be sufficient to supply immediate needs; there is no room here for *a volonta*. What remains in the "manger which is not clean" represents a waste of money.

If we ignore the effects of the various degrees of incorporation into the market, we may then view self-produced and bought-in feed as simple substitutes. Figure 2.8 gives a twofold substitution line: Yn represents a high level and Yn–1 a low level of feed. The relation between purchased and self-produced feed is projected in segments over the substitution lines. With a low degree of incorporation into feed markets (segment L) most dairy enterprises on the plain appear to give a high level of feed. Some 50% of the enterprises in this segment realize a GVP/AA (per adult animal) of more than 1.5 million lire. If one

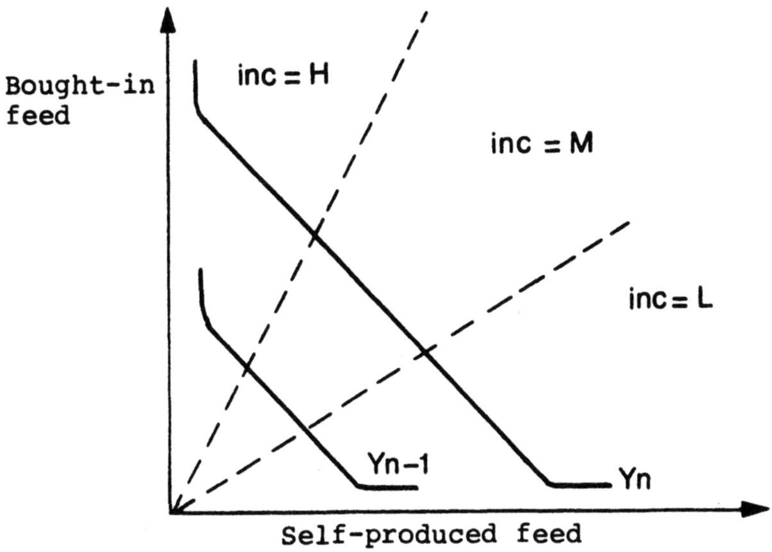

Figure 2.8 Feed levels and incorporation

then goes to the segment which represents a medium degree of incorporation into feed markets (M), this percentage drops to 34%. If incorporation is high, segment H, then the feed input per cow veers more strongly towards the lowest substitution line. Only 16% then achieve a GVP/AA higher than 1.5 million lire.

A higher degree of incorporation into the feed market leads to a falling level of feed input per cow. The coefficient of correlation is 0.42 ($p=0.001$). This decreasing level of feed input produces an improvement in the cost/benefit ratio. The ratio rises from 3.86 to 6.73 as one goes from highest to lowest feed levels. At the same time the earnings per cow drop. This drop is then compensated for by enlargement of scale. The flow diagram in Figure 2.9 corroborates the relationship suggested by the E-calculus: a lower level of feed is linked to raising stock density per hectare. Therefore the feed saved is distributed to more cows. Figure 2.9 illustrates how the underlying calculi work in practice. The clockwise flow represents the structure of the I-calculus perfectly—a low degree of incorporation through high impegno (seen in kind as a high feed level) and high production. The counter-clockwise flow appears to be the embodiment of the E-calculus. It depends on a high degree of incorporation and shows a specific ordering of enterprise interrelations—high in terms of scale (seen as cattle density) but with a low feed level per cow (producing an improvement in the cost/benefit ratio). In short, what in the current view appears only incidental, and primarily due to differences in entrepreneurship (Heady and Jensen,

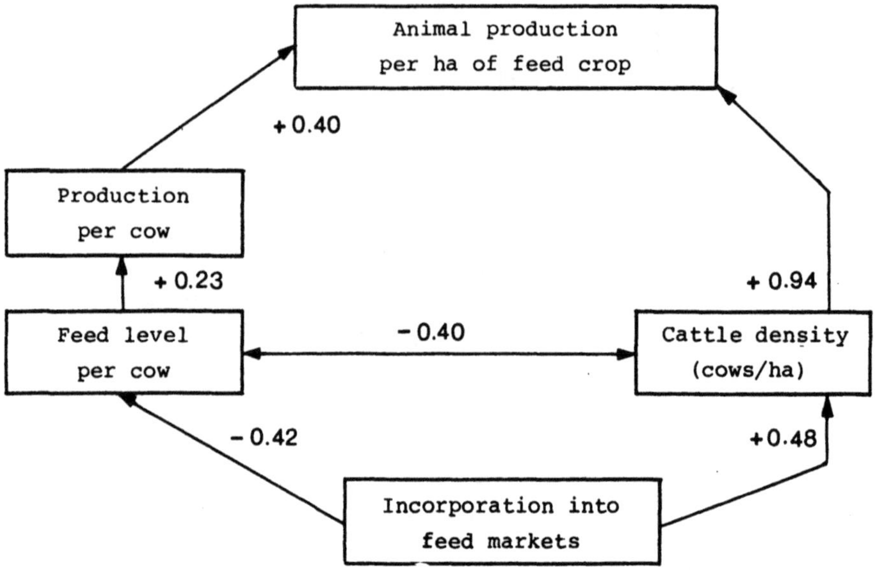

Figure 2.9 Relation between cattle density, feed level, production per cow, and degree of incorporation into feed markets (ERSA/plain)

1954:468)—is better understood when seen as the result of different calculi.

The E-Calculus and Its Link to a High Degree of Incorporation

With increasing incorporation, the structure of the process of reproduction drastically changes from autonomous historically guaranteed reproduction to market-dependent reproduction. If in the former the production factors and non-factor inputs are more or less given, in the latter they are variable, even in the short run. If in the schema of autonomous historically guaranteed reproduction progress is primarily reached through raising technical efficiency, in market dependent reproduction it depends first and foremost on keeping a watchful eye on the relationship between monetary costs and benefits.[8] That is the central tenet of the E-calculus, the logic upon which farmers who opt for scale enlargement and relative extensification base their arguments.

The group of eighteen dairy farmers who formed the Parma sample were asked which situation they thought more desirable—self-sufficiency or market-dependency? The question was asked per production factor (and thus per supply market). With respect to the labor market, for example, they were asked which situation is more favorable—the farm on which three brothers work, or the farm where one entrepreneur and two laborers work. The same was asked, with appropriate examples, for the other markets. There are clear differences in the choices of I- and E-farmers. Only 32% of the former thought that in general incorporation was the most desirable, in clear contrast to the 51% of E-farmers who thought incorporation desirable. The greatest differences, however, are to be found in relation to labor, feed and capital markets. Application of the method to the ERSA sample indicated that the differences encountered were statistically significant. In fact the differences are even greater than the statistics show. It became obvious during questioning that the real significance was not to be found in the statistics but in the reasons given for the choices. For example, 86% of E-farmers and 78% of I-farmers said that they preferred incorporation into the market for long-term loans rather than relying purely on private saving. The following argument was used by 66% of the E-farmers:

> "Naturally the farmer who borrows is much better off. He can then use his own savings to invest elsewhere where he will get better returns, in real estate or good shares."
>
> "It's always better to take an interest subsidy. If necessary you can steer it back into the bank and get a higher interest on it and pocket the difference."

Only 11% of the I-farmers found such reasoning legitimate. Instead, they argued quite differently about the value of a certain degree of incorporation for long-term loans.

> "Say you have 50 million lire of your own saved, then it would be crazy to reject a loan of 30 million if you get it under reasonable conditions. It is better to invest 80 million than only 50."

If you can increase impegno (and thus cura and produzione) then a long-term loan is attractive. Personal savings from the previous season are seen as a means to improve work and working conditions or to create la bell'azienda. Loans supplement personal savings. E-farmers, however, tend to see loans as a substitute for savings. One might go so far as to say that they perceive the use of personal savings, as it were, through the market. The word "shares" in one of the above quotes is interesting. Personal savings are projected into the sphere of circulation and judged by the exchange value they get there. Exchange value dominates over use value. An increasing incorporation changes the perception of the goods in question to quite a considerable degree: with such an increase, savings become indeed a commodity whose use is governed by the logic of markets. Again it should be stressed that this is a differential phenomenon. It emerges with a high degree of incorporation and within the E-calculus. Quite a different view is present in I-farms: savings do not function as a commodity, nor are they seen as such. Savings are, in this case, just another means to develop the bell'azienda. This implies that savings do not go the way commodities go, i.e., towards the highest profit.

The Parma farmers were also asked in which situation produzione would be the higher—on the farm which was highly market dependent, or on the farm that was self-sufficient. Or would there be no difference? Table 2.6 summarizes the answers. This table supports the hypothesis that a high level of market incorporation is not congruent with a calculus which gives high produzione a central place. I-farmers see incorporation or market-dependency as an obstacle and think production will especially fall with incorporation into markets for cattle feed and working capital. Their comments speak for themselves:

> "Look, the self-sufficient farmer is far better off. He doesn't suffer the ups and downs of the market. He is in a much better position to continue farming in a linear and stable way."

And in relation to working capital:

> "With the majority of farmers I think the system which puts them most in the red, is the bank. It's no problem for the bank, they even encourage it, but it's a silly situation to get yourself into, financing your running costs with commercial loans at 24%. It's the same when you take short term credit with the Conzorcio. You can easily become unstuck through such credit. In theory you need a product that renders 24% and no product does that. And the worst is you can no longer raise your production. You have to stay at a very modest level, because the high costs of loans are not economically viable and far too risky. It's different with the farmers who pay all their running costs themselves from the previous year's savings. They can keep their produzione up because they are beholden to nothing and no-one."

Craftsmanship as a Specific Structuration of the Farm Labor Process

The calculi outlined contain clear references to the operational mechanisms for achieving certain options or goals. I wish to discuss two of these mechanisms: craftsmanship and entrepreneurship. Craftsmanship refers to the practical capacity to optimize the productive results per labor object, both in the short and long term. Craftsmanship is ex-

Table 2.6. Percentage of I- and E-Farmers Who Think Incorporation Leads to a Drop in Production

	Ground Market	Labor Market	Market for Long-Term Capital	Market for Working Capital	Market for Feed and Fodder	Market for Machine Services	Average for All Markets
I-farmers	22%	78%	22%	67%	67%	56%	52%
E-farmers	14%	57%	0%	0%	14%	72%	26%
Difference	(-8%)	(-21%)	(-22%)	(-67%)	(-53%)	(+16%)	(-26%)

pressed in a particular organization of the relations between object of labor, means, and labor force as well as in a series of specific norms with which the organization is established, evaluated and further developed. It is by no means a subjective "residual factor."[9] Craftsmanship is the outcome of a specific arrangement of farm labor, understood as the permanent interaction between intellectual and manual work. It is the operational mechanism with which I-farmers develop their farms, and it produces a particular style of farming. In the I-calculus we came across craftsmanship as cura. Its central place in the calculus serves to emphasize how important craftsmanship is in the I-logic.

Entrepreneurship implies a willingness to permanently tune one's enterprise to external market and price relationships. Like craftsmanship, it also presupposes a certain arrangement of farm labor, in this case one which is functional to dominant market and price relations. The need to organize farm labor so that it relates to the market and prices depends on the degree to which commodity relations penetrate the farm. With a high level of incorporation they will penetrate to the heart of the labor process and will then condition it in a direct way. The willingness to attune one's enterprise permanently to these commodity relations requires a set of norms and a "goal" to which such tuning can be directed as well as a yardstick for judging the (ever changing) results of this process. These are provided by the two central tenets of the E-calculus—cost/benefit ratios and scale. In short, craftsmanship is the vehicle of the I-calculus and entrepreneurship is the vehicle of the E-calculus.

In the agricultural economics and agrarian sociology literature it is usually assumed that craftsmanship and entrepreneurship are aspects of one and the same thing. They are seen as extensions of each other. At times both may be present (in the good and modern farmer) and at others both absent. However, if we consider craftsmanship and entrepreneurship as different "operational mechanisms," and further, if we consider the relationship between specific options and operational mechanisms not as coincidental but as a consciously constructed rational link, then it emerges that the two may often stand in a negative relationship to each other. In this sense the empirical research provides an important test for one of the central assumptions of modern "agricultural management" theories. While in the current theories craftsmanship and entrepreneurship are conceptualized as an expression of individual capabilities, and therefore as often being in line with each other, I will consider, in the rest of this chapter, both craftsmanship and entrepreneurship as specific structurations of the labor process in farming, each conditioned by specific social relations of production. Special attention will be given in this respect to interrelations between the farming unit and markets and market agencies. Commoditization and institutionalization play a decisive role in structuring the farm labor process, and thus, they condition the qualities developed in this process. The differential impact of commoditization and institutionalization elucidates why craftsmanship and entrepreneurship are more often than not at odds with each other instead of being in line as is so often hypothesized in current theory.

Three interconnecting threads can be discerned in the production process in dairy farming:

1. breeding,
2. fodder production and cattle feeding, and
3. mechanization in both the fields and the cow shed.

Together these tasks form the theater in which farm labor is acted out. Each thread can in turn be subdivided into a number of elements, each representing an important practical aspect of farm labor. Figure 2.10 summarizes some of the most important elements. In the Parma research each element was presented to the farmers interviewed in the form of a question, such as: How do you do that? Why do you do it that way? Have you always done it like that? Why is this better than that?

By thus exploring, we learned that farmers who opt for intensification work differently from their counterparts on virtually all fronts. They feed, breed and select their animals differently. Mechanization inside and outside of the cow shed is such that there is more time available for each animal. Fodder production and conservation too are organized in quite a different way. Muck spreading is also different as is the period of time devoted to mowing and the choice of rotation scheme. In short, dozens of small contrasts taken together create a level of dissimilarity sufficient to warrant an assertion that I- and E-farmers work differently.

The most productive method for each element was identified by a panel of technicians. Their opinion of fodder production and cattle feeding was reasonably homogeneous. However, with respect to mechanization such an agreement was not possible. (For a detailed description of the panel study, see van der Ploeg and Bolhuis, 1983.) Subsequently, the methods explored were coded in terms of the normative schema proposed by the panel: in this way each farmer was given a craftsmanship score. The simple adding of these scores resulted in a figure representing a farmer's craftsmanship in relation to questions of selection and feeding (including food production). It was not possible

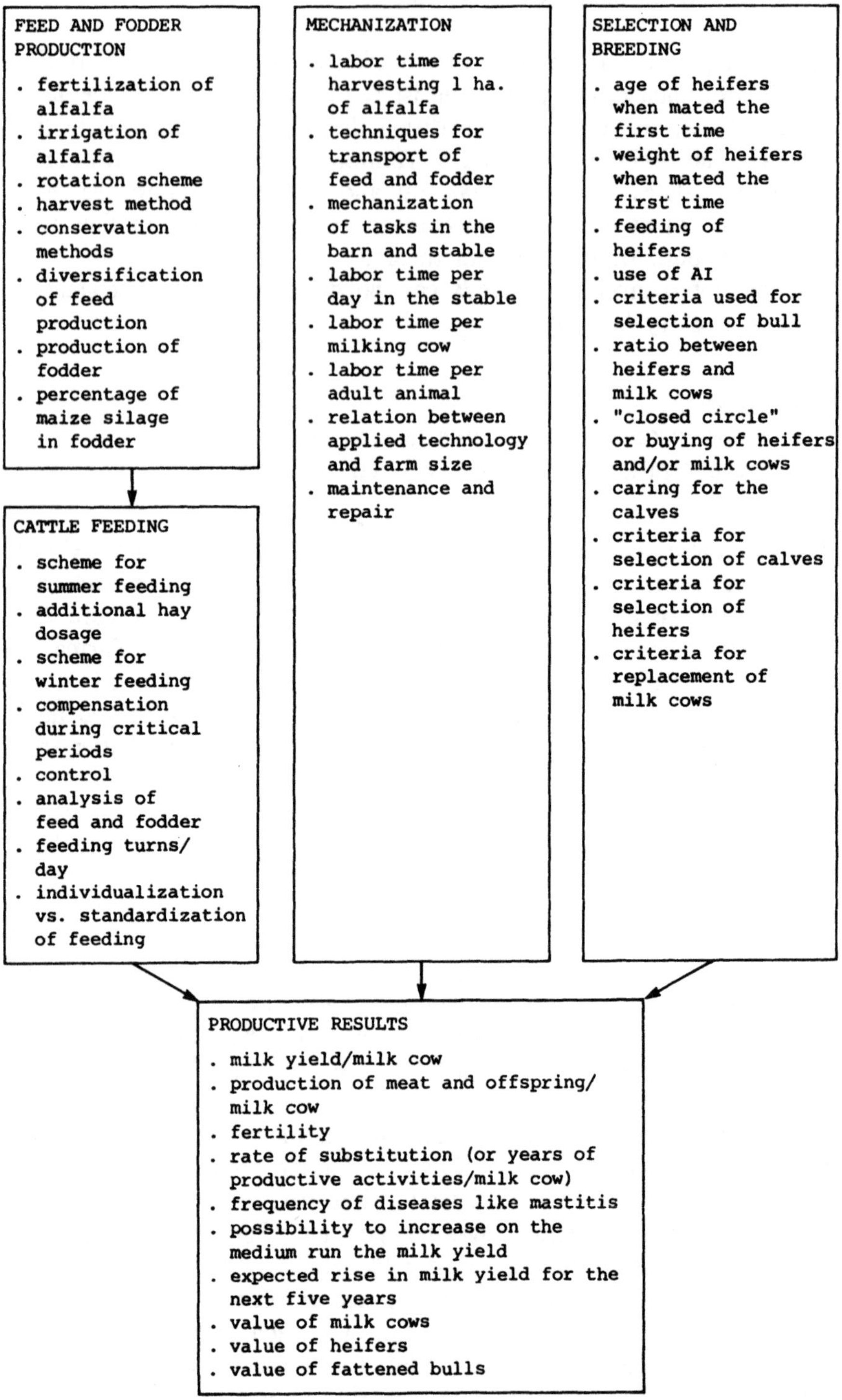

Figure 2.10 Work tasks in a dairy farm and some indices for productive results

Table 2.7. Indices of Craftsmanship for I- and E-Farmers (Parma/n=18)

	E-farmers (n=8)	I-farmers (n=10)
Craftsmanship in cattle selection and rearing	3.9	8.6
Craftsmanship in cattle feeding	2.0	6.7
Craftsmanship selection and feeding	5.9	15.3
Milk yield/milk cow	44.4 ql.	52.9 ql.
Rate of substitution	27.8 %	22.4%
Fertility	1.55	1.51

to measure craftsmanship in terms of mechanization because of the lack of agreement among the panel.

It appeared that there was indeed a high positive link between the calculated indices of craftsmanship and milk yield (considered as an indication of productive results per object of labor). Furthermore, and this is more interesting, it appeared that craftsmanship as here measured was not randomly distributed over the sample of eighteen dairy farmers. There are among these eighteen some who show a high level of craftsmanship over the whole spectrum and others who show a low level on practically every count. This result highlights a fundamental point. Craftsmanship is not simply the art of performing separated tasks in a more genial way. Craftsmanship is also the capacity to coordinate and integrate in a coherent way all the many tasks to be carried out in the field and in the cow shed. Craftsmanship refers to farm labor as an integrated whole. It is not a residual attribute.

If we finally relate indices of craftsmanship to the different I- and E-options, then the picture summarized in Table 2.7 emerges. Information on craftsmanship as a specific way of structuring farm labor is also available for the ERSA sample. I will limit myself to two elements: to the time spent on each cow and to breeding and selection. Working well takes time. To milk restfully instead of rushing the animals through, to use individual or block feeding, instead of standardized or even uncontrolled feeding, to permit three instead of two feed rounds, to inspect regularly—these are all aspects of cura, of craftsmanship. The time factor on a farm is not, of course, just a matter of taking a few extra minutes over a job or of working slower. It is a question of whether the relationship between applied technology, size of herd and available labor is organized in such a way that there is indeed time to care properly for each animal. (The type of cow shed and the techniques it employs for transporting milk, fodder and manure are important here.)

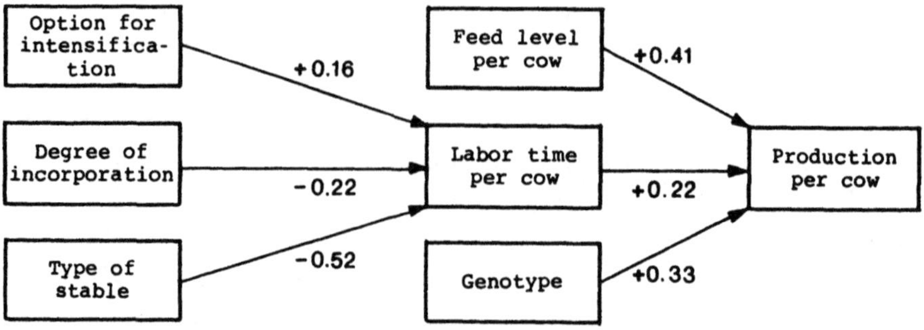

Figure 2.11 Labor time per cow (ERSA/mountains/n=59)

The ERSA questionnaire asked farmers how long they spent on milking and on feeding and cleaning out, and how much time was spent on young animals, etc. As the size and composition of the herd was known, work time per cow could be calculated for each farm. The time per cow varies considerably. It appears that by holding constant feed level and cow value (a rough indication of the genotype), labor time has a positive and significant influence on production per cow. Labor time also naturally depends on the type of cow shed (see Figure 2.11). If we hold this factor constant, then it appears that labor time is dependent on the goals which normatize labor and on the degree of incorporation. If intensification is opted for, labor is organized so that there is more effective labor time per cow available. This approach results in "more productive results per object of labor," i.e., in a higher production per cow. On the other hand, a higher level of incorporation leads to "rushing," a farmer expression to indicate that labor time per cow is reduced as much as possible.

A second highly fascinating part of craftsmanship-in-practice is cattle improvement. This is what one of the I-farmers from the Parma sample had to say about his *selezzione,* his efforts to selectively improve his stock:

> "In 1972 we decided to specialize entirely in dairying. Before that we had a lot of beets and tomatoes. In 1972 our average milk yield was 40 ql per cow. That has now risen to 58–59 ql. We are of course proud of that. All the cura that we put into improving the herd has paid off. It is the crowning achievement of our work. We follow the system of the *ciclo chiuso* (closed circle). . . . We hold on to all the heifers (one- to two-year-olds that have not yet calved). They will all be sired. We have our own bull though we sometimes use artificial insemination, especially if there are certain characteristics that must be corrected. If it appears that a heifer is not going to make the grade then she has to go, even before she has young, but that rarely happens. The remaining heifers join the

cow shed when they have calved so that we can get acquainted with the animal as a milk cow, with her milk yield, and with how that will develop. We are also in a position to know then what kind of calves she produces. This practice of holding on to heifers as long as possible is the basis of our selection. It's a relatively expensive system, you have a lot of cattle standing in the stalls which are not yet productive. But the advantage is that we can select the very best of the heifers to replace the older cows and those whose milk yield is falling."

Milk cows have to be replaced after a number of years. The productive period for a cow varies. Some can be milked for more than ten years and produce up to ten calves, while others are more or less finished after two years. The yardstick for such differences is the rate of replacement, i.e, the ratio between the number of cows that must be replaced and the total number of cows. If the replacement ratio is for example 0.20, then that means that in a herd of 100 cows, 20 will need to be replaced per year. It follows therefore that the average productive period for a milk cow is five years. If rate of replacement rises to 0.33, then the average productive life during which a cow can calve and be milked is three years. The replacement rate is not so much a result of the individual characteristics of the cows as it is the result of the cura practiced in the cow shed. Breeding and selection play an important role in this cura and give rise to quite complex relations. With good breeding and selection the productive life of a cow can be prolonged. The replacement rate will therefore be lower, and this, in turn, makes selection easier. Thus a self-reinforcing system develops. The practice of selection consists, among other things, in the organization of a specific relation between heifers and cows so that a material basis develops for allowing one to hang on to the best heifers and to sell off the others after calving and first milking. But as the farmer quoted above said, this is a relatively expensive system because it puts pressure on stall space; many calves and heifers have to be fed and the income gained from their sale has to be postponed. These are precisely the reasons why E-farmers reject the practice:

"I sell as many calves as soon as possible, usually at one go. To have to continue feeding and managing them for a further two years is really too expensive and takes a lot of work. No I am better off buying a couple of heifers that are in calf if I have to replace cows. . . . No, I don't buy really good ones, with a certificate, because that is also very expensive."

"Yes, I should maybe hold on to more heifers but that costs me too much. You can have a producing cow standing where you would have to put the heifer."

The I-farmer sees the system as relatively costly but nevertheless justifiable because it is the basis for high yields in the future. Such costs are not justifiable in the eyes of the E-farmer, to whom "expensive" means "too expensive." The explanation for the differences in their responses is that they judge the situation from two different perspectives, from two different calculi.

Cattle improvement, however, entails many other factors. Again, the farmer previously quoted takes up the point:

> "First calving is very important. I allow my heifers to conceive for the first time when they are 36 months old. They are sired when they weigh about 400 kilos. It is not possible to be more precise than that, because naturally it depends on more than weight. You have to be able to judge the animal as a whole and that means knowing it well. Above all you need to know the animal's development. . . . I have experimented with bringing first calving forward but it was totally unsuccessful. It isn't good for the milk yield, you get a greater propensity to sickness, conception is more difficult, the calf is uglier."

E-farmers think otherwise on this point:

> "On my farm I have the heifer mated to conceive when it's about two and a half years old. If you postpone that until it's about three, well I think you do get a heavier beast and a higher milk yield, but that's debatable. I have experimented for myself and there was a noticeable difference, but the test was also about costs and it is too expensive to wait so long, and meanwhile don't forget, you have to go on feeding, and that is feed with no financial offset. So I shall carry on with early mating."

> "I have brought forward the calving date from three to two years. I have also given up letting them conceive only in winter. Now they calve all year round. That is much wiser, as you have calves when the prices are high. But that aside, my heifers conceive at between 24 and 27 months old. They are milked for 200 days and are then allowed to conceive again. In this way I invest much less capital and have quicker returns."

> "Let the heifers mate at a later date? Well maybe that is better but it is ruled out because then you have to wait three years before you know whether its milk production is high or not. That is too expensive. Suppose it turns out to be no good? It's better to mate them as early as possible . . . and whether that produces a better calf or not . . . I don't know about that . . . but it doesn't interest me either."

Again one sees the rejection of a particular practice because it is too expensive. It is combined with the assumption that the only returns that make sense are the immediate ones. Costs must be covered by returns as quickly as possible. Time is judged differently in the I-calculus, where the improvement of the herd, something that can take years, is also seen in terms of returns.

A third important element, which has a direct bearing on the previous point, is the manner in which heifers are fed. This element is related to the question of earlier calving because the young animal develops better and more quickly if it is fed well. An evaluation of this theme, i.e., the feeding of heifers, leads to quite different solutions.

> ". . . of course concentrates and a substantial portion of the best hay: the heifers must learn to eat so that their system develops in such a way as to quickly produce at a maximum."

But reasoning according to the E-calculus,

> "No, that doesn't have to be so good, the heifers get what's left over from the cows. I am not going to make ridiculous costs."

They know each other well of course, these I- and E-farmers. They are often neighbors. And E-farmers know precisely what practices are possible for herd improvement. But whether such practices are deemed faulty or not depends on their convictions.

> "Yes, even among the younger modern farmers there are those who remain stuck at lower levels of production. They can't be bothered with improvement. They are people who live more by the day, who put in the minimum of work and think they have done well. We, on the other hand, are permanently busy with improving and getting ahead."

> "Yes, those others who never want to spend money on buying a good bull, who want to make fast and easy money, who want quickly produced calves, they damage their herds. But they are not farmers. I would never do such a thing."

And an E-farmer has this to say about the intensive farmer:

> "I know them very well. They are the farmers who go to the cattle market with 3 million lire and come home with two cows. They throw their money around . . . they are obsessed with their cow shed, in love with their cows, and have a hole in their pockets."

But his neighbor says:

"Quality comes first. That's worth paying extra for. I needed two cows. I bought good animals for one and a half million lire each. My colleague from next door, he needed four. Four bags of bone he bought, that together cost less than one of mine. Who is right? Time will tell, but I know it already."

We have now spoken of three aspects of cattle improvement:

- the relationship between heifers and replaceable cows,
- judging the calving age for heifers, and
- the manner in which heifers are fed.

There are countless other aspects: bull choice, making the right choice of artificial insemination, the outer appearance of cows, the use of ciclo chiuso, and many others. The problem is, however, that such aspects are not always easy to measure, and in themselves they say little. They derive their meaning only in relation to the whole, and according to the Parma data, improvement as a whole is usually organized in such a way that most aspects are either geared to raising future production or to keeping down costs and bringing in immediate returns. It makes little sense to use a pedigree bull or artificial insemination on poorly fed cattle which are themselves the offspring of low-quality animals which are mated early. A bull which "jumps without problems" and is not expensive is good enough for that. Artificial insemination is also of little value in such a case, for it is primarily useful for various corrections and improvements. Certainly, if it is more expensive than "that cheap bull," then the value of it is quickly denigrated as *tapibucchi* ("just filling holes"), and so on.

However, an insight into the three elements mentioned above is sufficient to judge the state of affairs regarding cattle improvement. What is cardinal regarding the three elements is that they do not refer to a special talent of any one farmer, nor do they assume any special biological knowledge of the laws of inheritance: they refer, in a simple way, to a specific ordering of relations between means, objects of labor and direct producer. The amount of labor per heifer, feed, the length of time involved and the composition of the herd are not only all-decisive for the quality of the herd (and thus for production) but are at the same time manipulable by the farmer. In principle a farmer can work in any manner, but in practice what he does will depend on what makes sense to him, and this will depend on the calculus that he works with. In the I-calculus, the logical way to do things is to organize the farm labor process in such a way that maximum production is obtained

Table 2.8. The Way in Which I- and E-Farmers Organize Cattle Improvement

	I-farmers (n=10)	E-farmers (n=8)
Feeding of heifers		
- as good as possible	63%	20%
- with leftovers and industrial wastages	37%	80%
Moment of calving		
- accelerated (24 months)	12%	40%
- anticipated (25-30 months)	50%	60%
- traditional (later than 30 months)	38%	--
Relation between heifers and cows that are to be replaced	1.96	1.31
Closed circle (for the reproduction of stock)	88%	0%
Use of artificial insemination		
- for all mating	50%	--
- sometimes AI, sometimes the bull	38%	20%
- mostly a bull, seldom AI	12%	40%
- only bull	--	40%
"A slow but self controlled selection is preferable to buying in a moment a completely new herd"	100% yes	0%

both in the long and short term. Not so in the E-calculus. From this comparison we get the relations sketched in Table 2.8.

A certain number of the variables mentioned above were taken up in the ERSA questionnaire, including the manner in which young animals were bred, the rate of replacement, the degree to which feed levels are stable or vary with the fluctuation in prices and the degree to which feeding a volonta or the clean manger is opted for. The rise in production expected over a five-year period and the degree to which fodder production per hectare was expected to rise were also included. We assumed that high expectations for future levels of productivity would principally be an outcome of craftsmanship and would therefore be an indicator of such. In retrospect, that assumption appears to be inaccurate. A high degree of craftsmanship results indeed in a modest expectation, but the lack of craftsmanship is linked to far greater expectations. The lower the craftsmanship the higher the expectations. A factor analysis of the six variables, applied to mountain and plains data (ERSA) is summarized in Table 2.9.

Table 2.9. Results of Factor Analysis Applied to Craftsmanship Variables (oblique) ERSA/Plain and Mountains/n-75 and n-59

	Plain			Mountains		
	CRAFTp1	CRAFTp2	CRAFTp3	CRAFTm1	CRAFTm2	CRAFTm3
The rearing of heifers	.21	.70	-.01	.00	.74	.15
The substitution rate	+.32	.28	.61	.76	-.15	.19
Stability of feed levels	-.21	-.21	.79	-.06	.00	.89
Norms applied in cattle feeding	.72	-.01	.16	.36	.41	.34
Expected rise in milk yield/cow	.33	-.81	.06	.81	.05	-.37
Expected rise in yield of fodder production	.74	-.10	-.28	-.13	.71	-.25
Variance explained	24%	21%	17%	23%	22%	18%
Eigenvalues	1.43	1.27	1.01	1.41	1.29	1.09

CRAFTp2 is a variable which refers to the plain and summarizes some aspects of craftsmanship, above all the careful rearing of young cattle. The variable "expected rise in milk yield/cow" over five years has typically a high negative loading on this factor. We have already discussed the OPTION factors which apply to the plain (see Table 2.3). Craftsmanship can therefore now be correlated with these goal factors. I limit myself to an illustration of the influence of goal factors on craftsmanship (CRAFTp2). The links are summarized in Figure 2.12. It shows that opting for intensification does indeed lead to craftsmanship and that opting for scale enlargement and relative extensification (OPTIONp2) blocks or impedes craftsmanship.

What is happening with those farmers who opt for scale enlargement? We have already seen that they know how craftsmanship is practiced on the typical I-farm. However, this form of organizing farm labor is expressly rejected by them. Their main arguments are that:

- it is too expensive,
- it takes too much labor, and
- it takes too long.

This last point is interesting: it refers again to an unmistakable shortening of time horizons, at least in the sphere of production. It is as if the velocity of circulation of capital must be raised as much as possible. Such an approach contrasts with an agricultural practice that

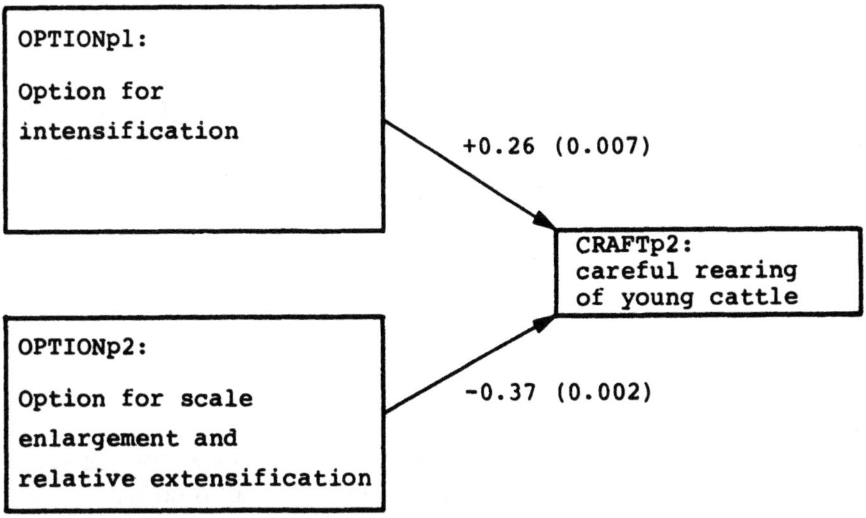

Figure 2.12 Correlation coefficients between options and craftsmanship in the field of animal breeding (ERSA/plain/n=75)

depends upon historically guaranteed autonomous reproduction, which not only produces a marketable surplus, but reproduces in each cycle some of the production factors used. When farmers speak of the importance of la bell'azienda, they are not just indulging in romanticism. They are expressing la bell'azienda's relationship to autonomous historically guaranteed reproduction. "To build a fine farm" is, in an analytical sense, the same as reproducing and producing inputs for future cycles. The relevant time horizon of the I-calculus, in other words, encompasses la bell'azienda of the future.[10] Increasing incorporation into feed, labor, and short-term loan markets implies that the costs incurred within a single cycle must be valorized. Thus future benefits are less relevant than immediate ones. Medium- and, to a lesser degree, long-term loans also lie within a very precise time span: the repayments and the yearly interest due require the realizing of immediate (monetary) returns, at least more than is the case for personal savings. The same applies to incorporation in land markets (via rent mechanisms):

> "The problem with rented land is that there is no certainty over the longer term. And that is a strong impediment. If, for example, you contemplate improving the herd, then you have to be able to think and plan long term, and that clashes with the rent contracts."

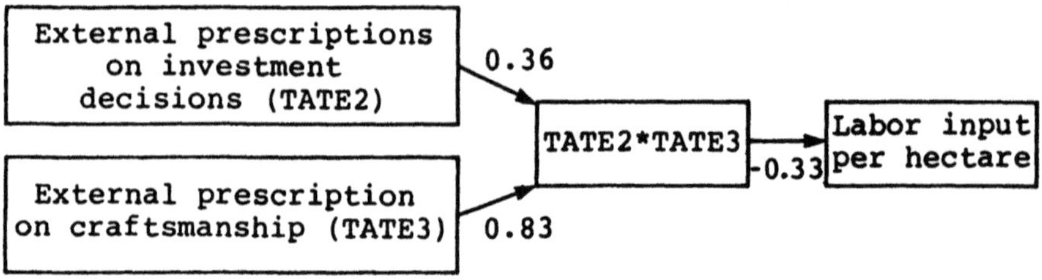

Figure 2.13

Craftsmanship and the Creation of a Frontier Function

If a simple linear production function is calculated for the dairy farming sector (on the ERSA sample n=134) we find: prod/ha = −497 + 0.42 labor/ha + 1.58 capital/ha + 2.38 inputs/ha (r^2 = 0.77). This result agrees with similar calculations made by others (Brugnoli et al., 1976; Messori, 1981). A loglinear function leads to better results, with the explained variance rising to 90%: prod/ha = 1.01 + 0.26 labor/ha + 0.07 capital/ha + 0.79 inputs/ha (r^2 = 0.90).

Incorporation has a twofold influence on the position of enterprises in the space defined by these functions. Incorporation has a substantial, negative and statistically significant effect on both the input of each separate production factor and the whole of the (summed) production factors.[11] As incorporation into markets and/or external prescriptions rise, the input of production factors and non-factor inputs per unit of land drops.

For the amount of labor input per hectare, for example, we find that a falling input of labor can make organizing labor as craftsmanship difficult (especially if labor input is reduced before technology can replace it) see Figure 2.13. This leaves little time "to farm well," in the words of local farmers. Take manure: Until recently this was left to heat for at least eight months and then spread the following spring. On farms where labor has been drastically cut back, however, this process is increasingly impossible to carry out. In the spring there is so much work to do that the carting and spreading of manure is mostly left undone. Its timing is therefore advanced to the winter months, which means that the heating process is foreshortened and the manure becomes increasingly a spreader of weeds. This compels the use of herbicides on the luzerne, which in turn shortens the vegetative cycle. In brief, labor reduction leads to a series of new problems and to negative effects on the quantity and particularly on the quality of fodder production.

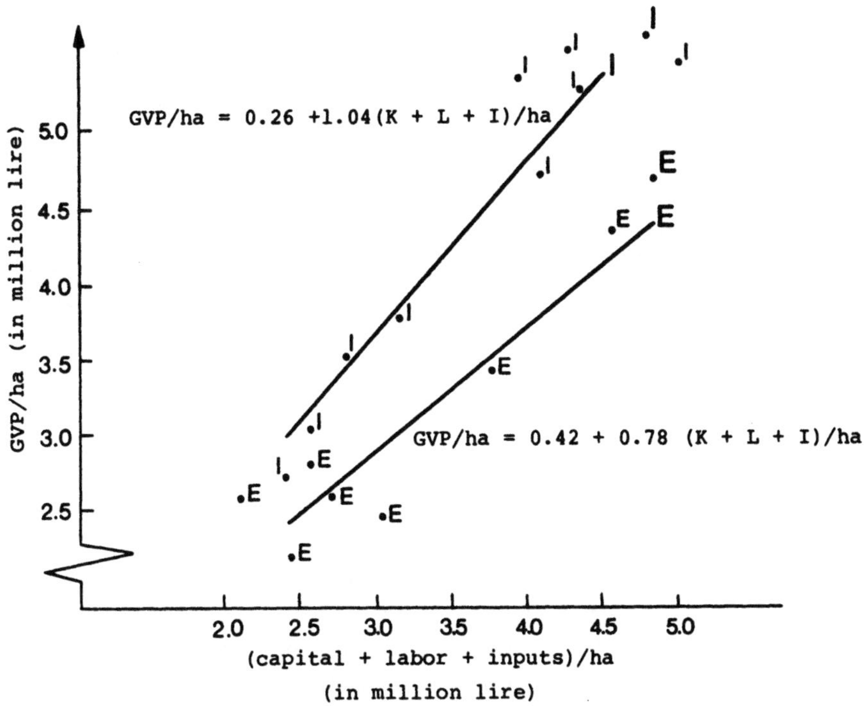

Figure 2.14 Production functions of I- and E-farmers (Parma data, 1979/n=18)

Craftsmanship, which is the structuring of labor leading to high production results per labor object, is thus made difficult and sometimes impossible. Thus we come from the input of production factors directly to their use. The degree to which production factors (once committed) are geared to obtaining high production results is measured with the help of the concept "technical efficiency." "A firm is considered more technically efficient than another if, given the same quantity of measurable inputs, it consistently produces a larger output" (Yotopoulos 1974:270). Timmer suggests that a firm is technically efficient "if the firm actually produces on the technical production function that yields the greatest output for any given set of inputs" (1970:99). In a graphic sense, one can imagine an increase in technical efficiency as the upward movement of the production function. Timmer expresses this as a "frontier function," that is, the production function of the most technically efficient firm. As craftsmanship is the means by which farmers achieve independent progress, then craftsmanship results in the creation of a frontier function. This is represented in a tentative way in Figure 2.14, constructed from the Parma data. Farmers who opt for intensification and who structure their labor as craftsmanship create a production function (given here as a linear link between totalled production

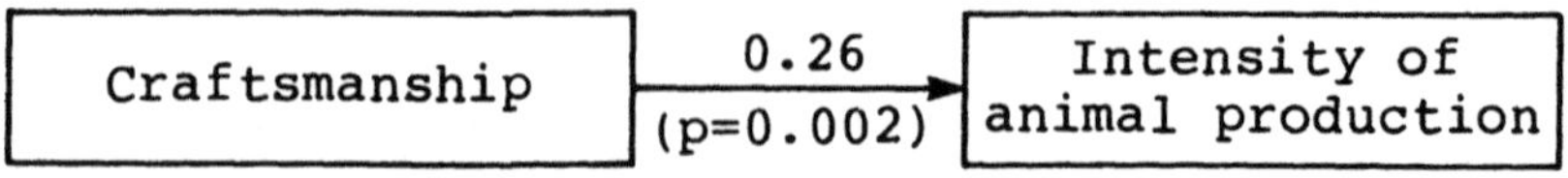

Figure 2.15

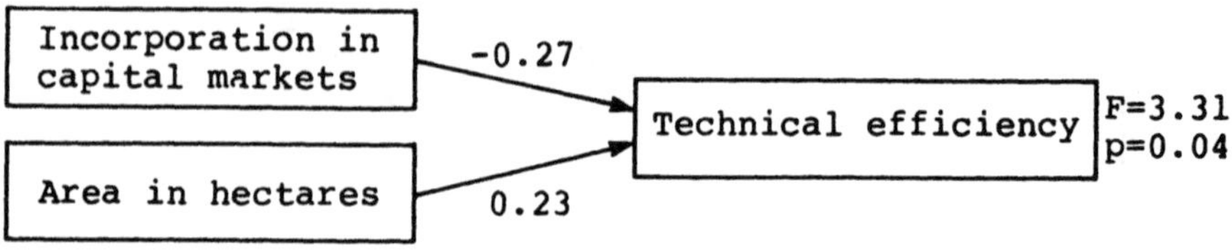

Figure 2.16

factors and production) which is higher than that of the E-farmers. The picture presented in Figure 2.14 highlights the meaning of craftsmanship.

In the larger ERSA sample (n=134) the influence of craftsmanship can be statistically tested. If we sum the earlier discovered craftsmanship factors (see Table 2.9), we find the relationship shown in Figure 2.15. If we formulate a production function to elucidate the gross value of production value per adult animal unit and give the labor per cow, feed-level and value of the cow as explanatory variables (see Figure 2.11), then it appears that the production function does indeed move upwards when craftmanship is added.

It can also be demonstrated statistically that with increasing commoditization and institutionalization, the E-calculus becomes dominant and that structuring labor as craftsmanship becomes increasingly difficult, if not impossible. If we combine the earlier production function (relating to the explanation of production/ha) with an index of incorporation,[12] then we find that, for the mountain, the plain and the total sample, the degree of incorporation has a negative effect on technical efficiency. The regression coefficient (not standardized) for the influence of incorporation for the mountains is −41.06 (F=2.53); for the plain −106.05 (F=2.13); and for the total sample −75.90 (F=2.10). Increasing incorporation makes the use of the I-calculus as a structuring principle of labor untenable; a downward movement of the production function (i.e., falling technical efficiency) is the consequence (see van der Ploeg and Bolhuis, 1983, for more details).

An additional and somewhat rough approximation is to operationalize technical efficiency as the relation between production and the input of production factors and to relate this term directly to incorporation. It then appears that incorporation into the capital markets has a significant and negative effect on technical efficiency (see Figure 2.16).

Interestingly, the effect of size (expressed in hectares) is positive. I will return to this particular point in a later section.

In synthesis, craftsmanship is a complex set of interdependent relations between producer, labor objects, and means. It is, in other words a specific way of organizing farm labor. Both labor and the relations between production unit and structural environment are molded by a strong orientation towards objects of labor and the optimal use of their productive potential as a means of further development.

Entrepreneurship as a Specific Structuration of the Farm Labor Process

Entrepreneurship tends to be the opposite of craftsmanship. Again it is not merely a matter of individual attributes. Insofar as entrepreneurship is a personal characteristic, it will nevertheless be the outcome of a specific complex of relations which not only lead to entrepreneurship but at the same time form the learning ground for it. If an orientation towards labor objects is crucial for the development of craftsmanship, then for entrepreneurship an orientation towards the market and market institutions is essential. Labor and production are organized around the relations, tendencies, expectations, prescriptions and recommendations operating in them. What is as often as possible outside of or even excluded from the framework of craftsmanship, is normative for entrepreneurship. However, a theoretical ordering of craftsmanship and entrepreneurship as an orientation to "production" and "markets," respectively, would be wrong. Craftsmanship would then be thought to relate merely to the "technical" talents of the farmer and entrepreneurship to the "economic" ones, and the coordination of both would then provide for a balanced complementarity. That, unfortunately, is indeed the assumption on which most theories of agrarian sociology and economics are based.

Of course, in craftsmanship, technical interrelations between producer, objects of labor, and means, are a continuing source of concern. One might even go further and say that the organization and planning of labor and production are to an important degree based on what appear to be purely technical arguments (raising milk yields, improving the quality of feed, bringing cattle rearing to a higher level, etc.); that produzione is not only a guiding principle but is also, in many fields, applied practice. However, one should not ignore the fact that the interrelations between producer, objects of labor, and means are not purely technical; they are also economic and social by nature. Even strictly technical arguments, as important as they are for structuring farm labor, are also important for defining and structuring economic and social relations. Cura and produzione reflect definite economic interests as the need for income, guadagno, and for an enterprise reproduced over time. Craftsmanship, therefore, contains the structuring of economic and social relations as well as technical relations and gives to economic and social relations a specific character. The same applies for entrepreneurship. Maybe entrepreneurship at first sight appears to concentrate simply on economics. However, the technical naturally becomes defined through the economic. If, for example, a herd of 50 cows per man is built up for economic reasons, then that has sweeping consequences on labor time, on the cura that can be given. If, for the same reasons, a farmer chooses to buy a combine harvester, then that will imply, often of necessity, a modification of plans in the technical sphere—making, for instance, the farmer more likely to pursue a monoculture of the crop for which the combine is appropriate. Only then can such a purchase be considered rational. One could go on at length with such examples. The main point is clear, however: with the definition of economic relations, technical relations in the labor process are also circumscribed, and vice versa. The economic and technical cannot be separated.

Before going on to discuss the way entrepreneurship is conceptualized in this present study, I would like to go into three aspects of current theories on entrepreneurship. A critical discussion of these provides a foundation for understanding the way the term is operationalized here.

1. In the first place, as Walters (1963:5) puts it, "there is no generally accepted cardinal measure of entrepreneurship," let alone, I would add, any valid theory on agrarian entrepreneurship (see also Hinken, 1974:27). No one can or will say what precisely entrepreneurship in agriculture is. As Benvenuti, Bussi and Satta (1983) maintain, it is a "phantom." But of course ghosts do have very important social functions. One could even argue that the very obscurity of the concept is one of the main prerequisites of entrepreneurship as social definition. Further on I will return to this point. Zachariasse, one of the foremost European scholars on entrepreneurship, gives an apparently clear definition when he writes that "the farmer must, if he follows purely economic principles, try to reach as high a net surplus as possible within the given circumstances of his business . . . with due regard to specific prescriptions for maintaining long-term profitability" (1972a). The problem, however, is that the crucial concept in this argumentation, i.e., *net surplus*, only recently emerged as a category in farming accountancy. Apart from that, the concept of net surplus represents, as we have demonstrated elsewhere (Frouws and van der Ploeg, 1973:105), more a specific interest than an objective, "universal" yardstick. Labor input and its remuneration, typically enough, fall outside the scope of this

particular concept. Thus Zachariasse's definition is very dependent on a specific location in time and space, to say the least. For Zachariasse, however, it is clear that the more the farmer strives for the highest possible net surplus, the more he demonstrates himself to be an entrepreneur. The question that next comes to mind is, how does an entrepreneur operate? What does he do that the non-entrepreneurial farmer neglects to do? Laboring goes with laborer, does then a business component called "entrepreneuring" go with entrepreneur? From whence does that "entrepreneuring" come? How is this striving for profit maximization achieved? Such questions remain unanswered.

The structure and mechanism of entrepreneurship—of entrepreneuring as an activity—is neither described nor researched. Zachariasse limits himself in his study to a scrupulously careful description of the cultivation techniques used by farmers for potatoes, sugar beets and wheat. Data about sowing, number of kilos of nitrogen per hectare, soil humidity and fifty-one other variables are detailed. It then appears that special cultivation methods go with higher yields per hectare, which correlate, in turn, with a higher net surplus. What is missing, however, is the construction of a link between data on sowing for example and "the striving for profit maximization." Does a farmer actually structure his decisions in such terms? Perhaps, and perhaps not, but Zachariasse does not make it clear.[13] For him, the personal experience, knowledge and insights of farmers do not count for much.[14] "Nevertheless the good cultivators among the farmers apparently know . . . how to come to decisions that will bring them a higher physical level of crop per hectare" (1974a:3). Thinking through this particular argument of Zachariasse, let us now suppose that farmers who get high yields, thanks to a certain method of cultivation, are good craftsmen instead of "good" entrepreneurs. Doesn't the whole story about entrepreneurship again hang in the air?

On reflection, there is little mystery attached to entrepreneuring as an activity. What does the farmer as entrepreneur actually do? He will manage his business in such a way (and this is a permanent process) that business organization and operation is congruent with market relations and tendencies. Zachariasse says the same in so many words: "the price of production factors, means and products will always be one of the most important guiding principles of his economic dealing" (1972a and 1974b). Without a continuing projection of market relations on the farm enterprise, such concepts as net surplus and profit are unthinkable, let alone quantifiable. Adjusting farm organization and operation to these relationships is a prerequisite for and a mechanism for achieving profit maximization. There-in lies the meaning of entrepreneurship.

That this meaning of "entrepreneuring" is never or seldom made explicit is not because of the supposed complication of the concept or the "insusceptibility" of the activity itself. The problem is just that the continual reorganizing of farm relations to the swing of prevailing market relations is linked to the goal of profit maximization in an *ex ante* way. Only in retrospect is it possible to determine the extent to which a farmer has been successful as an entrepreneur. The structural turbulence in the economic environment of farmers, to which not only markets but also agri-business contribute (Benvenuti, Bolhuis, van der Ploeg, 1982), is such that the congruence between ex ante projections and ex post facto results is not only a very uncertain matter but is sometimes completely missing. In other words, the logical legitimation for acting as an entrepreneur, as *homo economicus*, is increasingly swept away. Science solves this problem by means of a classic turnabout: legitimation (i.e., profit maximization) is made absolute, and the practice of entrepreneuring, as a guiding activity, is masked and hidden from view. The farmer as entrepreneur is thereby, to a large extent, an ideology. Although farmers do not appear in the strict sense to be entrepreneurs (so it appears from the follow-up study of Zachariasse [1979]), "the striving" to make farmers "better entrepreneurs . . . is undiminished." Given the structural turbulence which disconnects ex ante calculations and ex post facto results, entrepreneurship must indeed be represented as intrinsically good. A legitimation in terms of rationalizing entrepreneuring as an activity would undermine itself, given the economic environment in which it is enacted.

2. The second point I want to make is logically connected to the first. To the extent that current theories on agrarian entrepreneurship regard entrepreneuring as a social activity, individual attributes are given exclusive attention. An overview by Muggen (1969) discusses some seventy-three studies of "human factors and farm management." Muggen classifies the various facets of entrepreneurship derived from these studies using Nielson's model (1965), which is reproduced in Figure 2.17.

Under the heading "capabilities," Muggen reviews twenty-five variables, from "spatial insight" through "memory" to "verbal capacity." Age and training, which fall under "biography," emerge as the most studied variables. But here again, "no clear picture emerges from the results." In some studies, these and other variables are linked positively with "farm management performance," and in others they are linked negatively. Achievement motivation (sometimes measured in patently absurd ways) and the degree to which a man is "scientifically oriented" are the key concepts in the category "drives and motivations."

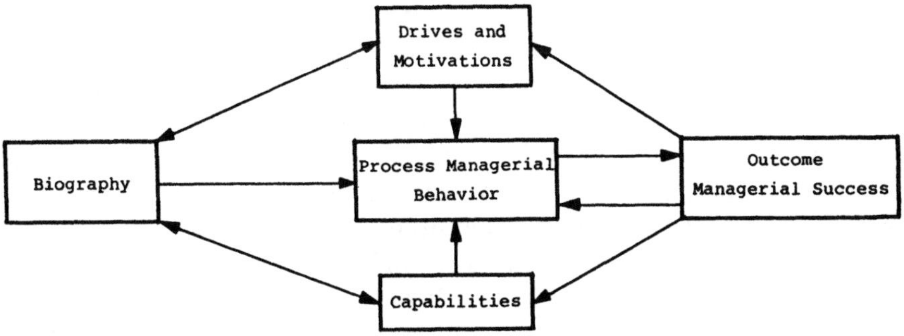

Figure 2.17 Nielson's model of entrepreneurship

The more interesting category for us might be "process managerial behavior." Muggen says that such studies are not yet very numerous. However, a closer inspection of the eleven studies noted makes it clear that attention is devoted again to the individual characteristics of the entrepreneur and not to the mechanisms of entrepreneuring. Muggen summarizes these characteristics as: "use of information sources, and use of the scientific method of decision making," and "decision-making ability." One is left to wonder what the underlying substance of such characteristics might be. Considered in this light, one might question what the difference would be between a secondary school pupil and a farmer, let alone what meaningful distinctions such criteria could make between the farmers themselves.

"Plans for the future" is another such category. Cole and Wolf (1974) describe in detail how one of their respondents and his son drove every year to a distant village to lay in huge stocks of good wine. His son drove on the way back so that father was already free to taste the wine. The father is content. He has enough wine laid in to supply his needs for the coming year. Are these "plans for the future"?

In synthesis, the current literature on agrarian entrepreneurship not only fails to tackle entrepreneuring as a conscious activity (as I made clear under point 1) but also obscures it behind a smoke screen of personal characteristics and attributes. These characteristics and attributes are, in their turn, often so ambiguous or unclear that it is really no surprise that, as Muggen concludes, "the magnitude of the correlations obtained (between attributes and characteristics, on the one hand, and outcomes, on the other) shows that our understanding of these factors is strictly limited."[15]

3. The third problem I wish to highlight is that currently, agrarian entrepreneurship is defined as being located in a kind of *tierra incognita*. Even Robinson Crusoe, who certainly had "plans for the future," could thus be considered an entrepreneur.

Considered sociologically, the concept of entrepreneurship refers to a role (see also Hinken, 1974:26). However, a role can never be defined in an isolated sense. One can only speak of a father if there is a son or daughter. A role emerges only in relation to others; it is the reciprocal role expectations and definitions as such which make up the role. If "farmer as entrepreneur" is a role or role definition, then this evokes, among others, the following quite simple questions: Who defines this role? On the basis of what interests? And who readjusts the "entrepreneur's role" (Benvenuti, 1975a and b)?

In short, the conscious activity of entrepreneuring (as defined under point 1 above) presupposes a social context (rather than a cluster of specific individual attributes, as discussed under point 2). In this context a network will be observable through which the farmer and other actors will negotiate and renegotiate the role enactment of the entrepreneur. Such a network will be discussed later in terms of the technological administrative task environment (TATE) of farming (Benvenuti, 1985b).

A number of items can be derived from the foregoing discussion. Together they form the profile of agrarian entrepreneurship as a specific social activity. They are the following:

1. The "farmer as entrepreneur" gives more importance to markets, to their development and interrelationships, than does his opposite, the craftsman, who is of the opinion that progress comes primarily from work. In the Parma data, indeed 86% of E-farmers (as against only 33% of I-farmers) were of the opinion that "the farmer who is always to be found in the marketplace is a better farmer than the one who never leaves the farm." This item was also presented to the ERSA respondents, where 90% were of the opinion that in general the farmer who is always at the market is the better farmer. This higher score could be influenced by the special situation of the mountains, where the marketplace is regarded as a meeting place. Men go once a week to the *piazza* to talk and swap experiences, but of the piazza as a marketplace, as a place to trade, there remains only the memory. They were also asked which farmer they most resembled. Only 77% (out of the 90%) replied that they were like the farmer who was always to be found at the market.[16] This difference between norm and behavior appears to be a good indication of the prescriptive power of "entrepreneurial ideology," "for a good entrepreneur goes to markets."
2. In an extension of the previous item, respondents were asked which of two factors was especially determining for income—

"prices, or the manner in which you work." It was prices for 71% of the E-farmers and 33% of the I-farmers.

3. The "farmer as entrepreneur" will be more inclined to adapt enterprise operations to market and price relations and their changes. Newby et al. (1978b) observed that there exist great empirical differences in market sensitivity, differences which, or so it appeared from their research, could not be linked to farm size. Also, in Italy, it appears (see for example, the five-year empirical study of Angeli and Omodei, 1981) that some farmers react with undue alertness to price changes while others do not. In the ERSA data the following item was presented to the respondents: "There are many farmers who are very sensitive to market prices, as for example, when milk prices fall, they lower their milk production, and rely more on the meat. Others, on the other hand, always continue in the same way as before. Who are you most like?" In the ERSA sample 42% of the farmers likened themselves to the adaptive farmer.
4. A precondition for all this is that the "farmer as entrepreneur" will, more than his hypothetical counterpart, perceive productive organization mostly in economic terms. The ratio of certain activities and relations lie not in their relation to produzione, but in their eventual correspondence with market relations. We asked, "Is it true that it is cheaper to feed a cow who produces 40 ql than to feed one that produces 50 ql?" The intention was to see how far farmers projected price relations into cattle feeding and considered feed as a cost item. In the ERSA data, 52% replied that it was cheaper to feed the cow that produced 40 ql.
5. The foregoing implies that the entrepreneur will be less likely to see the solutions to problems in terms of political intervention in the market ("price strife") but more likely to see them in terms of adjusting the individual enterprise to market relations (by a reorganization of the cost/benefit ratio). In the ERSA data, 85% opted for price strife, 15% for cost reduction. In the Parma data, 57% of the E-farmers opted for reducing cost against 22% of the I-farmers.
6. Originally we thought that the farmer as entrepreneur would be inclined to calculate each investment, each change, in terms of eventual profit improvement. We asked the following: "If you see a new machine which seems useful, what do you do then? Do you say, that machine will be useful to me and go and buy it, or do you calculate whether what the machine will save will come to more than the cost and depreciation?" Correlational analysis of various entrepreneurship items on the ERSA data

showed a high negative correlation between this item and all other items. The structure of "entrepreneurship factors," discussed later, also shows this. In retrospect, i.e., after re-reading the Parma interviews, that did not seem so surprising. What does an entrepreneur by nature say?

> "Look, if anything new comes onto the market, then you can be sure that the engineers and technicians have spent long enough brooding on it. After all they have to sell their products and that product will only go down well if the farmers wear it. They are smart enough about that, so with new products it is only a question of seeing it, and if you can get the credit for it well then, bang: you buy it."

This is a remarkable example of institutionalization: independent calculations and, in a certain sense, planning are given up and get delegated to external institutions. Thus technology becomes, as it were, normative. The "technique emerges as a language," according to Benvenuti (1982b:122). Insofar as a farmer still regards it as proper to calculate, the object of calculation shifts. It is no longer over technology, which has become the norm, but over the consequences of its applications.

> "For me a mechanized tomato harvester (he named the newest type) is without doubt the best. You have the lowest cost per hectare and you can harvest 70 hectares with it. But the problem for me is how I can arrive at extra land, to write the machine's costs off to make it pay for itself."

Those who do carefully weigh up the benefits of purchasing machines do so from another point of view, namely, the desire to maintain "a well-balanced business."

> "That means for example that you don't have a heavier tractor than is necessary for your farm; it means also that feed production and stock should be more or less equal to each other; that your rotation system is suitable for your hectaragen."

Only those who wished to keep a low level of incorporation gave serious thought to preventing technology (and the cost/benefit relations it entails) from becoming the factor which defined the scale and framework of their enterprise. In the ERSA data, 52% said you should "calculate carefully" while 48% said that it was not necessary to do so when buying new machinery.

7. Knowing that the market environment carries with it a certain degree of turbulence and that an enterprise geared to such an economic environment will, to a large extent, be exposed to this

turbulence means that the "farmer as entrepreneur" will be relatively more inclined to consider taking risks as acceptable. In the Parma data 71% of the E-farmers were of the opinion that risk taking was acceptable, as against 33% of the I-farmers.

8. The "farmer as entrepreneur" will assign more importance to investment relating to (future) income than to a continuing improvement of craftsmanship. This setting of priorities applies particularly if such investments bring about scale enlargement and/or cost reductions. Indeed, 86% of the E-farmers as against 45% of I-farmers from the Parma sample agreed that substantial investment is crucial.
9. Because heavy investment puts pressure on expendable income in the short term, the entrepreneur is more inclined to squeeze family income for the sake of investment. We asked respondents in the Parma sample, "if you have a net income of 10 million, what part do you spend on the family, what on the business?" The average family spending of the E-farmers was 3.6 million while that of the I-farmers 4.1 million.
10. A consequence of the previous point is that the entrepreneur will hold future income (over, say, a five-year period) to be a more important reference point than present income. The shortening of the time period in the direct organization of production (illustrated previously) urges a lengthening of perspective when it comes to income improvement. Income will depend less on business operation and more on business organization (in the eyes of the entrepreneur, at least), and the latter is only manipulable on a long-term basis. In the ERSA sample 52% thought "present income is more important than income over five years," 2% found both categories equally important, and the remaining 46% considered future income more important.

The majority of the above items were used in the Parma sample. The most discriminatory and theoretically interesting items (1, 3, 4, 5, 6 and 10) were taken up in the ERSA questionnaire. A factor analysis was applied to the material for both the plain and the mountains in order to be able to analyze some composite indices of entrepreneurship. Table 2.10 summarizes the results of this analysis.

These factors will be used in the multiple regression analysis of the following section. The first factor (ENTREPp1) relates to future income, which, on the plain, is considered to be a more important parameter than present income. The degree to which a farmer calculates, on the contrary, has a high, negative loading on this same factor. A similar, though weaker, structure is encountered in the mountain sample.

Table 2.10. Results of Factor Analysis Applied to Items of Entrepreneurship ERSA/plain/n=75 and mountains, n=59 (oblique rotation)

	Plain ENTREP			Mountains ENTREP		
	p1	p2	p3	m1	m2	m3
Feeding at a 40 ql level is more economic than at a 50 ql level	.06	-.27	.71	-.13	.65	-.31
An entrepreneur ought to calculate	-.87	.04	-.18	-.70	-.19	.35
It is necessary to frequent the markets	-.08	.90	.07	-.08	.09	.86
To adapt the farm continuously to the market is necessary	.28	.49	-.09	.02	.75	.20
A farmer ought to strive for cost reduction instead of engaging himself in price strive	-.01	.22	.84	.81	-.17	.14
Future income is a more important goal than present income	.75	.12	-.20	.29	.44	.31
Eigenvalues	1.66	1.14	1.11	1.42	1.21	.97
Variance explained	28%	19%	19%	24%	20%	16%

ENTREPp2 speaks for itself: market visiting in combination with the adapting of business operation to changing prices. Finally, ENTREPp3 represents entrepreneurship in "the immediate sense": operating is substantially in terms of perceived costs, and cost reduction is to a high degree normative over price-strife. One might suggest that ENTREPp1 is above all related to farm organization (where technology appears to be the guideline) and that ENTREPp3 is primarily expressed at the level of the daily labor process.

In the mountains something rather different arises, although the combinations are not different. ENTREPm1 combines cost reduction with the absence of any felt need for calculation. On first sight this may seem to be a rare combination, but it is not if one starts from the hypothesis which cropped up earlier around the normative working of the technology offered. ENTREPm2 applies to the conceptualization of feed as a cost, the impact of price changes, and future income as all being more important than present income. One sees that what were two different dimensions on the plain ("entrepreneurship over the long term," ENTREPp1, and "entrepreneurship in the immediate sense," ENTREPp3) begin here to converge to a certain degree. Finally, the structure of ENTREPm3 is typical for the mostly social function of

Table 2.11. Correlation Coefficients Between Option Factors and Factors for Entrepreneurship

	ENTREPp1	ENTREPp2	ENTREPp3		ENTREPm1	ENTREPm2	ENTREPm3
OPTIONp1	-0.23	ns	-0.34	OPTIONm1	+0.32	ns	ns
OPTIONp2	ns	ns	ns	OPTIONm2	-0.27	ns	ns
OPTIONp3	ns	+0.31	ns	OPTIONm3	ns	ns	ns

note: ns - not significant, with p - 0.05

the market or piazza, as a meeting place. Only the first variable distinguishes itself by a high positive loading. The other loadings are weak and contradictory.

Earlier we saw that craftsmanship was the operational mechanism for effecting the option for intensification. In the same way entrepreneurship is the means by which an E-option is realized. That link is not accidental; it is already entailed in the E-calculus.

Table 2.11 gives the Pearson correlation coefficients for the direct links between OPTION factors (derived from Table 2.3 and 2.4), and entrepreneurship factors (derived from Table 2.10). The result shows that in the mountains the "explicit opting for extensification and the priority given to labor and cost reductions" (OPTIONm1) relate positively and significantly with the first entrepreneurship factor ENTREPm1. On the plain, opting for intensification (OPTIONp1) correlates negatively and significantly with the first entrepreneurship factor (ENTREPp1).

The construction of more meaning-loaded patterns makes the relations clearer. If we imagine, for example, that ENTREPRENEURSHIPp = ENTREPp1 + ENTREPp2 + ENTREPp3, then this new combination expresses how high all the dimensions of entrepreneurship score. For the link between OPTIONp1 (opting for intensification) and this new variable, which comprises all the aspects of entrepreneurship, we find $r=-0.43$ ($p=0.0001$)! A specific combination of OPTION factors (E-OPTIONp = − OPTIONp1 + OPTIONp2 + OPTIONp3) expressing the extent to which scale enlargement and relative extensification is thought to be normative and effectively translated into farm planning is likewise related in a significant and positive way with all separate entrepreneurship factors. Finally, correlation of this E-OPTIONp with ENTREPRENEURSHIPp underlines once more the importance of pattern forming on both levels: $r=0.30$ ($p=0.008$). In short, entrepreneurship can indeed be considered as the concrete mechanism for the realization of the E-option.

Table 2.12. The Analysis of Farm Performance in a Typical E-Farm

Industrial concentrates	35.2 kg	9.100 lire/ql milk	25%
Bought-in Feed	15 kg	3.880 lire/ql milk	11%
Medicines, veterinarian		675 lire/ql milk	2%
Milking time	70 minutes		
Various tasks in the stable	90 minutes		

On TATE: The Technological-Administrative Task Environment

Although entrepreneurship contains a number of clear criteria on which to base activities in the cow shed and field, projecting market and price relations on these activities does not necessarily result in a specific design for organizing the labor process in practice. Market and price relations are abstract entities with respect to the labor process. They can be used—in the framework of entrepreneurship—to calculate the profitability of early as against late calving. The idea itself of early or late calving (or of using silage methods instead of hay drying techniques, etc.) cannot be derived from market and price relations. But once such alternatives are developed and recognized, then entrepreneurship can indeed function as an adequate mechanism for choosing them. The designing of new alternatives in the development of farming seldom springs from entrepreneurship as such.

This is particularly the case in highly developed agricultural systems, and above all in farms which have substantially enlarged their scale. Again—within the normative framework of entrepreneurship—a farmer can readily deduce from general price and market relationships that in order to increase the number of cows and the income per unit of labor, the labor input per cow must be minimized. Labor time per cow is then also one of the criteria for measuring enterprise progress for those farmers who define themselves as entrepreneurs. Table 2.12 records a small fragment of a self-styled bookkeeping system used by a typical E-farmer of the Parma sample. It gives us a glimpse of what the craftsmen among the farmers mean when they say "just look there, he doesn't take the trouble to care properly for his cows." It is only the time spent that counts.

Notwithstanding the fact that many small adaptations in the daily operations of the farm are derived from this way of thinking,[17] it is also clear that a major breakthrough in the organization of time is dependent on technological renewal. Milking machines and modern cow stalls with machines for automatically supplying concentrates and carrying away the milk are innovations which at a stroke reduced labor time from around 50 minutes per cow per day to less than 5 minutes. One can go further and say that such innovations make possible other adaptations, which taken together, can radically alter the way a farm operates. Letting a bull loose, for example, is unthinkable in a traditional cow shed but quite possible in a modern cow stall, where a "bull on the loose" can again save some time. Farmers who organize their labor within the normative framework of entrepreneurship will tend to be more receptive to the dominant technology offered.[18] Technological development provides them with those elements crucial for farm development that they cannot sufficiently develop themselves; namely, a continually changing gamut of techniques to be subsequently judged in terms of profitability.

There is a clear line running from entrepreneurship to a certain dependence on new technology, especially since the propaganda that goes with new technology is often in terms of improved cost/benefit ratios. This leads to a situation where independent calculation becomes superfluous: "The technicians and engineers have been brooding over it for long enough, it must be good."

In most agricultural systems a specific technological and administrative task environment (TATE) can be discerned. This concept was developed by Benvenuti (1975a, 1982a, 1985b) to describe the network of market-agencies and associated institutions to which farmers are tied both economically and technically (agricultural industries, banks, trade consortia, extension services, etc.). It is from within such a network that the concrete organization of the farm labor process—sometimes directly, sometimes indirectly—is prescribed and eventually sanctioned. It is from TATE that the farmer obtains those elements which are necessary but which he cannot independently or fully develop himself. TATE therefore forms the embryo of a specific division of labor between head and hand (i.e., TATE expresses the separation of what in craftsmanship, to a large extent, still forms a unified whole). A strong externalization of some of the tasks from the broad spectrum of farm labor leads not only to a substantial increase of the commodity relations in which the farm is involved, as already demonstrated, but also to the creation and reproduction of technicological-administrative relations. If milk cows are no longer reproduced within the enterprise but purchased from specialized breeders and if concentrates are no longer grown, milled, mixed and enriched on the farm but instead come from industry, then the classic unity of past and present labor disappears as a guideline for carrying out tasks. The coordination of some tasks is partly synchronic but above all diachronic. The insights and experience gained from previous cycles form the detailed empirical knowledge crucial for the next cycle. Perhaps that seems self-evident; but let one of the Parma respondents tell the story.

> "There lies the art . . . in all those thousand small things which the outsider does not see, and which earns us the name of dumb farmer. But if you don't know from what lot the feed came, how much humidity was in it, what the composition of the field was, or whether it has enough protein, enough roughage . . . then you also won't know why some cows crap badly, or drop behind in their milk yield. You will see them crap well or badly, but you won't know what it means. That is why you have to follow everything so closely and be forever trying to understand. Even if you're a modern farmer. The real art is in being able to get out of the situation what lies behind it, and make something fine of it. You don't get that from the technician or the market. Standard things, standard recipes, well you need to know them of course, but a good farmer goes a step further. That's why I speak of art."

One can add little to this except to say that this art is so important that when it is ignored or excluded, new technological designs often fail. For instance, a complete automation of cattle feeding (both roughage and concentrates) seems impossible, despite all efforts to achieve it, simply because this unity of past and present labor cannot be integrated into the prevailing software. At least where an optimal intensification is pursued, time cannot be reduced to a discrete variable. However, when externalization is vigorously pursued, this unity of tasks—part synchronic, part diachronic—is eroded. When this breaks down, the "art of the specificity" has to be replaced by external "directions for use" (Benvenuti and Mommaas, 1985) or "mode d'emploi" (Lacroix, 1981). Local knowledge then is increasingly substituted by a complete package of technical advice in which the how, when, why, where, and how much, the sequence and the required combinations, must all be specified.

Such directions for use are naturally not limited to the examples given here. They arise on all sides—between farm and bank when it is a question of external financing, and between farm and industry when the latter has taken over specific tasks such as cheese- and salami-making, slaughtering, etc. Farms must then meet the processing industry's requirements for quality, delivery procedures, etc. In the case of contract farming, such prescriptions reach a high point. But external prescriptions and sanctions from a technological-administrative task environment (TATE) are not limited only to contract farming. They are much more widely distributed, as Benvenuti and Mommaas (1985) have thoroughly documented.

In Emilia Romagna a clearly defined network of industry, government institutions, extension services (also charged with effectuating supranational policy programmes such as those of the EEC), and banks can be identified. This network provides farmers with a huge range of technological artifacts, organizational schemes, with several specific "opportunities" and an overwhelming, though sometimes confusing, array of "directions for use." However, the same network, although relatively efficient, is distinguished by a poor level of centralization and a concomitant inability to explicitly sanction.

The cheese-making factories (*caseifici*) are the most important institutional junctions with which every farmer has daily dealings. The milk he delivers has to be "good cheesemilk." The labor process in these cheese factories must therefore be coordinated with that on the farms. And the reverse is so to an even greater extent. Such coordination is channeled through technical and administrative rules. For many reasons these *caseifici* are very numerous and all of them quite small.[19] Most of them rely on only about thirty farmers for their supply. In such a cheese factory only one cheese maker works, usually with the help of an assistant (Bussi and Rizzi, 1974; Brugnoli, 1980). These small *caseifici* are all linked to a consortium. The influence of decisions made by this consortium are not difficult to trace on the farm. A part of the working practice is even explicitly prescribed. "The good farmer," they advise, "will limit himself to milking and leave breeding to the experts!" (Consorzio, 1980:17). The *Regolamento per la Produzione del Latte* (Consorzio, 1973) prescribes very precisely which feed elements are forbidden, the minimum and maximum amounts for other feed stuffs, and the qualitative and quantitative requirements for the composition of feed as a whole (see Figure 2.18). However, their ability to effectively sanction is almost nonexistent (Alvisi, 1980:16). The same holds true for subcontracting industries, firms that build cow sheds, and to a lesser extent the banks. There can be no question of monopolization or of high-level centralization, and even less of any categorical imposing of one-sided sanctions, which is often the case elsewhere in Europe and in Italy (Benvenuti, Bolhuis and van der Ploeg, 1982). If an external prescription is not to the farmer's liking, then he can switch to another institute or try one way or another to translate a particular prescription into its opposite. This does not mean that there is no institutionalization, it simply implies that institutionalization cannot be seen in terms of fixed, formal and unilineal relations. Institutionalization in this situation must be measured much more in terms of dual relations: as the degree to which the farmer systematically does or does not go along with the external prescription, i.e., the degree to which the farmer adjusts his own "criteria for farming" with that which is implicitly or

Tab. 3 - Frammenti del «Regolamento per la produzione del latte» edito nel 1973 dal Consorzio del formaggio Parmigiano Reggiano

GLI ALIMENTI:
ASPETTO QUALITATIVO E QUANTITATIVO

Per la produzione del formaggio parmigiano-reggiano occorre impiegare latte di vacche alimentate con *foraggi locali* (*) adatti ed integrati con idonei mangimi concentrati.
Gli alimenti che nel corso dell'anno si rendono disponibili e che possono essere impiegati nel razionamento delle vacche che producono latte destinato alla produzione di formaggio parmigiano-reggiano sono riportati nell'allegato calendario.
Si ritiene opportuno sottolineare il fatto che per effettuare una razionale alimentazione delle vacche da latte è indispensabile osservare, per ognuno degli alimenti considerati, dei limiti che possono interessare o la quantità massima o quella minima da impiegare giornalmente.

A) I FORAGGI

Sono da ritenersi adatti per l'alimentazione i sottoelencati foraggi che debbono costituire l'alimento di base e prevalente delle vacche in lattazione.

3) **Fieno di medica:**
— minimo: 4 kg capo/giorno (concomitante con l'impiego di elevate quantità di erba);
— massimo: a volontà.
4

4) **Fieno di prato naturale:**
— minimo: 5 kg capo/giorno (concomitante con l'impiego di elevate quantità di erba);
— massimo: a volontà.
5

5) **Erba di trifoglio:**
— minimo: nessun limite;
— massimo: 25 kg capo/giorno.
25

6) **Erbai di loiessa, segale, avena, orzo e misti:**
— minimo: nessun limite;
— massimo: 30 kg capo/giorno.
30

7) **Granturchino (*):**
— minimo: nessun limite;
— massimo: 25 kg capo/giorno.
25

8) **Sorgo ibrido** (da ricaccio) (*): falciato a sufficiente distanza dal suolo ed al raggiungimento dell'altezza di almeno un metro:
— minimo: nessun limite;
— massimo: 20 kg capo/giorno.
20

ALIMENTI IL CUI IMPIEGO È VIETATO

FORAGGI FRESCHI E CONSERVATI

1 - *Insilati di ogni genere;*
2 - *Erbai di sorgo zuccherino a maturazione estiva;*
3 - *Erbai di sorgo ibrido di altezza inferiore al metro;*
4 - *Colza, ravizzone, senape, fieno greco, foglie di piante da frutto e non, aglio selvatico, coriandolo;*
5 - *Ortaggi in genere (cavoli, rape, patate, piselli, pomodori) ivi compresi scarti, cascami e sottoprodotti vari;*
....
....
....
....

M A N G I M I

1 - *Farine proteiche di origine animale (pesce, carne, sangue, penne, sottoprodotti vari di macellazione, idrolizzati, solubili, ecc.) nonché i sottoprodotti freschi ed essiccati del latte (siero, latticello, farine lattee, ecc.) i grassi e gli olii di qualsiasi natura;*
2 - *Semi di diversa origine (veccia, svecciature, fieno greco, colza, ravizzone, lupini, fagioli, lenticchie) ed i seguenti sottoprodotti della lavorazione del riso: lolla, pula, puletta, farinaccio, gemma e granaverde;*
....
....
....
....

Figure 2.18 A fragment of the Regolamento: An example of "directions for use" as sent out by TATE. Reprinted by permission from Consorzio (1973).

explicitly prescribed. Thus I take the view that the more individual criteria for performance coincide with the "directions for use" as prescribed by TATE, the more a certain dependence on external institutions and markets is internalized by the farm (both in a material and symbolic sense). The farm becomes then so structured that a gradual appeal to external institutions will become necessary.

In more theoretical terms this means that: (1) an analysis of TATE dependency cannot ignore the dialectic between external forces and internal responses, but that (2) once established such patterns of technical administrative dependency can dramatically change the balance between both. Long is right to criticize the "tendency to interpret the restructuring of agrarian systems and farming enterprise as resulting basically from the penetration of external forces. . . . All forms of external intervention necessarily enter the existing life worlds of the individuals and social groups affected and thus, as it were, pass through certain social and cultural filters. In this way, external forces are both mediated and transformed by internal structures" (Long, 1984:17). Indeed the creation of technical administrative relations which link farms to a network of external agencies implies numerous actors who work from a variety of meaning systems. That is the reason why in this study the measure of external prescriptions is operationalized not by means of the formal prescriptions articulated by this network but by the degree to which farmers enter into and identify with these external criteria. The point is that the interrelation between "external forces" and "internal responses" is not to be analyzed in ontological terms which place "determinism" and "voluntarism" as opposites. As I have argued before "the question is basically a matter of social practice" (van der Ploeg, 1985:20). As Giddens states (1981:55), domination and power as structural properties of societal systems do not only entail sets of rules and their definition, but "also sets of resources and their allocation." In modern agriculture the development, allocation and control over resources become externalized quite rapidly. Related to this process is the emergence of a new type of "domination machinery" as shown by Benvenuti (1982a), Eizner (1985), Rambaud (1983) and many others. The E-logic then can be considered as the nexus of this process at the farm level. It is the vehicle by which the subordination of farming to outside agencies and conditions takes place. The result is a certain expropriation of the "design" dimension of farm labor. Although such a result cannot be considered as "necessary"—given the example of the I-farms—holly or partially "expropriated" farm labor undergoes a change of analytical status from being an independent or "explanatory factor" (as it is in the I-logic) to becoming increasingly a dependent or determined variable (as is the case in the E-logic). This means that

TATE, as a dual relation, can clearly result in a continual narrowing of space for the farmer to maneuvre in; i.e., "external forces" become a dominant guideline for the labor process of at least some farmers.

Within the conceptualization of technical administrative relations as in essence dual, a daily stream of information can be observed from TATE agencies to the farmers. The forms are many: newspapers, radio courses, the agricultural press, the regular visit of extensionists and advisers to the various farms. This information flow relates only in part to recommended items as such. An innovation, be it a new cow shed, machine or medicine, only makes sense when used in a particular, prescribed way, as argued more generally by Latour (1983). And vice versa, it is that particular practice (often promoted by the information flow from TATE) that makes the new innovation, so to speak, necessary. The representative of a firm that delivers farming machinery described it to me at a *fiera* (market) as follows:

> "You are selling not so much a machine, you don't get anywhere by shouting about the few plus points of a new thing. You have to first introduce a new way of thinking and working, making it reasonable for farming to be organized in a new way. . . . When that works then these men will buy your machines . . . and without any problem. Then you will find adverts are superfluous for men decide for themselves that buying is necessary."

Through such techniques agribusiness has a substantial and dynamizing influence on farm enterprises. In a note from the Italian branch of Unilever (1978), the redesigning of farming using industrial criteria is unmistakable:

> It is simply necessary to make the cost structure of agriculture the main theme of our policy. . . . This requires a renewed approach, where the dynamizing role of industry is placed central for the lowering of agricultural costs. . . . Industry must turn its sights to that field which is most her own, that of efficiency and of organizing production factors with the aim of accomplishing a minimalization of costs; a strong concern for agronomic themes is thus indispensable. Industry is in the best position to find the most adequate solutions, to provide agriculture with the necessary means, through technical extension in particular, so that production will be more appropriate for industrial transformation and allocation to the market.

An OECD document (1979:91) concludes, with some justification, that the food industry (along with industry that produces production factors) definitely cannot be seen as a simple extension of agrarian production, but rather "as a domineering pole." In relation to Italy, Galizzi observes that "the actions of the food processing industries have always been characterized by efforts to hold down the price of agrarian staples" (1980). Corazza (1980:1–14) argues that the regulation of quality and amount delivered, time for delivery, etc., are also increasingly important in relations between the farm enterprise and the food industry; they become the object of detailed prescriptions and sanctions. With this and with the continual efforts to reduce the price of staples, the structuring of labor on the farm is naturally at stake (Benvenuti, 1980). One of the effects will be standardization. Autonomous control of the producer over production and reproduction, and the continual experimentation in order to further develop an enterprise, will be then less relevant than a uniform role-enactment conforming to general and centrally defined prescriptions. In that respect, it might be accepted that the E-calculus, in which conscious references to cura and *autosuficienza* are absent, will reasonably dovetail with an increasing measure of external prescriptions and sanctions over farm labor. Naturally that applies a fortiori to those elements which are central to this calculus: the cost/benefit relation and the scale coincide at a micro-level with the macro terms indicated above (price reductions in the primary agrarian sector and quantitative increase of output per enterprise).

TATE1: Information

The flow of information from TATE to farm enterprises is voluminous and diffuse. This flow, which increasingly concerns, directly or indirectly, the organization of labor and production, can involve the following questions: to what extent is the farmer inclined to accept the content of information from TATE institutions as guidelines, and to what degree does the farmer try to counterbalance this information by turning to other information and experiences? This issue concerns acceptance versus counterbalance, TATE information taken as normative or approached critically.

A number of items were taken up in the ERSA questionnaire to quantify this relation between farmer and TATE agencies:

1. "If you have to build a new cow shed, do you then take one of the models offered by the firms specializing in them, or do you yourself make a number of changes in the models they introduce?" Of the total sample (n=134) 24% said they would use the standard model while 76% said they would make alterations.

2. The following item measured relations to the technology offered in general. Is the "newest" that is offered normative and direction giving of itself or should one appraise it critically? Of the sample, 21% were of the definite opinion that "you can buy all new concerns with a quiet mind, because the technicians who develop them take care that they are alright"; 35% more or less agreed with this statement; while the remaining 44% disagreed.
3. In relation to the bank: "Do you think that the bank . . . works with criteria which, generally speaking, are those used by the farmers, or is there in your opinion a certain discrepancy between the different criteria?" Results showed that 30% thought there was no discrepancy; 70% thought there was.
4. The extent to which farmers were inclined to accept the advice of technicians from the consortia selling cattle feed or to approach it critically was also measured. "If the concentrate specialist . . . gives you advice, how often does this agree with what you had thought yourself?" The answers were as follows: 3% said "always"; 31% said "often"; 58% said "sometimes"; and we were not without the pigheaded 8% who said "never."
5. Giving an example from fruit cultivation, we measured whether there was a willingness to delegate the decision-making concerning the labor process and development of production to external agencies in order to reach "better results": 44% said that in principle they were willing to do so.
6. The last item measured opinions about the so-called *libro genealogico*, an important instrument of institutions which control trade in genetic material.

The given items were summarized in a scale, TATE1, rating the degree to which TATE information is accepted or approached critically.

TATE2: Investment Decisions

Important investment decisions will fix the organization of labor and production on the farm for a long time period. A high degree of institutionalization is possible in these investment decisions, and in the context studied by us this was no exception. Cow sheds, techniques for manure conservation and transportation, and cattle density all had to comply with regional, national and supra-national planning regulations. Plans for specialization and cropping had to correspond with regional development plans. Frequently a farm development plan or financial scheme will have to be drafted in such a way as to not only meet the advice of but often to receive the definitive approval of

Table 2.13. Importance Accredited to Advice from Varying Sources by Considering Investment Decisions ERSA/n=134

	Without any importance	Of minor importance	Important	Extremely important
Wife	8%	7%	35%	49%
Members of family	3%	11%	30%	56%
Successor	40%	11%	17%	25%
Farming friends	17%	40%	34%	6%
Very experienced farmers	5%	25%	43%	25%
Technicians of the consortia	29%	27%	30%	14%
Technicians of farmers' organizations	15%	18%	41%	26%
The bank's consultant	43%	21%	29%	4%
Agricultural press	41%	26%	25%	4%

various experts from the different external institutions. The latter applies a fortiori if the level of dependence on banks, i.e., on capital markets, is high.

With the help of a series of questions, it was possible to examine whose advice was sought about investment decisions, and what weight was given to such advice. Table 2.13 gives an overview of the answers. By dividing the total weight accredited to TATE by the weight given to wife, family members, etc., a scale was developed, TATE2, which gives the relative weight of TATE in investment decisions.

TATE3: Learning the Craft

Each farmer has his own specific way of organizing labor and production, his own "system." If we project these various ways of working on our previous definition of craftsmanship (structuring labor in such a way that the productive potential of labor objects is used optimally and at the same time further developed), then it is possible to speak of "better" or "worse" craftsmanship. Each level of craftsmanship assumes a conscious learning process. Different actors, from the father figure to diverse TATE institutions, can play a role in this system. We asked the following question: "From whom or what have you learnt most in acquiring the craftsmanship you now possess?" Table 2.14 summarizes the answers. The information, as summarized in this table, was again used to construct a scale, TATE3, which expresses the relative weighting of TATE instances in the development of personal craftsmanship.

At the request of regional farmer organizations, we investigated the relationship between the extent to which TATE information is simply

Table 2.14. The Weight Given to the Influence of Diverse "Agencies" in the Development of Personal Craftsmanship (ERSA/n-134)

		"Very much"	"Little"	"Nothing"
Father		58%	33%	8%
Own experience		96%	2%	2%
Colleagues		16%	73%	11%
TATE	Agricultural school	11%	16%	73%
	Technical courses	21%	33%	46%
	Individual advice of technicians	25%	37%	19%
	Agricultural press	10%	37%	53%

Table 2.15. TATE and Craftsmanship (ERSA/plain/n-75)

	"Objective" level of Craftsmanship (CRAFTp1 + CRAFTp2 + CRAFTp3)			
	low	medium	high	
Uncritical acceptance of external information and advice (TATE1 - high)	**57%**	**13%**	**30%**	**(n-26)**
Intermediary position (TATE1 - medium)	**19%**	**38%**	**43%**	**(n-24)**
Critical attitude towards external information and advice (TATE1 - low)	**27%**	**27%**	**46%**	**(n-25)**
Difference between acceptance and critical attitude	**+30%**	**-14%**	**-16%**	**(n-75)**

accepted, or approached critically in the light of personal experience, and the extent to which farmers developed the capacity to independently optimize the productive potential of labor objects.[20] The relations between TATE and craftsmanship appeared *grosso modo* to be negative, as shown in Table 2.15. As one of the leaders of the farmer organization exclaimed when confronted with these data: "It is indeed not TATE which makes the best farmers."

Social Relations of Production and the Farm Labor Process

The foregoing, mainly qualitative explorations suggested that the E-logic becomes dominant in farming when high levels of market incorporation and institutionalization occur. Increasing commoditization shortens time horizons in the farm labor process, introduces new uncertainties concerning prices, increases the risks involved, and affects

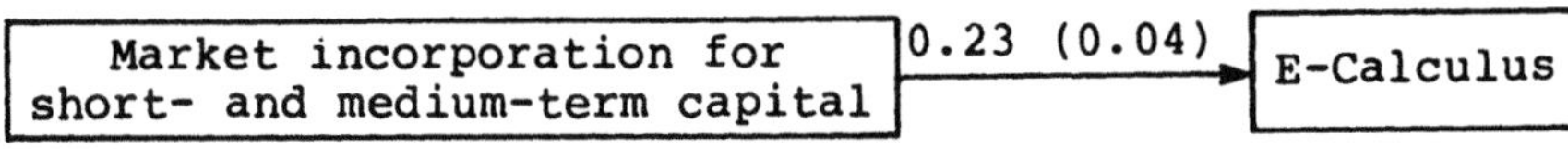

Figure 2.19

especially the quality of the means of production. Beyond that, it introduces a new logic for the organization and planning of the labor process: the E-calculus. All this leads in practice to a reorganization of the labor process to fit in with all-pervasive commodity relations. The E-calculus can be considered as the main vehicle for achieving this reorganization. In this respect the E-calculus represents a clear case of functional rationality; it subordinates the labor process to prevailing commodity relations. Apart from the empirical demonstration, it can also be assumed on theoretical grounds that this functional rationality will occur earlier and more widely where system-integration is highest (Gouldner, 1970:213–216).

"*La banca non scherza mica*," "the banks are not joking," said one of the Parma respondents, who wished to convey that taking credit is a serious matter and that the consequences have to be taken equally seriously. Within the ERSA sample, the E-calculus could be formulated as an unambiguous statistical variable, namely, as the synergic presence (in statistical terms: the interacting) of the E-option and a high degree of entrepreneurship. In short, if a farmer mainly considers himself as an entrepreneur and acts accordingly, and strives at the same time for scale enlargement instead of intensification, then it can be assumed that his thinking and behavior are structured by the E-calculus. So, if we now relate the degree to which a farmer is dependent on the bank[21] with the thus operationalized E-calculus, then we find that the ERSA/plain sample follows the model shown in Figure 2.19. There is, at least for farmers, a basic difference between handling one's "own capital" and dealing with "credit." Credit implies, as every other commodity, a complex series of commodity relations as well as specific technical administrative (TATE) relations vis-à-vis the bank. One has to act in accordance with these different kinds of relations, simply because the "banks aren't joking." So the E-calculus emerges as a guideline for structuring the labor process, thus ensuring the necessary correspondence between labor process and capital market incorporation.

One could study in a similar way all the direct links between the E-calculus and incorporation factors on the one hand, and between that and scales of institutionalization on the other, but for an overall view, it is quicker and easier, and above all more interesting theoretically, to investigate the interactive effects of incorporation and institutionalization on the E-calculus. The path diagram shown in Figure 2.20[22] gives

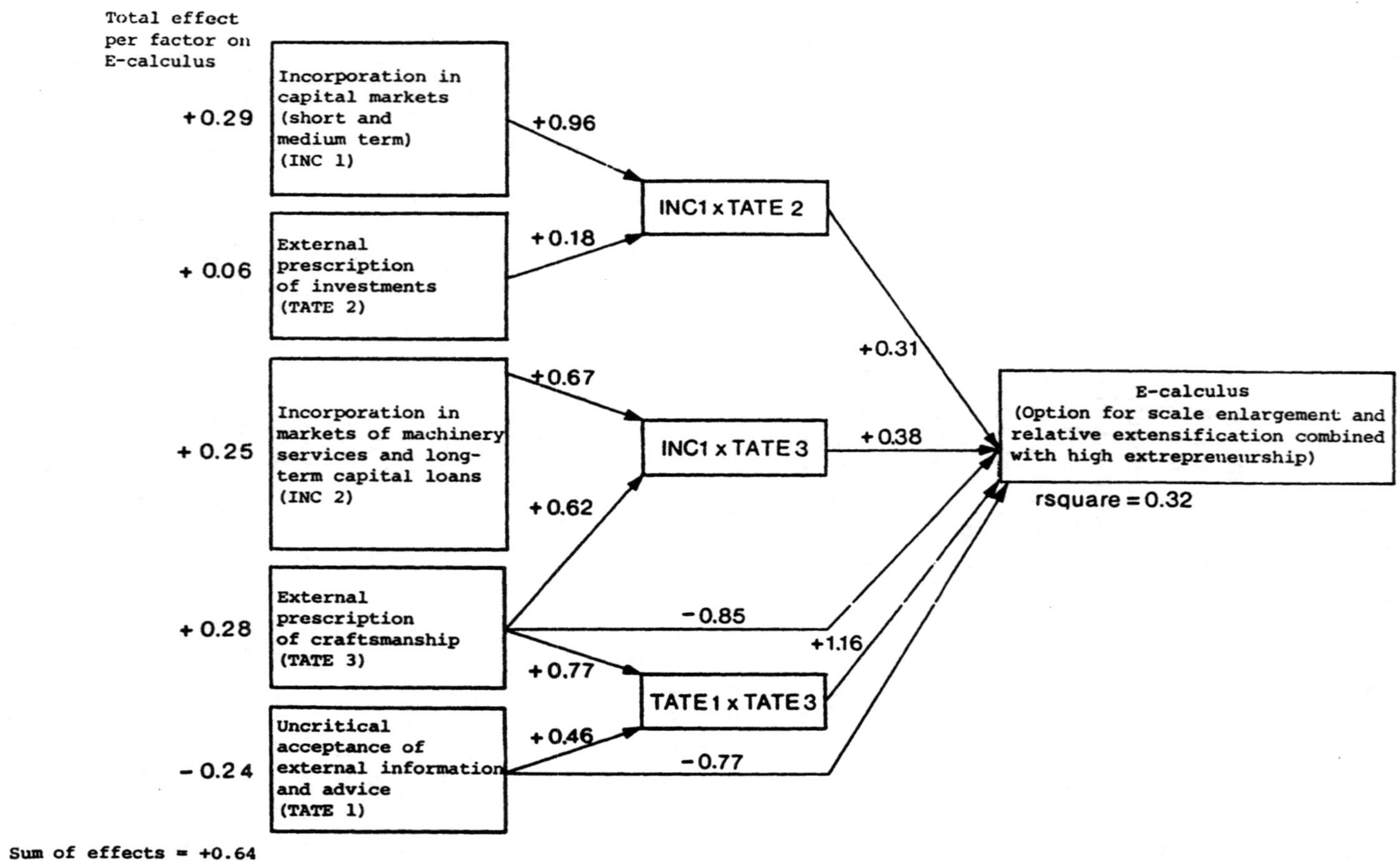

Figure 2.20 Path diagram showing the direct and indirect effects of incorporation and institutionalization on the E-calculus (ERSA/plain/n=75)

such an overview of the direct and indirect effects of incorporation and institutionalization on the structuring of the labor process. It shows several positive and significant links. If all the aspects of incorporation and institutionalization in the model were to increase with a unity similar to their respective standard deviation, then the dependent variable, i.e., the E-calculus, would rise +0.64.

The model also offers a view of the relative weighting of each of the separate forms through which incorporation and institutionalization manifest themselves. Isolated from other independent variables, "external prescription of craftsmanship" (TATE3) exercises a negative effect on the E-calculus. Combined, however, with "an uncritical acceptance of external information and advice" (TATE1) and/or "incorporation into markets for machine services and long-term loans" (INC2), such strong indirect and positive effects occur that the total effect of TATE3 becomes positive, 0.28, almost the same as the factor with the heaviest weighting, "incorporation into capital markets (INC1)," which has a positive effect of +0.29 on the E-calculus.

This brings us to an exciting aspect of the model, the synergism or mutual strengthening of separate aspects of incorporation and institutionalization. Incorporation into capital markets is one matter, external prescription of investment decisions (TATE2) is another. But it is their mutual interaction which forms the meaningful situation. One cannot discount the effect of one on the other; both are important, especially where they strengthen each other. It is not so much incorporation as such, but commoditization of increasing parts of the labor process together with the ongoing articulation of the "rules" that must be followed (TATE), which produces significant changes in farming.[23] Other interaction moments also point to meaningful patterns. Incorporation into machine markets (INC2) implies that important parts of the labor process are sometimes contracted out. In general this can be extremely functional for lowering costs, but, on the other hand, it makes the optimizing of production potential difficult. Add to this a high level of "external prescription of craftsmanship," and the effect on the E-calculus is then positive. What is already potentially determined by the contracting out of certain activities is completed by the diminishing relevance of personal experience necessary in developing craftsmanship. The E-calculus, in which there is no place for craftsmanship, then becomes dominant.

The same is true for "external prescription of craftsmanship" (TATE3) and the "uncritical acceptance of external information and advice" (TATE1). Alone they correlate negatively with the E-calculus, but together (TATE1 * TATE3) they have a strong positive effect. In short, it is the interaction, the coming together of different relations in a systematic network in which each of the separate links is strengthened by the others, that the relevant whole of the E-calculus is formed.

This principle implies that commodity relations as well as technical-administrative relations (as symbolized by the different TATE scales) emerge as dominant social relations of production. As Poulantzas (1974, 1976) would argue, they effectively "constitute" the farm labor process. They give the labor process a specific form and structure, as symbolized by the E-calculus. At the same time one must stress that this is in no way a universal phenomenon. Commodity relations and the technical administrative relations with which they interact only emerge as social relations of production in the context of high systems interweaving (see also Figure 1.4 in Chapter 1). It is not simply the market economy and its institutional superstructure as such which have this particular impact. The logic of farming, the strategy consciously used and reproduced by the farmers concerned, is equally important. For, in this same setting of markets and market agencies, another empirical reality can be observed, i.e., the reality that the farm labor process is not determined in the same unilinear way by commodity and TATE relations.

The I-Calculus and Its Relation to Incorporation and Institutionalization

Farm labor structured according to the I-calculus can also be defined in statistical terms. The key terms in the I-calculus, as mentioned before, are produzione and cura, which go hand in hand with intensification and the practice of craftsmanship. Only when intensification and craftmanship are taken together does a meaningful pattern emerge. In other words, we can define the I-calculus in statistical terms as the interaction between opting for intensification and the structuring of farm labor as craftsmanship.

The E-calculus emerges, as we saw, in a context characterized by a systematic interweaving of the farm and the farmer's thinking with markets and relevant institutions. However, for the I-calculus, it can be assumed that a certain level of functional autonomy (operationalized as a low level of incorporation and a low degree of institutionalization) is an essential prerequisite. Table 2.16 shows that, indeed, the E-calculus goes with a high degree of incorporation and institutionalization ($\Sigma=1.04$), and the I-calculus with a low level ($\Sigma=0.06$), taking average values for various aspects of incorporation and institutionalization. Identifiable differences occur with incorporation into capital markets (INC1), with incorporation into markets for machine services and long term loans (INC2) and with external prescription of craftsmanship (TATE3) (=0.69, 0.52 and 0.47, respectively).

Table 2.16. Mean values of Incorporation and Institutionalization When I- and E-calculi Are Respectively High

	INC1	INC2	INC3	TATE1	TATE2	TATE3	Σ
High I-calculus (n=25)	-0.21	-0.06	+0.15	+0.21	+0.05	-0.08	+0.06
High E-calculus (n=16)	+0.48	+0.46	+0.21	-0.37	-0.13	+0.39	+1.04
Δ	+0.69	+0.52	+0.06	-0.58	-0.18	+0.47	

INC1 = Incorporation into capital markets (for short- and medium-term loans)
INC2 = Incorporation into markets for machinery services and long-term loans
INC3 = Incorporation into markets for labor, feed and fodder

TATE1 = Uncritical acceptance of external information and advice
TATE2 = External prescription of investment decisions
TATE3 = External prescription of craftsmanship

Figure 2.21 consists of a path diagram which summarizes the effects of incorporation and institutionalization on the I-calculus. Several relations are demonstrable between farm labor structured according to the I-calculus and incorporation and institutionalization. Some are negative, some positive. It is remarkable, however, that these interrelations are so structured that the joint net effect of incorporation and institutionalization on the I-calculus is virtually nil (–0.04). If all aspects of incorporation and institutionalization were increased by 1.00 (the SD), then the I-calculus still wouldn't alter much. They are not mutually reinforcing but have a counterbalancing effect on each other. Comparing Figure 2.20 (the effects on the E-calculus) with Figure 2.21 (the effects on the I-calculus) one detects a shift as one moves from high systems interweaving to functional autonomy. The sociological importance of the constellation in Figure 2.21 is that it highlights the fact that farm labor, typified by the I-calculus, does not exist by the grace of some Utopian economic autarky. Both forms of labor are "real" in the sense that they manifest and reproduce themselves in a market economy. Alvisi (1980:1,2), building on the work of others, argues that farming as an activity is characterized by two fundamental aspects: production and marketing. The first aspect assumes that the entrepreneur orientates everything he does towards maximum technical efficiency, and the second assumes that he "subordinates production choices to adapt . . . to price and cost relationships." He subsequently places both aspects in a certain chronology: "the production aspect was dominant up till the end of the second world war on account of the self provisioning nature of agricultural enterprises and the promotion of this through agrarian policy." He argues that "subsequently economic de-

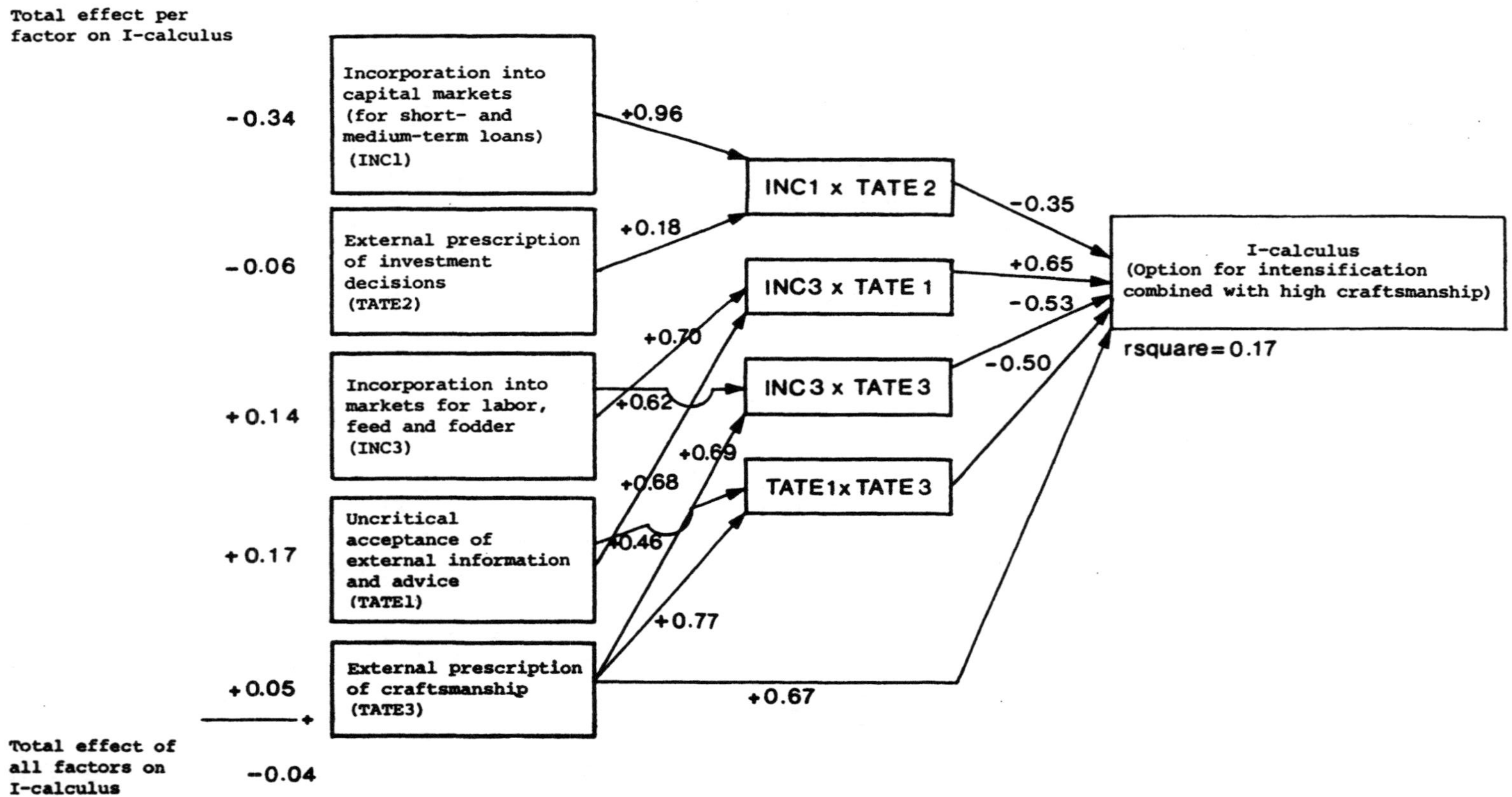

Figure 2.21 Path diagram showing the effects of incorporation and institutionalization on the I-calculus (ERSA/plain/n=75)

velopment led to a rapid integration of agricultural practice into markets." With this, the second aspect then became dominant. Newby (1978a:17) follows a similar reasoning.

Although not per se incorrect, the suggested sequence is nevertheless deceptive since it suggests that an emphatic orientation to production (ref. the I-calculus) is to be seen as an anachronism, as a phenomenon that only goes with a stage already passed, i.e., that of economic self-sufficiency. However, farm labor structured by the I-calculus is still an empirical reality within *present day* markets and market institutions, or so it appears from the path diagram. It assumes no autarky, no absence of links and relationships with the economic and institutional environment, but simply a field of activity that is actively controled by the farmer himself; a constellation in which relations with market and TATE are of a multifarious and mutually opposing nature. The characteristic of systems interweaving so typical for the patterning in which the E-calculus functions is missing. It is much more a question of a certain degree of functional autonomy, a situation of opposing relations which assume that actors make conscious choices.

The path diagrams, in which the impact of the economic and institutional environment on the specific structuring of farm labor were analyzed, point to diverse constellations which cannot be interpreted in a deterministic sense. A continuing interaction between the farmer and his environment forms the basis of both constellations. In the one case, this interaction tends towards a reproduction of systems interweaving; in the other, to the maintenance of a certain level of functional autonomy. Elements of this interaction were touched upon several times in the earlier text. Thus, where farm labor is structured by the E-calculus, the reliance on TATE, on external prescriptions for the carrying out of various tasks, will often be necessary. An ongoing incorporation will then emerge quite often as a logical step. Also, as we have seen, where intensification is explicitly opted for, increasing incorporation is considered undesirable.

A high level of system interweaving is both a condition for and a consequence of farm labor structured along E-calculus lines. With the advance of incorporation and institutionalization, a network of commodity and institutional relationships emerge which penetrate the farm at all levels. The ratios existing in and between markets (and their many explicit voicings by TATE agencies) function to increase and mutually reinforce each other. This implies for the individual farmer a certain reduction of space to make decisions; but, more important, it implies that this space is increasingly hemmed in by parameters which (a) are irreversible, (b) show a high degree of mutual correspondence, and finally, (c) where no other decisions are possible except those that fit in with these parameters.

Thus decision making becomes increasingly a question of applying a logic which lies embedded within an entirety of external parameters. The farmer loses his space to maneuver as well as his control over the labor process: consequently, labor undergoes a certain expropriation and tailorization.

Mountain Farmers and Intensification

So far the analysis has been based on dairy farming on the plain. The ERSA sample included a number of mountain dairy farmers (n=59). The differences between the samples are not purely ecological. Systematic differences occur also in the levels of incorporation and institutionalization. The mountains are characterized by a lower and much less consistent pattern of these indicators. Generally speaking, the differences boil down to differences in the norms and structures for organizing farm labor. Mountain farmers are more likely to opt for intensification than farmers on the plain. Scores for craftsmanship are correspondingly higher for the mountain farmer, while the average score for entrepreneurship is lower. Similar differences in the geographical distribution of the degree of incorporation, options, etc., have also been shown by the recent research of Sauda and Antonello (1983).

The respondents in the mountains answered the same questionnaire as their colleagues on the plain, and the same analysis was applied to their answers on the various questions concerning goals, craftsmanship and entrepreneurship. The analysis confirms the relationship earlier observed on the plain, namely, that an increasing commoditization and institutionalization leads to a restructuring of the labor process; the I-calculus gives way to the E-calculus as the structuring principle.[24] But there are important differences as well. To begin with, incorporation and TATE indices in the mountains strongly reflect the nature of state interventions. Government and EEC programs for the mountains (considered as marginal areas that deserve special help) are primarily based on the provision of credit, which must be used to buy "productive" breeds of stock and to build "modern" cow sheds. The care of these new breeds, however, constitutes a noticeable break with current practices for local stock. Hence TATE information becomes necessary. Consequently statistical analysis shows the development of clear chains: incorporation into both capital and cattle markets and a high dependency on TATE forming a clear interdependent cluster.[25]

A second difference between mountain and plain is strongly tied to this finding. The distribution of farm enterprises along a continuum of low to high systems interweaving is more bipolar. They are either strongly or only slightly interwoven with markets and market agencies. This difference is substantially related to the planned nature that various institutions try to give to the integration process, and it is this which entails that the boundary between "inclusion" and "exclusion" is much sharper. The manner and degree to which relations between farm enterprises and various markets and market agencies are created can be less circumscribed by the farmers themselves, since the creation of these relations are the goals of state intervention itself.

This can be vividly illustrated by local events, which were much debated by the farmers themselves during the period of our research. Farmers in the mountains are often faced with lengthy and arduous tasks. Producing animal feed from steep hillsides is one such task, and many farmers (particularly female farmers) will gladly invest in techniques which lighten or shorten their work load. At first sight the State seems to give substantial support and loans for this. However, the loans are each linked to a series of conditions. One of these is that after "modernization," the enterprise is expected to produce an income on a par with that of city workers. Meeting this requirement often entails a considerable expansion of the herd. To fulfill a second condition—namely that the density of cattle should not be too high—means buying extra land. In order to compensate for the financial burdens that this imposes, herds have to be further expanded.

One of the most probable results then is that certain tasks are indeed lightened but that the total work time remains the same or is even lengthened. Thus, the planned nature of the integration process confronts farmers with chains of consequences which in the end go against their interests. For farmers this means either making use of the available opportunities, following the prescriptions and accepting the consequences, or remaining on the margin. There is little room for a middle way. And when farmers create their own solutions to problems, such solutions are usually perceived as being undesirable by the planning agencies and sometimes made impossible by the accompanying regulations.

A third striking difference between the mountains and the plain is that the explicit option for scale enlargement and extensification (the E-option) is less frequent (20% versus 43%). Mountain farmers reason that this difference is closely related to the ecological setting and to the social construction of time associated with it. One can well cultivate less intensively in the short term, and this can even be highly remunerative, but to do so over the longer term leads to typical forms of soil erosion, such as the *frana*, to the uncontrollable growth of brush wood and weeds in the meadows, and to the rapid deterioration of the quality of fodder. So in the long term, extensification of the labor process becomes strongly counterproductive. Thus the ecological setting, or more precisely, the tension between long- and short-term rationality, precludes as ready or as widespread an acceptance of the E-calculus in the mountains.

Finally, let me broach a point that can be discerned both in the mountains and on the plain but which is seen much more clearly in the mountains—the conflicting relationship between craftsmanship and entrepreneurship. In the abstract, one might endorse the proposition of Newby et al. that "a conflict between the two does not necessarily need to exist" (1978b:181).[26] However, in reality they each appear to take strategic positions in qualitatively different systems of logic, which makes the relationship between them problematic. The statistical analysis shows a negative correlation, $-.25$ ($p=.05$), between the I-calculus and entrepreneurship, and $-.34$ ($p=.008$) between the E-calculus and craftsmanship. This suggests that where craftsmanship is dominant, there is no "structural room" for entrepreneurship. And vice versa, where production is dominated by market and price relations, there remains little room for craftsmanship. The relationship between craftsmanship and entrepreneurship on the one hand, and between craftsmanship and TATE on the other, strengthens this suggestion. The correlation between TATE and craftsmanship is negative ($r=-.29$; $p=0.02$); the relations between TATE and entrepreneurship positive ($r=.27$; $p=0.03$).

Styles of Farming as Social Constructions

A first impression of heterogeneity in Emilian dairy farming was given in Chapter 1 (see especially Table 1.2). Here I wish to use statistical data of a larger number of farm characteristics[27] to show that styles of farming can only be properly understood when viewed as social constructions. They cannot be understood when perceived as simple (or even complex) derivations of prevailing commodity relations, nor when seen as determined by technological or institutional factors. That is not to say that these factors are irrelevant, but that the impact of commodity and TATE relations is a differential one and that their influence is mediated by farmers as conscious actors operating with different logics of farming.

One of the points I hope to establish in the following analysis is what in recent Italian debates on agriculture has been rightly called "the centrality of labor." A style of farming is the outcome of a particular labor process guided by certain options, structured in a specific way by a corresponding "logic," and conditioned by particular social relations of production. Through the farm labor process both the social relations of production and the style of farming are reproduced. In the heterogeneity manifest in Emilian agriculture, clear patterns or "styles" are identifiable which are constructed by farmers. Factor analysis shows that more than 80% of the variation can be explained in terms of six factors. Three are directly related to intensity of production, a fourth describes scale and labor input, a fifth the density of cattle to land, and a sixth, capital input per hectare (see Table 2.17).

Precisely the same factors appear from a similar analysis of the mountain material (summarized in Table 2.18). More recent research has shown that such factors are also evident elsewhere in Italy. The six factors can now be seen to be six dimensions within which each specific farm structure can be described. If we take an ideal typical example of an intensive dairy farm, it will be first characterized by a high score on the factor "intensity of animal production" (the produzione of the I-calculus), and likewise a high score on "intensity of feed production." Land and animals are treated with "cura" resulting in a high yield of feed per hectare and a high yield per cow. Finally, to achieve such "cura," a high "impegno" is necessary (input of labor per labor object must be high). This means that the score on the "scale" dimension will be low. With such a high inset of labor a relatively large number of other resources is necessary: the input of capital does not so much substitute for labor in this model of intensive farming as complement it. Thus the score of "capital input" will also be high.

A similar exercise is also possible for the typical opposite: for large scale, relatively extensive farm enterprises. Such an enterprise will be characterized by a high score on the scale and animal density dimensions. "Intensity of fodder production" as also "intensity of animal production" will be characterized by low factor scores.

Essential for the subsequent analysis is the assumption that, for individual enterprises, scores on the dimensions shown are the result of goal conscious activity and that the different options, as well as the operational mechanisms such as craftsmanship and entrepreneurship, play a decisive role in this activity. The farming enterprise, in other words, is a result of "the purposeful action of the farm operator" (Crouch, 1972). This first assumption leads to a second—that a whole profile of scores on several dimensions represents more meaning than one score on one isolated dimension.[28] A high score on the scale dimension is naturally important, but it is only in combination with other scores that one can rightly speak of a large-scale, relatively extensive enterprise. This argument can be elucidated by a simple example. As we saw earlier, high incorporation into fodder markets clearly affects the organization of the production unit: yield per cow

Table 2.17. Summary of Factor Analysis Applied to Farm Characteristics ERSA/plain/n=75

Variables	Factor 1 "cattle density"	Factor 2 "scale"	Factor 3 "intensity of plant production" and level of cattle feeding	Factor 4 "intensity of animal production	Factor 5 "intensity feed production"	Factor 6 "capital input/ha"
Cattle density	.91	.14	.03	-.08	.01	.10
Total production/ha	.76	.09	.47	.29	.19	.13
Surplus/ha	.87	.02	-.11	.25	.01	.02
Benefit/cost ratio	.83	-.12	-.18	.67	.21	-.05
Labor input/ha	.40	-.77	.28	-.14	-.02	-.07
Acreage in ha	-.15	.77	-.16	-.07	.04	-.21
Number of adult animals	.22	.81	.02	-.20	.11	-.18
Total input of production factors per adult animal	-.35	-.74	-.02	-.01	.14	-.14
Costs of plant prod./ha	-.16	-.10	.82	.14	-.05	.00
Yields in plant prod./ha	-.06	-.19	.79	.28	.19	.15
Variable costs/ha	.26	.26	.66	-.14	.35	.36
Total input production factors/ha	.48	-.40	.60	-.18	.19	.25
Production/milk cow	.32	-.16	.17	.80	.39	-.07
Surplus per adult animal	.62	-.14	.01	.69	-.03	-.04
Physical milk yield/ milk cow	.01	-.01	.06	.67	-.35	.22
Production of feed and fodder/ha	.15	.06	.06	-.07	.91	.04
Feed level/adult animal	-.48	-.06	.10	.07	.83	.00
Value of stock and machinery/ha	-.14	-.11	-.12	.09	.00	.77
Ammortizations/ha	.41	-.06	.18	.03	.03	.77
Income/unit of labor force	.59	.45	.08	.57	.01	.09
Eigenvalues	5.78	3.43	2.61	1.69	1.47	1.39
Variance explained	29%	17%	13%	8%	7%	7%

Table 2.18. Summary of Factor Analysis Applied to Farm Characteristics ERSA/mountains/n=59

Variables	Factor 1 "cattle density"	Factor 2 "intensity feed production and level of feeding"	Factor 3 "size of stock"	Factor 4 "intensity of animal production"	Factor 5 "capital input/ha"	Factor 6 "intensity of plant production"
Cattle density	.92	-.05	.31	.01	.05	-.01
Labor input per ha	.78	+.11	-.52	-.09	.09	.00
Total input of capital, labor and inputs per ha	.88	+.17	-.31	.18	.16	-.02
Variable costs per ha	.62	+.51	+.11	.49	.18	-.02
Acreage in ha	-.78	+.27	+.34	-.04	-.07	-.13
GPV per ha	.72	+.31	.06	.51	.25	-.02
Production of feed and fodder per ha	.50	.65	.04	.43	.19	-.07
Feed level/adult animal	.12	.77	-.06	.56	.19	-.05
Surplus per ha	.38	-.90	.00	.10	-.01	-.03
Surplus per adult animal	-.03	-.95	-.20	.13	.02	-.01
Benefit/cost ratio	-.14	-.68	-.26	-.21	-.09	-.12
Income per unit of labor force	.02	-.93	+.21	.08	-.00	.02
Number of adult animal units	-.15	+.26	+.83	-.01	.07	.03
Input of capital, labor and inputs per adult animal	.04	+.20	-.82	.37	.13	-.06
Physical milk yield/ milk cow	.06	.02	-.05	.80	-.04	.40
GPV of animal production/ adult animal	.14	-.13	-.32	.86	.25	-.08
Value of stock and machinery per ha	-.02	.11	-.10	.23	.89	.09
Ammortizations/ha	.35	.05	.07	-.02	.86	-.09
Costs of plant prod/ha	.03	.20	-.14	.08	-.03	.80
Yields of plant prod/ha	-.04	-.17	.25	.05	.10	.76
Eigenvalues	6.38	4.32	2.16	1.89	1.81	1.07
Variance explained (cum.)	32%	54%	64%	74%	81%	87%

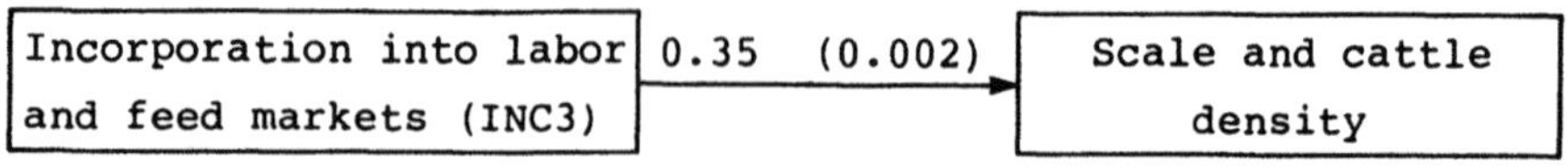

Figure 2.22

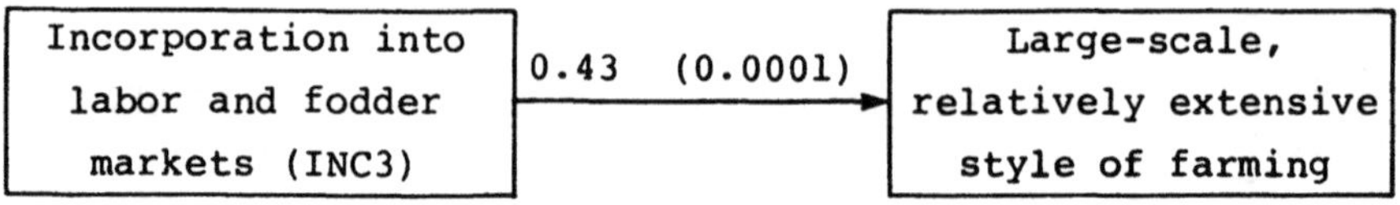

Figure 2.23

drops and cattle density is raised. Likewise with the labor market: the more labor is mobilized on the market, the lower the input of labor per hectare and per cow. The regression coefficients of INC3—the factor which stands for incorporation into labor and feed markets—on cattle density and on scale confirm this relationship (r=0.25 [0.02] and 0.25 [0.02], respectively). If we now combine "scale" and "cattle density" additively then we find the relationship shown in Figure 2.22.

The explained variation rises: i.e., the combination of factors is more meaningful and highlights fundamental relationships better than do factors in isolation. One can, however, go a step further and broaden the definition to the combination of scale enlargement and relative extensification: "intensity of animal production" and "intensity of fodder production" together refer to an intensity applying over the whole range. The inverse of this term refers to a dairy farm with relatively extensive practices. We can thus define a large-scale, relatively extensive farm as: (cattle density) + (scale) − (intensity of animal production) − (intensity of fodder production). If we now relate this composite term to incorporation into the labor and fodder markets, then we find the relationship shown in Figure 2.23. The example points to the importance of an integral definition of farming "style." The equation also indicates the statistical relationship between high incorporation and the tendency for farming to become large scale and relatively extensive. However, in the following analysis we will take a different course.

The direct relationship between incorporation and style of farming is less interesting than the indirect relationship which goes via farm labor. Instead of removing the farmer "from a system for which he has been responsible and within which farm practices are an integral part" (Crouch 1972), it is essential to place him at the center for the why's of a particular relationship to become clear. Why is there a regressive effect of incorporation and institutionalization on the style of farming? Why do scale enlargement and relative extensification become dominant? The qualitative research suggested that the answers to such

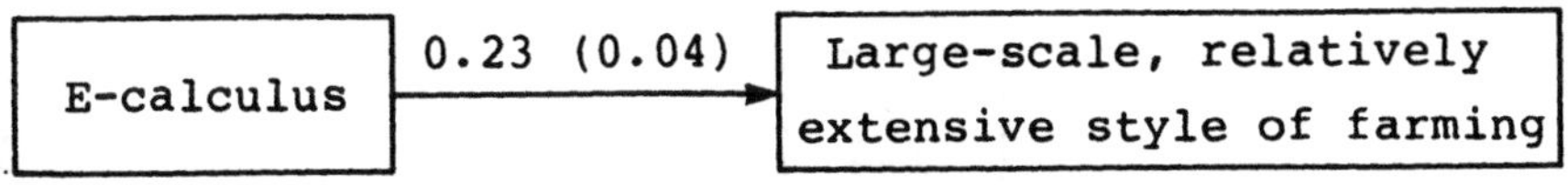

Figure 2.24

questions are to be found in the increasing development of an E-calculus associated with incorporation and institutionalization. This hypothesis can now be tested with the help of the terms and techniques described. To begin with, we examine farm labor normalized and structured via the E-logic, the dominant logic of high systems integration. The direct link between it and a "large-scale, relatively extensive style of agricultural practice" is shown in Figure 2.24.

Figure 2.25 shows the total effect (direct and via farm labor) of incorporation and institutionalization on the style of agricultural practice. With an increase of incorporation and institutionalization and the associated domination of the E-calculus, the farm becomes modeled along the lines of scale enlargement and relative extensification.

It is worth emphasizing that the essential difference between intensive and large scale relatively extensive farms lies in their level of incorporation and institutionalization. Age and training of the farm head, family composition or whether an heir is available make no significant difference.[29] Only the point of view of the head tends to make a difference: intensive farmers are more orientated to the rural world and their own families than are extensive farmers; intensive farmers are more often than their colleagues the sons of *mezzadri*, of sharecroppers who managed after the war to acquire their own farms through collective and individual effort and sacrifice. Their sons, as well as inheriting the farms, also inherit the emancipation aspirations of their fathers to farm better than had been "allowed."

Whether a farm is large or small in terms of area, it can be extensively or intensively farmed. The average farm area of extensively operated farms is 24.1 ha, as against 11.0 for intensively operated farms, but the standard deviations are such (16.1 and 6.8, respectively) that the range overlaps. A large area is a favorable condition for raising the scale, but it is not essential. A large area likewise can be intensively farmed. The reason for the development of one or the other style is again the structure of the labor process and the corresponding degree of integration into the politicoeconomic system.

Back to the Mountains

While on the whole the patterns are again similar for the mountains, there are some differences which are of great importance. In Chapter

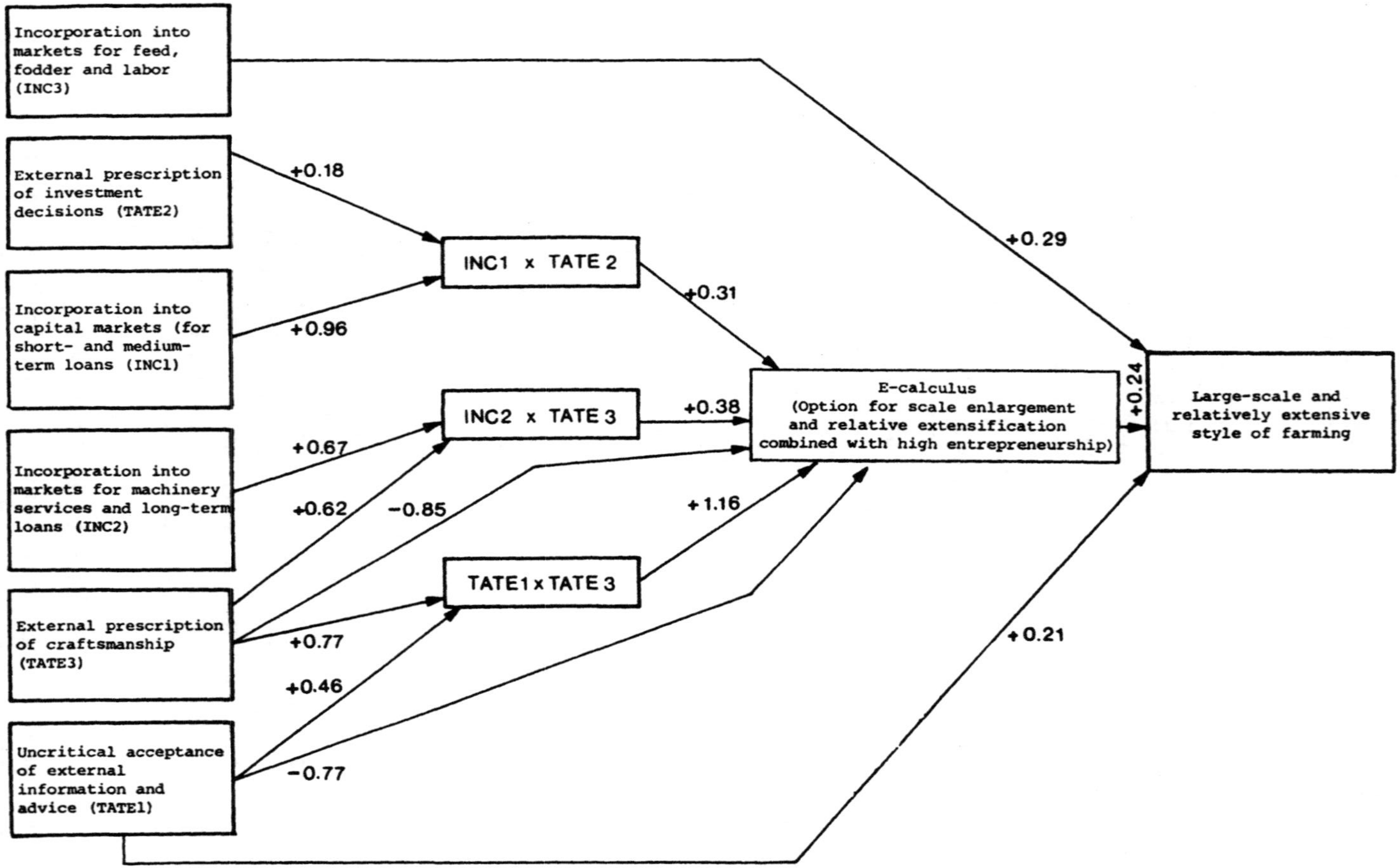

Figure 2.25 Path diagram showing the direct and indirect effects of incorporation and institutionalization on scale and intensity of farming (ERSA/plain/n=75)

5 where the differences within and between Italy and Peru are discussed, I will explore these differences further. Limiting the discussion here to a comparison of the mountains and the plain in Emilia Romagna, two essential differences emerge.

On the plain, increasing cattle density was seen to be a typical mechanism for scale enlargement. The greater the density, the lower the intensity of fodder production. More fodder was bought in, and the amount given per cow was less. Statistically there is a positive and significant relationship between the E-calculus and cattle density. In the mountains, however, increasing cattle density appears to go with the I-calculus. Thus, cattle density appears there as part of a strategy oriented towards intensification. In the mountains, the raising of density coincides with the raising of fodder production per hectare. Consequently the buying of supplementary feed is not needed.

In a theoretical sense this means that no universal relationships between calculi and farm styles can be postulated. One must conclude therefore that the way in which the I- and E-logics take concrete form depends on the setting in which they are realized. Ecological, political, economic and historical circumstances may be essential to this process.

A second important difference between the mountains and the plain concerns the intensity of animal production. As was constructed in Figure 2.25 for extensive farming, an explanatory model can also be constructed for intensity of animal production. For both mountain and plain, a positive and significant relationship emerges between an I-logic and the intensity of animal production. The reverse calculation also applies; a negative relationship exists between the E-logic and intensity. However, the relationship is much stronger in the mountains than on the plain. There are recognizable correcting mechanisms at work in the plain. In both the mountains and on the plain incorporation and institutionalization exercise a negative effect on the intensity of animal production. But the dequalification of craftsmanship, at least on the plain, can be partly compensated for by buying and using certain technological innovations and prescriptions. Indeed, incorporation and institutionalization accomplish by means of the emerging dominance of the E-calculus a certain stagnation in the development of productive forces; but where external TATE networks succeed in producing highly productive innovations—such as uni-feed systems, Holstein cattle, automated feeding, etc.—such stagnation can be broken, presumably to subsequently reappear at a higher level. The continual development and implementation of external innovations which raise production levels then replace farm labor (structured by the I-calculus) as the driving force of agrarian development. Intensification based on the continual improvement of quantity and quality (impegno and cura) of farm labor is superseded by an intensification based on technological development. The "making" of intensity becomes externalized, and the phenomenon of an "*intensification scientifique*," as the French say, then becomes dominant (Capelle, 1986). However, apart from its potentially counterproductive effects (Ullrich, 1979), such a new phenomenon implies another important effect; that is, its selectivity, or geographical bias. It often provokes a marginalization of production areas which are characterized by conditions that do not fit with the premises of technological design. And that is exactly what is shown by the statistical interrelations already referred to. The negative effects of the E-logic on intensity can be remedied to a degree on the plain by specific technological developments. In the mountains, however, the applicability of such an intensification technique is much lower or nonexistent. Anyway, when discussing another chain of mountains, the Peruvian Andes, in the next chapter, I will focus on this problem again.

Concerning Scale and Social Relations of Production: A Supplementary Argument

Although it cannot be reduced to pure tradition, the I-pattern represents, more than the E-pattern, a historical continuity. The language is sometimes literally that of the earlier *insegnamenti*—"teachings"—(circa 1700) (Spaggiari, 1964) and *lezioni*—"lessons"—(1771) (Barigazzi, 1980). The "father figure" is less a subject of taboo, and the reference group is more rural than urban. And even where the I-pattern appears in a new guise—in some of the young, well-trained farmers who openly criticize the emergence of the E-logic and opt for an intensive way of farming—the ratio of thought and actions, structurally speaking, is the same as that which agricultural development has taken through the ages, which depends as far as is possible on an autonomous reproduction cycle.

However, at the moment that, historically speaking, the conditions arose for a complete development of this ratio,[30] a new rationality, the E-calculus, was introduced in agricultural practice through the advance of commoditization and institutionalization.[31] Its rapid spread can be inferred from several studies relating to the 60s. At the beginning of the 1960s—when specialized dairy farms barely existed—the standard deviation (expressed as a percentage of the gross value of production per hectare) for mixed dairy farms fluctuated between 9% and 25% (for the grain-wine-dairy complex and for the wine-dairy complex, respectively). The mean standard deviation was 17% (INEA, 1962). A similar index, calculated for fifty-five dairy farms seven years later, all located in Emilia Romagna—ten specialized, forty-five mixed, but all with

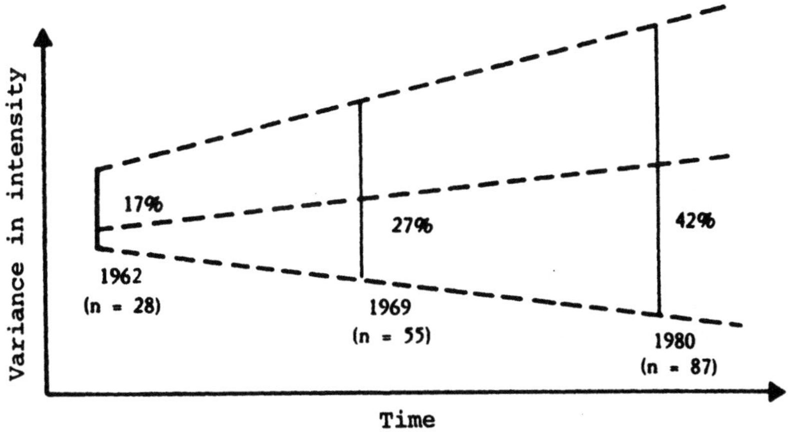

Note: For each data set the standard deviation of intensity is expressed as a percentage of the mean intensity.

Figure 2.26 Increasing heterogeneity in Italian farming (1960–1980)

dairying as the mainstay—showed that the standard deviation had risen to 27% (INEA, 1969). At the end of the 70s and beginning of the 80s, it had risen to 42% and practically all the farms were completely specialized dairy farms. This is an intriguing development. The shrinking of the zones to which the data refer, from northern Italy via Emilia Romagna to the provinces of Reggio Emilia and Parma, the increasing specialization, and the rising random sample size should all lean to a falling standard deviation. But instead it more than doubled. Figure 2.26 graphically illustrates this process. The variance sharply rose, which points to the rise and gradual spread of the E-pattern. Of course one cannot suppose that the growing divergence will continue to reproduce itself. There might well emerge factors which block the reproduction of the I- or the E-pattern. For the period considered, however, the growing divergence, which contains the crystallization of the E-pattern as a *new* phenomenon in agriculture, is an empirical fact.

On the basis of Turbati's work (1971), it is possible to reconstruct the pattern of development for thirty-four partly specialized, partly mixed dairy farms for the five consecutive years from 1965 to 1969. The results can be compared with those of the Parma sample relating to the 1970s, which showed that alongside a continuing intensification, scale enlargement and relative extensification also led to a considerable rise in income. One looks in vain for such a pattern in the 1960s. Where there was falling production per hectare, this did not accompany rising incomes. The world was still simple: "bad" farming and "good" earnings were not yet associated. The detailed documentation clearly

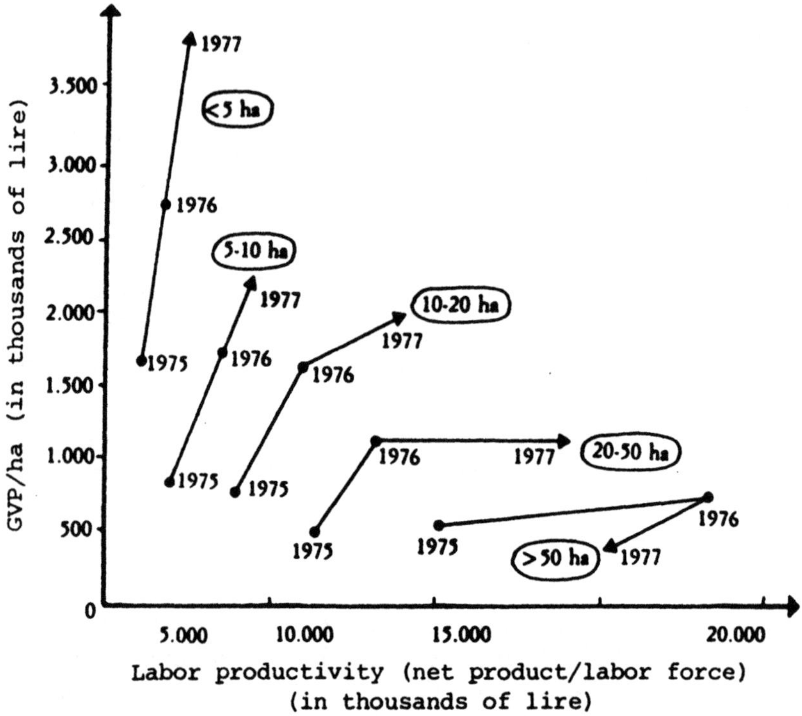

Figure 2.27 Developmental patterns for size groups (derived from INEA, 1975, 1976, 1977)

shows that where a drop or stagnation of produzione occurred, this was usually due to hail; to sicknesses such as brucellosis; to frost on the vines, etc., and not to any conscious choice. The complete absence of scale enlargement and relative extensification as a valid pattern at the time, at least in the overview of Turbati, can again be interpreted as demonstrating the relatively recent nature of this E-pattern.

INEA has published economic results ("*risultati economici*") for the years 1975, 1976, and 1977 based on a very large (though not constant) number of farms. All specializations are present. The farms are divided into broad bands. Analysis shows that the smaller farms (less than 10 hectares) are distinguished by significant increases in "ground productivity" while labor productivity remains constant. For the large farms, the opposite—i.e., stagnating or falling hectare yields with a large rise in labor productivity—is the case (see Figure 2.27). There is, in short, a marked variation in farm development patterns, which in these studies primarily appears to be a function of farm size. Such figures, which seem to support the current hypothesis on "scale advantages," can also be constructed for the ERSA sample. The input of production factors

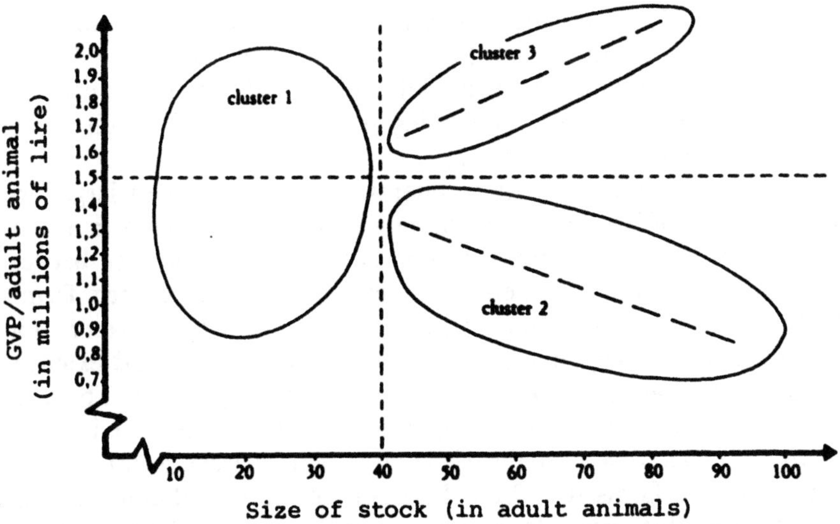

Figure 2.28 Intensity and size of stock in dairy farming: A cluster analysis (ERSA/plain/n=75)

per hectare is negatively correlated to the size of farm in hectares: r=−.43 on the plain, r=−.70 in the mountains. The GVP/ha also correlates negatively with area: −.19 and −.49, respectively. The level of incorporation, however, correlates positively with farm size: .35 and .47, respectively.

And therein lies the problem. Is intensity a function of farm size? Or is such a relation a spurious one caused by a third variable, the level of incorporation? Although the question can already be answered in principle on the basis of the previous analysis, we will here take it up once more. For this it is necessary to make a strict distinction between the concepts of scale and size. In an absolute sense, area and/or stock are indicators of size, and the relation between input of labor and area or stock is an indicator of scale.

The range over the variables "size of stock" and "production per adult animal" in dairy farms on the plain is considerable. With a cluster analysis, three subgroups become discernable in this mosaic. A first group (see Figure 2.28) is composed of smaller farms (less than 40 adult animals). There is a wide variation of production per adult animal in this subgroup and no direct relationship can be demonstrated between the number of adult animals and the GVP per adult animal. A second group comprises the larger farms (more than 40 adult animals), which achieve a low GVP per animal (less than 1.5 million lire per adult animal). Within this group an increase in livestock goes with a decrease in production per animal: GPV/adult animal = 1.13 − 0.36

* stock (r=.22). The most interesting is the third and numerically smallest cluster comprised of the larger farms which achieve a GVP of more than 1.5 million lire per adult animal. Within this cluster, an increase of livestock goes with rising productivity: GVP/adult animal = 1.33 + 5.98 livestock (r=.72).

In other words high levels of intensity are possible with large farm size (in terms of livestock)[32] if and only if labor is structured to achieve such a positive relationship. And that is exactly the difference between the second and third clusters (both of which are large-sized farms). Scale in the intensive group (i.e., the relationship between labor input and amount of stock) differs clearly from that of the second extensive cluster. Large farms do not need per se to be large scale. There are significant differences in incorporation levels also. For the large intensive farms, incorporation into the short-term loan market is nil. For the large extensive farms it is 9.2%. Incorporation into medium- and long-term loan markets presents even greater differences: 2.8% versus 45.9%. Incorporation into the labor, fodder and cow markets also show significant differences. Thus, although a high level of intensity is quite possible for large farms, the high levels of incorporation often to be found in them precludes intensive production and extensive large-scale production becomes dominant. And to the extent that large intensive farms become exceptional and strongly incorporated relatively extensive farms become the rule, the commonsense notion that "large farms are just more extensive" appears empirically at least to be cogent.

Scale: Capitalist and Cooperative Farms

Besides family farms, one encounters in Emilia Romagna a number of very large farms that are either capitalist-organized or agrarian cooperatives. A farm area of 1,000 hectares and a herd of more than 500 milk cows is more the rule than the exception in these categories. The differences in the organization of production and in production results are, however, astounding. The aims and structural conditions differ so greatly between the capitalistic and cooperative farms that the labor process (and thus, scale as the quantitative relation between labor object and direct producer) is structured in completely different ways.

The cooperative Bracciantile, of Novellara, is one of the biggest production cooperatives of Emilia Romagna. It has 970 hectares of land and 250 workers/members, half of them men and half women. This fifty-fifty division does not actually coincide with marriage patterns. There are in fact only thirteen "couples" among the members. A striking number of workers are under forty years old, which is remarkable considering the aging population of the countryside. Agricultural plans for 1979 and 1980 were very diversified. Each year more than ten crops were grown, besides dairy and beef cattle, pigs, etc.

> "We organize the agricultural cycle and other activities so that the labor peaks follow each other neatly. That is the reason for starting the production of beet seeds again now. We are also thinking of using the transplanting technique again for tomato plants. That way the peak harvest time is more favorable to us and production per hectare is noticeably higher. Also that way the land is available earlier for a second crop. . . . We are continually busy experimenting, also with a variety of techniques."

The level of mechanization is high—but not aimed at the substitution of labor. Raising the level of production is fundamental and continually kept under review. A high and specialized level of mechanization is one of the preconditions for this approach. The following anecdote about the purchase of the first tractor is typical:

> "We bought our first tractor in 1955. I remember it so well. To be able to buy it all the members put in a week's work with no pay, so they could earn the tractor. It was baptized with a feast."

The gross value of production per hectare in 1977 was 2.56 million lire, significantly higher than the GVP of some thirteen similar capitalistic farms. These averaged in the same year 1.99 million lire per hectare ($s=0.44$). One is conscious of the noticeable differences in work opportunities created by the intensity of farming on the cooperative farm.

> "Capitalist agriculture regards us as mining, as practicing *agricoltura di rapina* [literally, rapist agriculture], that is harvest without sowing. Obviously as a cooperative we have different aims: to produce and to work. We have to create as much work as possible and that can only be done through an intensive way of producing. That is our basic concept and that is why our production per hectare is so high. If this was a private concern there would not be 250 men and women working here but a meager 50. . . . There is a capitalist farm nearby here called La Fondiaria, and it is decidedly one of the better. It has about 300 hectares but there are only 25 men working it. Now and then you might see a lonely tractor in the fields but that is all. You know you can never farm well in that way. . . . To produce more and keep down cost is the way we work."

But if I provide a suitable counterargument, then the answer comes in a routine way as if the problem had been thought through many times:

> "You must view the economy in the right way. Feeding less and using less fertilizer is naturally out of the question. The produzione cannot be messed around with—that gets raised!"

Along with production cooperatives Emilia Romagna has several cooperative dairies (stalle sociale). The form is interesting: small farmers join together in order to form a communal dairy and in this way their private land and labor are freed for other activities. Livestock in the communal stall is cared for by a group of workers, some of whom are also members. A typical cooperative dairy will have some 300 milk cows and an average of five or six permanent workers and about eight part-timers. The average gross product per adult animal lies above the average for the capitalistic farms, i.e., 1,507 million lire per adult animal as against 1,346 million (Spaan, 1982).

Again we can conclude that farm size, in itself, bears no relationship to intensity of agricultural practice. Both capitalist and cooperative farms can rightly be called large. The production structure, however, is created from different goals and interests, and the concrete form of production factors is also very different. In capitalist farms, labor in itself is not the carrier of craftsmanship. In the above production cooperative of Novellara, just the opposite holds:

> "It is essential that everyone tries to work as well as possible, our internal organization stems from that. . . . Though it speaks for itself, it is actually in everyone's best interests!"

Both the owners and working managers of capitalist-organized farms (of the BOLKAP-sample, n=24)[33] were interviewed in Emilia Romagna. Some essential concepts figured in their opinions—concepts which together form a specific kind of the E-calculus described earlier. Craftsmanship as a decisive quality of labor was not only excluded; production was organized in such a way that it became superfluous.

> "Seeing the quality of my livestock, I can, without any anxiety, describe my milk yields as low. This low yield was a conscious choice which I have my reasons for. On a farm like this, with wage laborers, there is no structure for giving each cow an individual dose of feed. Between the best level of feed and the cow there is always the worker, and you know, it is risky to make the business too dependent on the workers. If I decided to give individual instead of standardized feeds to the cows then I would need far more workers. And apart from the fact that that would multiply the problems, it would be economically unsound."

Craftsmanship is being replaced by a rapid and wholesale adopting of externally developed innovations.

> "With beets, I harvest about 500 ql. per hectare. That is good, I mean look, ten years ago I was harvesting about the same, and since then there has not been a single innovation in beet cultivation."

Another manager/owner said:

> "Not bloody likely, I buy craftsmanship when I need it—I don't give myself problems about that!"

Craftsmanship has to do not only with direct production but also with the reproduction of production factors. An interesting comparison can be made about between a capitalist farm and an adjoining production cooperative where both had to contend with similar land. The owner of the capitalist farm related how his low harvest yields were due to the soil type, saying "it is now such that the land here is very bad: it is *terreno forte* [heavy soil]." But the president of the production cooperative who has to cope with the same heavy soil said:

> "When we finally managed to acquire the land, it was not much more than pure misery. The land had been stripped to the bone, there was nothing decent to be gained from it. The *latifundista* [capitalist farmer] had sucked it completely dry and ruined it for us. . . . Ai, the work that that cost us. But we have succeeded, we have transformed the land. The cartloads of manure that it took to rescue it [the cooperative produces outstanding compost manure, while the capitalist farm has closed its sheds], and the number of workings over! . . . Don't ask me to set everything on the line for you because it would make me tired all over again. But we made it, and now just take a look at how our beet fields produce!"

But say the capitalist farmers:

> "You can't compare the two, because on family farms and cooperatives, labor stands for income; while with us its pure cost."

In capitalist farming enterprises there is no central place for the practice of craftsmanship in the structuring of the production process. The following factors are fundamentally against this:

- high mobility of production factors,

Table 2.19. Differences in Productive Results Between Family Farms and Capitalist Farms (ERSA and BOLKAP/n=75 + 18)

	Family farms		Capitalist farms
	I-Group	E-Group	
GPV per adult animal in 1979	1.60 (0.18)	1.33 (0.12)	1.12 (0.20)
GPV per adult animal in 1980	1.67 (0.25)	1.31 (0.27)	1.35 (0.53)

- the planning and evaluation of the production cycle in terms of market and price relationships: in these capitalist farms entrepreneurship is given *tout court* as it were. The capitalist farmer is the agricultural entrepreneur par excellence,
- a strong orientation towards profitability.

Criteria such as produzione, so central to an I-logic, are explicitly absent.

> "Look, you don't need to make such hard work of aims and production and so on. All it is, is a question of keeping down costs and making sure there is a profit left over. That is why bookkeeping is so important for me. Your accounts will give you information that you will never get from a cow or a fruit tree; they will tell you whether things are going well or not. . . . Normally you tend to know only what comes in, what you receive in cash. The costs tend to slip through your fingers, and you forget them, and then you make false estimates. You have to be forever calculating."

In short, a clear E-calculus can be seen in the reasoning of the managers and owners of capitalist-organized farms. From here, it can be proposed that the E-calculus, insofar as it is generalized in the family farm sector, introduces from the beginning and for structural reasons those guiding elements which have applied over time in capitalist-organized farming: entrepreneurship tout court, a highly manipulable scale and a labor force from which craftsmanship is eliminated. In terms of production structure, those farms which are structured by the E-calculus veer strongly then towards those organized along capitalistic lines. Table 2.19 summarizes some comparisons. For 1979 they are based on the Parma and BOLKAP samples, and for 1980 they are based on the ERSA sample.

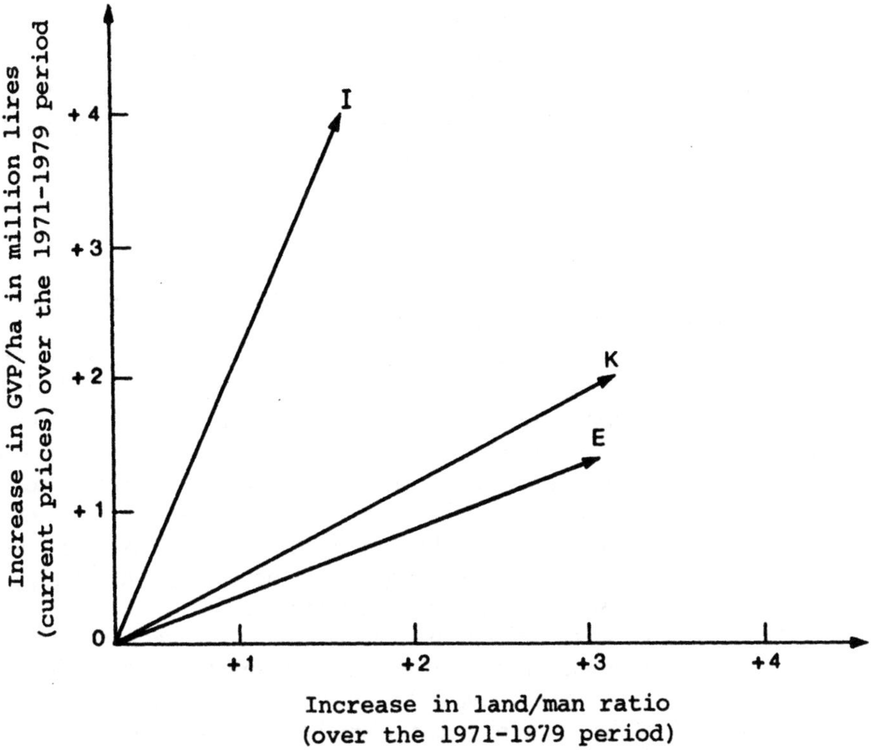

Figure 2.29 Developmental patterns for capitalist farms (K) and family farms (divided into I- and E-farms) (Parma and BOLKAP/1971–1979/n=18+18/dairy farms)

An analysis of farm development patterns over a ten-year period leads to the same conclusion (see Figure 2.29). Family farms organized according to the E-calculus, develop in a similar way to those farms organized along capitalistic lines, i.e., via a combination of scale enlargement and relative extensification. The difference between intensive and extensive family farms is greater than between extensive family farms and extensive capitalist-organized farms.

In the debate over the "agrarian question," the presence, absence or reappearance of the capitalist farm are questions rightly hedged with theoretical and political meaning. However, the reasoning developed above gives nuance to this debate because it demonstrates that the interaction between capitalist relations of production and agrarian production cannot be reduced to a simple presence or absence of capitalist agricultural enterprises. Such a question can even less be resolved by resort to conceptualizations such as "structural dualism" (de Benedictis and Consentino, 1979; Gorgoni, 1973, 1977; Koning, 1982).[34] Relations of production which are structurally present in capitalist agricultural enterprises are becoming increasingly real for a growing segment of family farms. Hence, the growing dominance of an E-calculus as the structuring principle of farm labor is a sign of the generalization of capitalistic relations in agriculture through the advance of incorporation, differential commoditization, and institutionalization.

Notes

1. In the ERSA data, different developmental patterns at farm level can be identified. If this data is trichotomized along the GVP/AA (per adult animal), three subgroups emerge. The first represents a growth in GVP/AA of only 2.5% over four years. Given the rate of inflation, this is an absolute drawback. The second group is characterized by a growth in GVP/AA of 25.8% over four years (which more or less equals inflation), while the third group shows a growth of 61.2%. The first two groups showed, on the other hand, a scale increase (adult animals per unit of labor force) of 21%, while in the third group this was only 7%. Income increased by 44% in both first and second groups as against 38% in the third group. So again two opposed growth strategies emerge at farm level: scale enlargement vs. intensification.

2. Not only as far as the different patterns are concerned. Both sets of data are also quite appropriate for an analysis of mean trends, as shown in the work of Cantarelli and Salgetti (1983). Of course, there remains the question of the significance of these "mean" trends, when there is in fact such a difference behind the "mean" picture.

3. This technique was developed and checked in an exploratory phase of the research, oriented to a better understanding of the elements and interrelations of the so-called "goal-function" of different farmers. The methodological inspiration was derived from an "actor oriented approach" (Long, 1977:117; and Galeski, 1972:12). It was especially important that no respondents defined the presented examples as being irrelevant or as being inadequate for symbolizing their own situation. Nor did any of them propose that scale and intensity ought to be combined in other modes (e.g. a farmer having 20 cows each producing 40 ql as symbolizing in some way a "marginal" farm, or 30 cows each producing 50 ql as being an indication of an "industrialized" farm). I am aware that this particular absence is due to the specific characteristics of the farming sector in Emilia Romagna that produces milk for cheese, where marginalization does not exist and industrialization is impossible (for a further discussion, see van der Ploeg, 1987).

4. The same interrelation between types of farming and varying hectarage expansion was also found in Newby et al. (1978b:185). We also registered the prices farmers were willing to pay (and actually paid) for land. Again there was a remarkable difference between I- and E-farmers, the latter being prepared to pay considerably higher prices.

5. Deriving from the classical work of Bishop and Toussaint (1958:45). In more recent agro-economic literature the same exposition is encountered (de Benedictis and Consentino, 1979:211, and Cramer and Jensen, 1982:89). The

production function presented in Figure 2.6 can be interpreted in different ways, e.g., as representing relations between input and output per labor object, as well as representing the farm as a whole. In the text we follow the first interpretation.

6. The validity of production functions based on the assumption of decreasing returns is increasingly questioned. Lipton (1968) discusses its use for peripheral agriculture, while de Benedictis (1984) does the same for "modern" agriculture. Theoretical agronomy now increasingly proposes that constant or even increasing returns should replace the notion of "decreasing returns" (De Wit, 1976, 1981 and Rabbinge, 1979:149). Their arguments are supported by historical research (Slicher van Bath, 1960; Ishikawa, 1981).

7. A general discussion is to be found in Messori (1984). Marasi and Salghetti (1980) demonstrate the same relation for the Emilia Romagna mountains. For the ERSA data used in this study the same relation was also calculated. Thus "costs" were defined as all monetary expenses, "benefits" as all monetary income derived from sales. Then the cost/benefit ratio was calculated for each farm. This ratio then fits well the "folk concept" ("*la margine*") as used by E-farmers.

There is of course a positive and significant relation between "costs" as defined above and "benefits" ($r=0.65$, $p=0.0001$). However, the relation between "costs" and the "cost/benefit ratio" is negative ($r=-0.32$, $p=0.005$), while the relation between "benefits" and the "cost/benefit ratio" (with a constant level of "costs") is insignificant. This implies that reduction of "costs" improves the "cost/benefit ratio." The same reduction of "costs" will produce an extensification.

I-farmers can also change the ratio between costs and benefits on their farms, but for them, such a change is never a goal in itself. Second, they follow a different strategy to get such a change, as spelled out earlier in the text.

8. On an abstract level it could be argued that in the end there is no difference between economic and technical efficiency, or that when the cost/benefit ratio is high technical efficiency is high and vice versa. On a conceptual level, however, it is useful to distinguish between *input* of production factors and *technical efficiency*. Only then can the position in the input-output space be adequately described. Timmer (1970:99) makes such a distinction when he discusses "allocative decisions" (regarding input levels) and "technical efficiency." He adds that "only recently" have economists started to make this necessary distinction in their analysis.

The cost/benefit ratio can be seen as the vector of two possible movements: both input and technical efficiency can change. Theoretically as well as empirically (Messori, 1984) it is possible that a decrease in the input of some or all production factors combined with a related decrease in technical efficiency leads to an increase in the cost/benefit ratio.

9. As can be derived, indirectly, from the fact that in industry it was strongly related to control over the process of production and therefore became an object of long struggles between management and labor (Braverman, 1974).

10. With this notion a long time perspective and the necessity of sustainability are introduced. Hence, one of the essential functions of an I-calculus is

that all relevant practical activities are "assigned a time—i.e., a moment, a tempo, a duration—which is relatively independent of external necessities, of climate, technique or economy" (Bourdieu, 1982:162).

11. The interaction of INC2 (see note 21 for a detailed description) and TATE2 has a negative effect on the input of capital per hectare (r=0.29). INC3 and the interaction term INC1*INC2*INC3 have the following partial regression coefficients on use of non-factor inputs per ha: +.39 and −.90!

If labor, capital and non-factor inputs per hectare are summed in an additive term, then this total input of production factors and non-factor inputs per hectare is negatively influenced by INC2 (−.23) and TATE2*TATE3 (−.32).

12. The Z-score used in the calculation is one of the composite indices for incorporation. Each partial degree of incorporation was standardized and then the eight Z-scores were summed up. There is a high correlation between this specific composite index and another one (represented as well in Table 1.4 of Chapter 1). The latter implied an addition of the monetary value (following current prices) of all the production factors and inputs mobilized through the markets. This total amount then was divided by the total value of *all* the factors of production and inputs used (including therefore the factors of production and inputs reproduced within the farm). Graphically this is illustrated in Figure 1.5 of Chapter 1. Both operationalizations imply certain biases and problems.

Apart from that, farmers have a certain room for maneuver vis-à-vis the different markets. They can choose, for instance, between dependency on capital markets for buying necessary machinery, or dependency on markets for machine services. For all these reasons the different dimensions of incorporation indicated by factor-analysis were used in the statistical analysis later in this chapter (see note 21).

13. Neither theoretically nor analytically. The problem is that Zachariasse applied factor-analysis to the *whole* of the dependent and independent variables, deriving from the loading all kinds of conclusions regarding the causal relations and statistical correlation between variables. This is definitely wrong (Thurstone, 1950; Brand-Koolen, 1972).

14. The underlying problem is that Zachariasse ignored the importance and validity of "l'art de la localité": "it is uncertain whether every farmer makes a correct evaluation of his own experience, it is questionable whether he is even capable of doing so" (1974b:67 and 3).

15. The only solution then to this problem is to halt traditional research into attitudes (nearly all research on entrepreneurship belongs to this category) and start research on the actual structuration of the labor process (of which the management of external relations is an integral part) from an actor-oriented point of view.

16. These and the following data are presented in the first place to give an impression of their variance; the intention is not to suggest whatever association. The interlinking of entrepreneurship variables with other factors will be discussed further on.

17. This is not to deny all the minor adaptations for raising the cost/benefit ratio that are invented and implemented by the typical entrepreneurs themselves.

For the important breakthrough however, they remain dependent on external innovations. A good example might be the delegation of calf and heifer rearing to other farmers (who are then paid for their work) versus the externalization of the reproduction of cattle to specialized institutes which manipulate Holstein material. The former can and was effectively done by typical entrepreneurs on their own account (till the introduction of the milk quota system); the latter depends crucially on external technological and institutional developments.

18. This aspect is spelled out in van der Ploeg and Bolhuis, 1983.

19. These caseifici or cheese factories, remained small (and therefore widely distributed all over the Emilian countryside) for two reasons. The first is that making Parmesan cheese is a craft which until now has proved unsuited to any form of industrialization. Scale enlargement, standardization and division of labor are impossible. The second reason is of a political nature. Since cheese-making implies a careful coordination between the caseificio and the farms delivering the milk, the cheese-maker can indeed exert a considerable influence on the farms and farmers concerned. Hence farmers prefer small, cooperative cheese factories which allow them to counterbalance this influence and to exert, in turn, some control over the cheese-maker.

20. The craftsmanship index here is based on objective criteria regarding the organization of the labor process; that is, craftsmanship is here operationalized as an additive term of the CRAFTp factors presented earlier.

21. The data on incorporation presented in Chapter 1 were submitted to factor analysis. The factors which emerged were then used in the calculations presented here and in the following path diagrams. Table 2.20 summarizes the results of the factor analysis applied to incorporation variables on the plain and mountains.

22. This and the following path diagrams are constructed using the Goldberger approach (Goldberger, 1970). This method was later described by Duncan (1971:122) as a "path-analysis [that] amounts to a sequence of convential regression analyses." This applies if certain conditions such as recursiveness and closedness (Blalock and Blalock, 1968; Blalock, 1971) are fulfilled. This is the case.

The "E-calculus" = (OPTIONp2 + OPTIONp3 − OPTIONp1) * (ENTREPp1 + ENTREPp2 + ENTREPp3). Several other combinations were tried out as well (for a detailed description and discussion of the results see Bolhuis and van der Ploeg, 1985, Chapter 3, note 76). Some of them generate much higher partial regression coefficients. However, I prefer to work here with the most composite and theoretically relevant operationalization. Additive instead of interactional models were also tried out for the operationalization of the E-calculus (since interaction might disturb the requirement of a normal distribution). Comparison, however, shows that the same results are generated in both cases (see Bolhuis and van der Ploeg, 1985, Chapter 3, note 73).

Finally it must be added that in this and in the following path-analyses, INC4 (see note 21) has been omitted from the analysis because of its ambiguous character.

23. Examining some of the debates (see among others Bernstein, 1986) one might well conclude that this crucial interaction between dimensions of incor-

Table 2.20.

	INC1	INC2	INC3	INC4
I3: incorporation into market for short-term capital	.93	-.04	-.06	.00
I4: incorporation into market for medium-term capital	.90	.02	-.13	-.05
I5: incorporation into market for long-term capital	.03	.70	-.22	.33
I2: incorporation into market for machinery services	-.05	.84	.19	-.26
I7: incorporation into markets for feed and fodder	-.09	.04	.84	.17
I1: incorporation into market for labor	-.20	-.02	.70	.01
I6: incorporation into market for land	-.09	-.05	-.05	-.84
I8: incorporation into market for genetic material	-.16	-.11	.15	.58
Eigenvalues	2.06	1.43	1.10	1.05
Variance explained	26%	18%	14%	13%

The same was done for the mountains:

	INCM1	INCM2	INCM3	INCM4
Medium-term capital	.90	.02	.13	-.05
Genetic material	.94	.01	.01	-.02
Long-term capital	-.04	.84	-.16	.30
Land	.11	.71	.16	-.19
Short-term capital	-.09	.11	.78	-.17
Feed and fodder	-.08	.04	-.74	-.45
Labor	-.04	-.04	.08	.80
Machinery services	.05	-.50	-.11	.19
Eigenvalues	1.99	1.56	1.07	.99
Variance explained	25%	19%	13%	12%

poration (or commoditization) and dimensions of TATE, shows that there is little that is "intrinsic" to simple commodity production. Its essence lies time and again in the specific *interaction* between simple commodity production and its "environment," i.e., in the concrete social formation of capitalism as defined by time and space. Without that interaction, SCP is really unthinkable. There is no "intrinsic" nature to SCP *outside* time and space, i.e., outside its interaction (or articulation) with other modes of production. It is instead the interaction which might be interpreted as essential. The same argument is developed by Mamdani (1986), who shows that what is really "intrinsic" for African simple commodity producers is their relations with the state, the political machinery and the economic circuits controlled through that machinery.

24. Readers interested in a complete statistical representation of all the material regarding the mountains are referred to van der Ploeg (1986, Chapters 7 and 8).

25. As can be derived from the data presented in note 21, the structure of incorporation factors in the mountains differs considerably from those on the plain. This mirrors state intervention in the mountains which is oriented to "development."

26. In applied research, however, most authors, such as Zachariasse (1974a and 1974b) go a step further: they include craftsmanship as part of an all-encompassing entrepreneurship. This is mainly done through the following two steps: (a) "entrepreneuring" is not analyzed as an activity, as a process, but essentially as an attitude (or set of attitudes) which by definition leads to and is associated with high profit levels; (b) craftsmanship is then operationalized within the same conceptual framework, i.e., as the capacity to produce a high net surplus per acre or per adult animal. This surplus is clearly associated with profits, so craftsmanship and entrepreneurship emerge in the end as being two sides of the same coin. However, this unilinear vision remains at odds with the overwhelming diversity and heterogeneity found in farming. Consequently, the image of "bad" entrepreneurs and "bad" craftsmanship as being something frequently found in the countryside is reproduced.

27. These characteristics are strictly limited to the technical and productive features of the production process. This is not to suggest that other domains, aspects and interrelations of farming (see Chapter 1) such as the family, gender relations, etc., are not relevant. The intention behind the selection of these technical aspects (derived from the ERSA bookkeeping records, and mainly analyzed to calculate farm income and profit levels) is to relate the specific technical structure of the farm to the labor process, and hence, to relevant economic, institutional and social relations.

28. The construction of these additive terms is theoretically grounded, since the farm labour process cannot be conceptualized, as was argued in Chapter 1, as the execution of isolated and separated tasks. Farm labor, above all, consists of the careful *coordination* of all tasks, thus constructing a "meaningful pattern."

29. The Italian version of this study (van der Ploeg, 1986, Chapter 9) includes an analysis which explores the similarities and differences in demographic structure, social reference groups, biographies, additional incomes, etc., between I- and E-farmers. No significant differences emerged.

30. From the end of the second world war onwards, the *mezzadria* system (a kind of share cropping) largely disappeared, and the political power of labor unions in the countryside became strong. They imposed a law which fixed labor input per hectare to a very high and constant level. All this spurred intensification which reached a peak in the 1950s. The first wave of mechani-

zation which then took place reinforced this process further, since it allowed farmers to dedicate more time (given because of the mechanization of burdensome tasks) to the cura of their fields and animals.

31. The market-oriented nature of farming was a constant feature. Far-reaching commoditization and institutionalization are, however, relatively new processes (see Benvenuti, 1985a; and Chapter 5 of this study).

32. It is the bigger herd that allows for better selection and improvement. However, whether such a possibility is put into practice depends primarily on the farmer's strategy.

33. The BOLKAP sample consists of twenty-six capitalist farms on which there exists structural and economic data gathered by the University of Bologna over a ten-year period. In 1981 all the managers and owners of these farms were interviewed. A detailed analysis is contained in Bolhuis and van der Ploeg, 1982.

34. This specific empirical constellation again highlights the stupidity of ascribing ahistorical, generic or intrinsic properties to simple commodity production in order to juxtapose it with (as done in all dualistic theories) the capitalist mode of production.

3

Potato Production in the Peruvian Highlands

In the southern highlands of Peru, 30 kilometers from Cuzco, in "the centre of the world," lies Anta Pampa. In earlier epochs it consisted of part morass and part lake, but it now forms a largely waterless plain some 3,000–3,500 meters above sea level. In the mountains which rise around it lie the *comunidades indigenas* "indigenous communities," or *comunidades campesinas* "peasant communities," as they are now called. During hacienda times the farmers of these communities were restricted to growing their crops on the rain-fed hillsides, often up to a height of 4,500 meters. However, during the turbulent years of the 1970s most of them also obtained plots of land on the pampa, where for the most part they grow potatoes and barley for the market and wheat, beans and maize for home consumption. Cattle raising is also important.

In and around Anta Pampa a style of agriculture is practiced which is representative of small-scale production throughout highland Peru, and of its problems. Some 70% of the economically active population of the highlands work in agriculture. In 1972, they contributed almost 42% of national agricultural production, mainly in the production of food crops. Unlike the rest of the country (coast and tropical lowlands), where about half of the cultivated area is devoted to industrial crops for export, agriculture in the highlands is almost 100% food production and is therefore of strategic importance for national food provision. Productivity in the highlands is low in comparison with that of other zones. On the coast, five times more is produced per unit of agricultural labor than in the highlands. Income levels are also low. As Caballero states (1981), "The farming population of the Peruvian sierra forms the largest component of the 'impoverished masses of Peru'; the average income of the sierra (between 130 and 160 dollars per capita in 1972) is comparable to that of the poorest Asian or African countries: the average farm income in the sierra amounts to about 50 dollars per capita."

This brief overview identifies a central theme of agrarian development, namely the need to increase food production and at the same time improve the incomes of the agrarian population. Further hectarage expansion is mostly precluded by the dominant economic and ecological relations. That holds a fortiori for any absolute reduction of those practicing farming. In other words, agrarian development in the Peruvian highlands is only conceivable as a continuing intensification of agricultural production. For various reasons, this theme has been noticeably absent from agrarian policy during recent decades, and technological and agronomic research has been minimal. However, the few studies that do exist on this problem highlight the great possibilities for intensifying agriculture in the highlands. Eguren (1977, 1978) has made a number of concrete suggestions for doing so. Caballero (1981) also emphasizes such possibilities and urges the need for rapid intensification, though he is sceptical of its practicality "if the global parameters for agrarian development in the sierra are not changed." He points out that private interest on investments in the highlands is unattractive compared with other economic or geographical sectors. As an allocation mechanism, the market is simply incapable of dynamizing agricultural technology or of alleviating the poverty of the farming people (Caballero, 1980).

Whatever the views as to the potential for intensification of highland agriculture, the trend in practice has been in the opposite direction. In an as yet unpublished study, Guillen has presented an historical analysis of agricultural production in the Department of Cuzco. On the basis of data from the Chamber of Commerce, he was able to estimate the gross value of production (GVP) for nine products for 1943 and also for the period 1950–1981. Of the nine products only three met the conditions for hectarage expansion: coffee, coca and cacao, cultivated on virgin land in the low-lying, semi-tropical zones (known as the *ceja de la selva*). The other six products are typical of the highland zone: potatoes, maize, beans, barley, wheat and wool. This distinction is significant because tropical and highland products manifest a completely different pattern of development. In Figure 3.1, the total gross value of production for the nine products is calculated using the 1960 value of the Peruvian currency, the *sol*. The abrupt fall in 1956, when production was halved, was due to the most serious drought of the century. The continuing fall in the 1960s was linked to the vehement and widespread peasant struggles taking place at the time in Valle de Convencion y Lares, a central area for the production of tropical crops (Fioravanti, 1974). After 1969 (the beginning of the Velasco land reform) a new growth period occurred during which small-scale peasant agriculture was consolidated and expanded. Thus, in the years following 1968, the

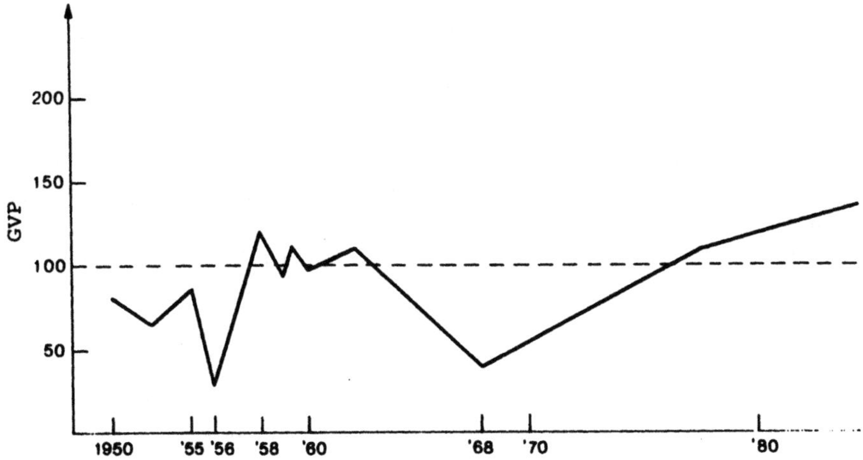

Figure 3.1 Development of the annual gross production value for nine agricultural products, Department of Cuzco (at 1960 exchange rate)

gross value of agricultural production for the Department of Cuzco grew by 4.9% per year. The average growth rate for the 1950–1981 period was 1.73% per year and measured from 1943 the growth rate was only 0.52%.

If, however, we compare the growth rates (1950–1981) for tropical and highland products, then the annual growth rate for tropical products is 2.83% as against a negative growth rate for highland products of −0.76%, and of −0.40% even after 1968. This decline was not equally distributed over all products. The GVP for potatoes fell annually by −1.73%, and maize and wheat also had negative indices, but, in contrast, barley (grown for beer production) rose. For the highland agricultural sector as a whole, however, in both the long and short term, only regression is to be observed. In fact output fell even more sharply than the figures imply, for over the same period, real prices rose by 1.75%. The small price elasticity for highland products compared to those for tropical products is also worthy of note. Indices for maize, potatoes, wheat and barley amounted respectively to −0.08, 0.00, −0.37 and −0.38; whereas for tropical products (coffee, cacao and coca) the figures were, respectively, 0.47, 0.30 and 0.27.

Balancing these data to some degree against the proposition that marked price increases ought to be able to act as a lever for agricultural development, then the stagnant and negative growth patterns elucidated by Guillen naturally require separate explanation. Caballero, taking potatoes, wheat, barley and maize together, arrives at the figure of 0.67% for average yearly growth in production (1964–1972), for the highland as a whole. In the Peruvian literature, authors often confine themselves with references to land shortage, the possibly declining fertility of the soils, the scarcity of capital, and the high levels of risk. However, these factors, taken either separately or together, do not appear to offer a sufficient explanation. During recent decades, commoditization in the highlands has undoubtedly made significant strides. Caballero (1981) writes of a "commercial revolution," while others, such as Villasanta (1982) and Ccori (1982) write of increasing "commercialization."

In the following discussion I examine the effects of this commoditization process on agricultural practice. Using a comparative analysis we shall see that increasing commoditization leads to extensification. This extensification offsets the growth effects of other factors so that, at a macro level, the net effect is to be seen in the pattern of stagnation already identified.

To investigate closely the relationship between commoditization and agricultural development one community out of the thirty or so communities of Anta Pampa was selected for detailed study. Chacán, the community chosen, shows a degree of internal socio-economic differentiation and has a relatively long experience of being integrated into markets, a view that is supported by peasants from the other communities. Data were collected on fifty-two farm enterprises of Chacán by a team consisting of a sociologist, an agricultural economist and two comuneros, sons of farmers who after training in agriculture, had returned to work on their fathers' farms in the community. Through frequent interviews and at a later stage through questionnaires a large quantity of agronomic, economic and sociological data were gathered. These data, which contain a bias which will be discussed later, form the basis of the analysis which follows. Before proceeding with the actual theme, however, a number of key questions must be answered. Who are these farmers? How do they farm? How do they perceive their enterprises? Into what relationships do they enter with surrounding markets and institutions? And how does the commercialization mentioned by so many Peruvian authors work out in practice?

The Farmers and Their Enterprises

Let us look first at the extent to which a man can be considered a farmer. It is known from research (Figueroa, 1982; Long, 1979; Long and Roberts, 1984; De Janvry, 1981) that only a minority in the highlands are "real" farmers in the sense that they limit their labor to working only their own fields (and thus enjoy no income other than that derived from their own farm enterprise). The average situation is much more complex. As well as being a farmer, the individual is an agricultural laborer, a wage laborer and a craftsman. Retail and wholesale trade and the possession of lorries for transport complicate the picture still further. Agricultural practice cannot be understood in isolation from these other activities. Occasionally agriculture is purely and only directed towards subsistence and monetary income is derived from other sources. This group of farmers in Chacán are described as "the poor." Then there are those involved in other activities in order to earn enough working capital to engage in market oriented agriculture. Agriculture and other activities cannot thus be separated. Secondary earnings are the cork on which agriculture floats. One also encounters examples of a certain complementarity: agricultural production is directed towards the market and income is supplemented by the other activities. Within this framework various combinations are possible. Thus one can imagine that where the other activities are more numerous and lucrative, they will directly compete for labor devoted to farming. Farm labor then is subjected to market-derived relations: "Where will I earn the most? Where will my labor or money render most value?" (see Fernandez, 1977).

In Chacán only a fraction of the farmers (9%) work solely on their own farms. A large group (45%) is involved in wage labor elsewhere, possibly combined with retail trading or weaving at home. This is the principal craft activity in Chacán where ponchos are woven for the tourist market in Cuzco. Twelve percent devote themselves entirely to this activity. There is a small group of chauffeurs—one owner-chauffeur and two who drive for the profit of others. Five farmers trade extensively (in potatoes and cattle). Finally, some 12% are involved in weaving and the retail trade (in running small shops).

Are these differences relevant to agriculture? Figure 3.2 gives the number of man days per topo (1/3 of a hectare) in potato cultivation for the various occupational categories. As we shall see later, the number of man-days per topo is a reasonable indicator of the intensity of potato growing. Those involved as intermediaries in wholesale trade "invest" noticeably less labor in agriculture than others. Evidently the high return obtainable from these other economic activities is in this case a decisive criterion for on-farm decision making. These farmers finance potato growing with large loans from the Agrarian Bank while they invest their own financial means in trade.

Caballero (1981) speaks of a *despachamamamización* in relation to the regression found in agricultural production in the highlands. *Pacha mama* is Quechua for "mother earth." The concept refers to the value that land has for the Indian population. Land must be "cared for" and "properly controlled." A clear relation between use and maintenance can be discerned in such a concept of land, which clashes with the

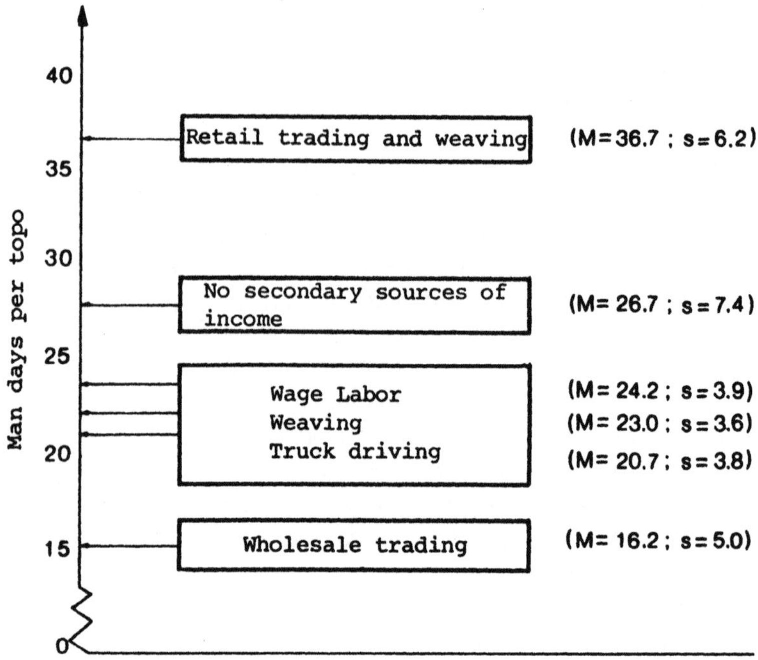

Figure 3.2 Relationship between man days/topo in potato cultivation and the nature and scope of secondary occupations

notion of simply accrediting land a purely commercial value. The ecology of the highlands urges a careful maintenance of soil fertility through careful farm management. Neglecting to fight against erosion, for example, is perhaps easier in the short term but will exact a heavy price in the future. The notion of "pacha mama" is thus that of the guardian of the future interests of the generations to come. The present commercialization of agricultural production (Villasanta, 1982; Ccori, 1982), and in particular the dependence of enterprise decisions on general market relations, leads to this "despachamamamización," and to the substitution of land as use value for land as exchange value.

This is quite clearly the case for those heavily integrated into general economic circuits, such as wholesale trade. They not only use less labor per unit of land (i.e., they extensify), but when asked which crop they would most prefer to grow they all opted for barley, the crop which in their words "costs the least, needs little labor and which earns the most."

A second observation concerns the high-flyers (see Figure 3.2), those who cultivate more intensively than is generally the case and who weave and engage in retail trading in addition to farming. They usually have at their disposal a small shop in the community. This combination

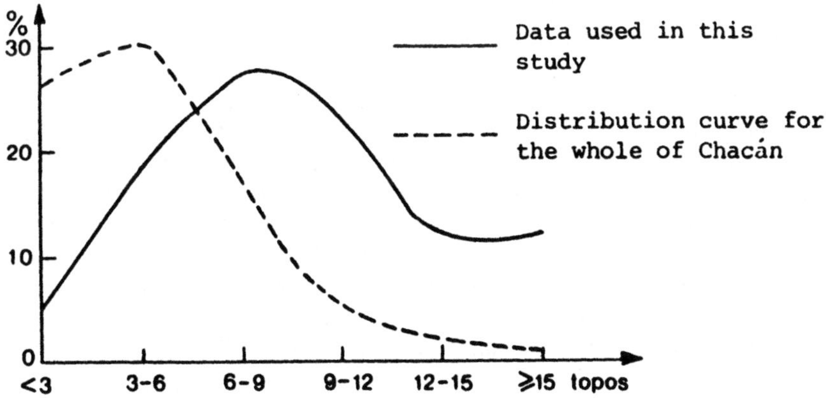

Figure 3.3 Distribution curve for the land available per farmer

reflects a certain level of prosperity, since they do not need—as do the poor—to rely on wage labor as a supplementary source of income. However, unlike those involved in wholesale trade, they plough their earnings back into further intensification of agriculture. For them, weaving and retailing offer few opportunities in themselves for accumulation or increasing investment. They invest therefore mainly in agriculture, and these investments result in a labor input per unit of land higher than for the other categories.

The hectarage that the fifty-two Chacán farmers work or at least have at their disposal for eventual working (fallow land forms a necessary part of the rotation system) again varies markedly. Figure 3.3 gives a picture of the land available. The frequency distribution for the whole of Chacán is also shown. From this one can see that the data collected by the team were biased in that the small-scale farmers are under-represented and the middle and large-scale farmers are over-represented. Despite this, land is clearly *not* a constant. Marriage and inheritance serve as mechanisms for the dividing and recombining of land, and then there are various mechanisms, partly based on the market, by which land can be mobilized. Although strictly speaking it is not possible, there is increasing talk of buying and selling land. Land can also be leased, and there is a *compania* mode of gaining access to land whereby a man works the land of another and divides the harvest in a manner previously agreed upon. In practice the rules for this type of arrangement can be extremely complex, as one farmer explained:

> "Next year I shall work in compania again. As you know there are lots of ways of doing that. Firstly, take barley, here in the village. The other man looks after the land, puts in the work and sees to the manure, and I pay for the insecticides, do the spraying and take care of the seeds.

> Come harvest time we harvest half of the *topo* each: one half of the topo is for me, one half for him. It's again different for potatoes. We harvest them together and then we carefully weigh them and take exactly half each. Sometimes for potatoes it also works like this, I pay for example for the tractor to break up the land, get the land ready for planting and look after the seedlings. The other pays for the fertilizer. We do the sowing and fertilizing together; each does half of the work and the costs are divided honestly. Also with the spraying . . . and like I said, we then harvest together and go and weigh them. The seed potatoes we divide and those for selling are also honestly shared. Naturally it is sometimes done otherwise; you must always reach some agreement about these things. . . . It's a question of understanding each other well."

The reciprocity stressed in this example can get lost in some patterns of cooperation. An asymmetrical relationship can arise which is experienced as a certain corrosion of the principle of reciprocity:

> "In compania the advantage quite often goes to the owner (of the land). Sometimes the land is the only thing they are willing to bring into it. You are expected to put in all the work and the harvest is then divided into two . . . imagine! . . . The problem these days is that many of the rich [i.e., those with much land and the capital to work it] are no longer willing to go along with the poor [in Spanish acompañar, hence compania] . . . unless it's according to unfair rules."

Despite this complaint, it is noteworthy that 70% of the farmers work, in addition to their own land, some land in compania. This pattern is encountered not only among the small-scale farmers but equally among the large-scale ones. In short, there is a network of compania relations that is mutually binding for farmers through which land and, according to our observations, also labor and tools are exchanged. Through these relationships farmers can mobilize a missing production factor or exchange it for another which they possess in sufficient measure. There are no abstract or general rules governing such exchanges; costs and benefits are worked out according to the individual case. Social relationships (e.g., kinship or symbolic kinship relations "*compadrazgo*") are often decisive for the exact conditions of the exchange.

The effectively worked land area is also, of course, dependent on shifts in the cropping system (see Table 3.1 for the average land area as well as the average cropping system). By reducing the amount of fallow land, for example, one can increase the effective area. However, there are some specific communal rules governing this.

Table 3.1. Average Area and Cropping System in Chacán (n=52)

Land use	Topos
Potatoes	2.72
Maize	0.85
Beans	0.97
Barley	2.05
Wheat	0.35
Fallow	1.45
Total area	8.39

Farmers were asked how much land they would work during the coming season compared to the amount worked during the previous season. The results underline the extent to which the amount of land is a variable rather than a constant factor. Hardly any farmers planned hectarage similar to that of the previous season: 58% wanted to reduce hectarage, 37% to increase it.

Finally it should be said that 44% thought they had enough land. Although, given the built-in bias in data collection, these data cannot be generalized, they are nevertheless interesting. They suggest that land expansion for the middle group and also for larger enterprises (note that the average farm consists of less than 3 hectares and the largest cultivates 6 hectares) is not the most crucial element for further development. It is the ability to mobilize sufficient capital and labor that is seen by these farmers as the most important means of developing their farms.

Labor

The average household in Chacán has about five mouths to feed. Size is only partially related to the age of the household head, since parents and younger brothers often live with the young farmer and older farmers frequently have married children living with them. On average, only one family member is available to help the head of household with farm work. This arrangement produces a skewed relationship, with a lot of mouths to feed and few hands to offset the amount of work this necessitates. The explanation for this problem lies in the high level of migration. Older children leave to earn cash elsewhere, and those who remain are too young to help much in the fields. Only after the death or invalidity of the father do one or more of the older children return "to help mother" and later to take over. This means that the available labor per family is insufficient to do all the work in the fields. The peak nature of a number of activities makes the problem even more acute and means that outside labor has to be recruited in one way or another. Once again we encounter the market and other social mechanisms through which additional labor may be secured.

The most important social mechanism for mobilizing labor is *ayni*, although in Chacán, where commoditization of labor is most advanced, this is less the case. Ayni in Chacán involves 70% of family production units, whereas in many communities it involves all such units (see also Hibon, 1981). Ayni is an exchange based on reciprocity; labor for labor, labor for traction, labor for other production factors (ploughs, saws, donkeys), and traction power for traction power. The form varies but the core concept remains that of reciprocity (Mayer and Zamalloa, 1974; see also Long and Roberts, 1978). The reciprocity principle is strictly adhered to, and by preference the exchange was for similar services: "I work one day with oxen for you (or I lend you my oxen), you lend me in return a similar service." One day of harvesting requires considerably less effort than a day's work in the *lampa* (these particular cultivation tasks will be described later). Thus one day's help of this kind is by preference exchanged for one day's help of like kind. If this cannot be arranged, then it is one day in the lampa for two in the harvest. Further (though this differs from community to community and even within communities), one or two day's labor is exchanged for the loan of oxen for one day; the loan of a plough is worth a day's labor and oxen are exchanged for oxen.

The system of ayni is closely interwoven with various social relationships operating within the community or family, friendship, compadrazgo, and *cofradia*-membership (i.e., of the semisecret religious fraternities). It is within these relationships that an ayni arrangement is discussed and implemented. Hence ayni is one of the most important binding tissues of the social organization of the community, involving service and return service, gratitude and obligation. In short, it entails mutual dependency structured by a socially controlled reciprocity. It fulfills, in addition, other functions. To begin with, the way it works is closely related to the nature of the agricultural production process. Certain activities, such as sowing, must be carried out within a short space of time because the soil dries out. Thus, being able to summon a large body of workers at any time is of strategic significance. But it is more than this. The most important function of ayni, now as in former times, is that by means of it the unequal distribution of land, tools, and traction are to some extent alleviated. It is through reciprocal exchange that missing resources can be mobilized.

The interweaving of ayni with relationships of respect and social affection is indispensable for its adequate functioning as an exchange mechanism. If a team of oxen (*yunta*) are given to someone to use (in exchange for a day's work), then one must make sure that the oxen get enough to eat and drink, are given rest when they need it, and are not overdriven or beaten to the extent that the team will be unmanageable in the weeks that follow. These relationships of affection and respect govern even the simple exchange of a day's labor, since a day's labor will be exchanged by labor of a similar kind, and they are consolidated during ayni by eating and drinking together. From this practice stems the comment heard time and again in interviews: "With ayni you can rest assured that your people will come; someone who only comes for a day's pay can just as easily stay away." The working of ayni is then closely linked to the pattern of stratification in the community, just as it is tied to different "phases" of the demographic family cycle (Chayanov, 1966).

Chacaínos order themselves by reference to three concepts: *ricos*, *medios* and *pobres*; that is, rich and poor farmers and a group in between. The rich are those who have at their disposal a lot of land, their own working capital, and one or more pairs of oxen, but have problems of insufficient labor. The poor do not have sufficient land at their disposal, nor do they have the necessary means of traction. They do have, however, a surplus of labor power. For the group in between, the medios, these factors are in theory balanced, although the nature of the production process implies that at certain times they will suffer shortages and at others will have a surplus of labor, tools and traction. This uneven pattern seems to entail some need for an intensive system of mutual exchange. In this case this is structured through ayni. Figure 3.4 provides a schematic representation of this.

An essential function of the structure outlined is that general market and price relationships are excluded as organizing principles in agricultural production. Tupayachi (1982) suggests that the following rule operated in the three communities in which he carried out detailed research: "one yunta for one day's labor." At local market rates, however, a day's labor was worth 250 soles a day as against 400 soles for hiring a team of oxen for a day. Could this then be described as "non-economic behaviour?" Certainly not. The most one can conclude is that the mobilization and organization of production factors are regulated by other relationships of scarcity and by a different mechanism of exchange from that operating in these "surrounding markets." In other words, ayni is a mechanism of exchange (which, in a manner of

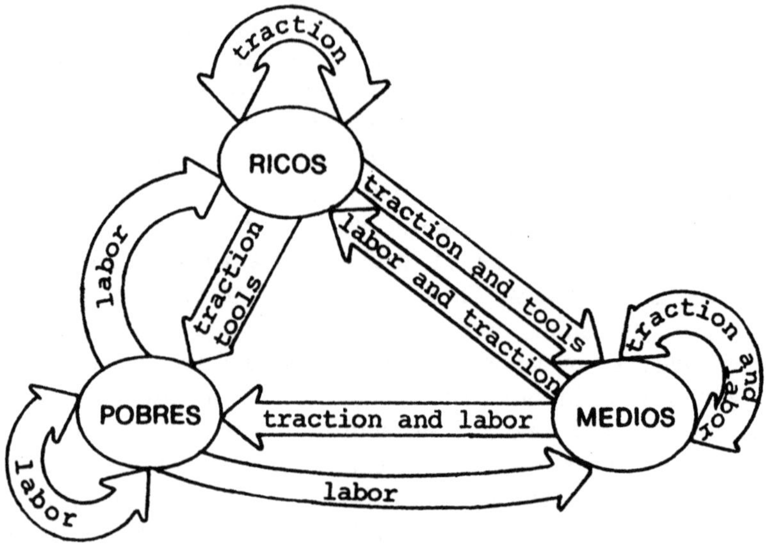

Figure 3.4 Mutual exchange between different social strata

speaking, creates a relatively autonomous market) that makes possible a high degree of adaptation to local conditions, needs, structures and relationships. Like compania described earlier, ayni keeps commoditization at bay, that is, it excludes a complete subordination of the agricultural labor process to general market relations. The regulating of exchange by social factors introduces boundaries which would gradually disappear if a complete commercialization of exchange relationships took place.[1] Then general scarcity and power relationships would become dominant, thus excluding particular adaptations through ayni and compania.

The subordination of local and relatively autonomous markets to national, supranational and even international markets, where the same price structure holds for ever greater geographical units, results in a "universal market," which Eisenstadt (1963) described as a market of "free floating resources not embedded within or committed beforehand to any primary ascriptive particularistic groups." Incorporation, or rather, "the direct attachment of local production, exchange and consumption to the national market-system" (Pearse, 1968), is the main process from which the development of such a "universal market" springs. With the partial superseding of ayni by the labor market, which, as shown, is beginning to take place in Chacán, socially regulated mobilization of production factors (in this case labor) will give way to market determinism. New relations of production may arise and the sphere of economics may become decisive in the organization of production.

In its first phases this process can exercise a noticeably unsettling effect on agricultural practice—certainly where there exists a strong discrepancy between locally operating exchange relationships, on the one hand, and the politico-economic relationships expressed in general market and price relationships, on the other. Polanyi (1957) argues that

> "Land and labour of course could not be transformed into commodities, as actually they were not produced for sale on the market. But as the organization of labour is only another word for the forms of life of the common people, this means that the development of the market system would be accompanied by a change in the organization of society itself. All along the line human society had become an accessory of the economic system. . . . But while production could theoretically be organized in this way, the commodity fiction disregarded the fact that leaving the fate of soil and people to the market would be tantamount to annihilating them."

Oxen, Livestock, and Dung

Practically all the farmers in our study had one team of oxen or yunta at their disposal. Only 15% had none, and 9% had the use of two. Under normal circumstances, 0.5 to 0.6 hectares of land can be worked in one day with one yunta. Thus, in terms of depending on ox teams, ayni is all but indispensable for agricultural practice, as the average hectarage in Chacán, as noted earlier, is about three hectares. As well as oxen, most farmers have some livestock, including a few cows and young steers, some sheep (often up to 20 or 30 head) and some pigs. Only 6% had no livestock. Apart from their value as a source of meat, milk and traction, livestock are also important for manure. Dung is carefully collected, dried, and stacked and later taken to the fields. Livestock are also important as a form of "security." After a bad year some animals can be sold to make it possible to purchase the necessary means of production for the following cycle.

Capital

The relative increase in various costs compared to prices, the impoverishment of the rural population, which often leads to the consumption of some of the means of production (such as potato seedlings), technological development, which requires an increasing use of inputs manufactured elsewhere—all these factors have resulted in an increase in the need for working capital to finance each stage of the production cycle. Working capital is not only used to purchase relatively expensive inputs only available on the market, such as chemical fertilizer, but is also increasingly used for obtaining labor, oxen and land. The ways in which capital is obtained and thus the relations it introduces into the labor process can differ markedly. One can distinguish three patterns, which at the same time reflect the social stratification of the community. Patterns of differential commoditization and the internal stratification of the peasantry seem here to be closely linked: the different mechanisms for mobilizing capital are strongly associated with the social definitions of ricos, medios and pobres.[2] Although from a theoretical point of view it may be surprising that mechanisms of economic accumulation define to such a degree the patterns of social stratification, the Chacán farmers themselves recognize this and use specific commoditization patterns to define social strata. They also use the argument in reverse when they explain a certain commoditization pattern by reference to a specific socio-economic position.

One must bear in mind in the following discussion that the concepts used to define the different strata are sharply demarcated, whereas empirically the boundaries are much more fluid and liable to change.

Los Pobres

The production and reproduction schemes of los pobres are in essence very simple. The harvest is divided into seed potatoes for the coming year and the rest for family consumption. Production is not marketed and production factors seldom pass through markets. If the household is not self-sufficient in land, labor and traction, they are mobilized through ayni and compania. The only working capital they have is in the form of their standing crop or the seed potatoes preserved in the house.

Agriculture as practiced by the poor does not interact with the market. This does not mean to say that the poor live a life of economic self-sufficiency. Far from it. Necessary cash income is earned in wage labor elsewhere and through the sale of home-produced craft products. The point is, however, that this form of earning cash is structurally separated from agricultural practice, and so agricultural practice is not commoditized. The president of the community at the time of the study summarized the position of the pobres as follows:

> "The poor farmers are those with very little land, a few topos, let's say less than two hectares. They are the ones who only plant for their own consumption, who harvest some ten bags of potatoes per topo, a poor harvest, and they eat from this, and you'll see no chemical fertilizer there. They are people who work elsewhere. . . . Actually they are laborers who grow their own food because it's cheaper than buying it. Moreover they don't earn every day . . . so they have to work their fields as well."

Following Figueroa (1982), Kervin (1982) argues that this stratum of the poor is unable to fulfill a meaningful role in any kind of agrarian development. Suppose for the sake of argument that the total income of such households is 100 soles and that the market value of the food grown for home consumption amounts to no more than 46% of total income (Figueroa, 1982), of which the potato share is no more than 20% (Equiplan, 1979). If a part of the potatoes were sold (in a situation of surplus), then this would imply only about 3% of the total income. Now suppose that new varieties are introduced which double the potato yield per hectare. Leaving aside the fact that the purchase of such varieties and the additional inputs of fertilizer and pesticide hardly fit with the production and reproduction scheme sketched above, the fact remains that the effect of such a doubling of monetary earnings would be minimal—according to Kervin, 15% at most. None of the authors reject improvement in agricultural methods per se but they conclude that the results would be manifested primarily at the level of improved nutrition. There will be little impact on agricultural growth. Furthermore, they contend that improving the lot of this group could best be achieved through raising wages; the virtual absence of any commoditization of agriculture as practiced by this group precludes their being manipulated by prices and market intervention.

Los Ricos

"Rich farmers," explained Chacán's president,

> "often have four or five hectares, or perhaps even more; sometimes of course less. And they don't need credit. They have their own capital and can buy what they need . . . just like that. . . . Mostly you see its the rich who buy most inputs, the most fertilizer, they also then produce the most, they till the land quite well. . . . Of course there are the traders too. They are also rich, but often they till the land badly. They borrow money from the bank even though they have enough money of their own. But we don't really consider the traders as members of the community, they are not comuneros, just folks who bring us down."

It is neither the amount of capital at a man's disposal nor simply how much land a man has that forms the decisive criterion for a definition of el rico, but the combination of land and having "the money with which to work it."

> "A farmer who has a lot of land but has no money of his own to work it and has to borrow from the bank, that is decidedly not a rich farmer."

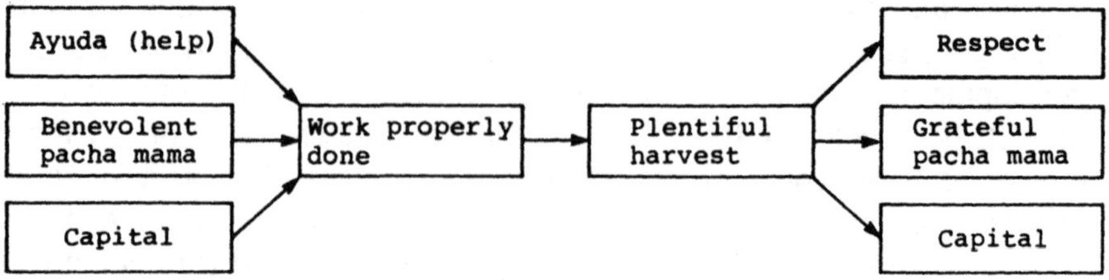

Figure 3.5 The calculus employed in Chacán

And the reverse, a little land and a lot of money?

> "Such a person is also poor because he can't earn a sou with his money, he will use up his money on the market (for buying the things he can't grow) and so in the end all that remains to him is the small bit of land."

Asked whether someone with a lot of land but without capital would be considered rich or poor, 58% answered that they would call him a pobre and 27% said that a fair description would be that he was a medio. This answer was motivated by the argument that in this situation credit could be obtained. Only 12% were of the opinion that a lot of land was sufficient to be considered a rico.

The production and reproduction scheme of the ricos is of a "historically guaranteed autonomous" nature. The pool of labor, work objects, and means are ensured by the preceding production cycles as well as by the cooperative links previously entered into (ayni, compania). Some inputs, such as fertilizer, must be purchased on the market. Their purchase is likewise paid for from the harvest of the previous cycle. Thus reproduction remains historically guaranteed. "*De la buena producción viene tu capital*" ("from good production comes one's capital"), say the people of Chacán. Good production (i.e., a good yield per hectare) forms the key, both for the maintenance and for the further development of production. "One's capital" is thereby seen not only as a result but at the same time as a prerequisite for being able to gain a good production.

Such a process is clearly evident in the logic which the farmers of Chacán draw upon. This logic is schematically summarized, using folk concepts, in Figure 3.5. A benevolent "pacha mama," sufficient capital (sometimes also called *la principal*) and *ayuda* (literally meaning "help" but used here to mean additional labor input ensured by the mechanism of ayni) are held to be the essential prerequisites for "working properly." The latter is a strongly held normative principle in all respects. The carrying out of all the specific tasks was repeatedly accompanied by some amplification such as "to do it properly," or "to do it the proper way." And asked which is better, "to do a good job," or "to get it done quickly," for the various tasks such as planting, weeding, and the like, virtually no one thought the second possibility a legitimate one. Working properly leads to a good harvest. The meaning of this cannot be simply reduced to the monetary value of the harvest when sold. A plentiful production of course results in the increase of one's capital, but it also increases the respect a person receives from the community. Such a person is seen as a capable, good farmer—an image which can be enhanced by giving small parts of the harvest as gifts or as help to others. This respect is in turn necessary for rallying ayni and compania relations in the following season. And finally, a good harvest will ensure for you a "grateful land." Because it is worked well, the land will provide a good beginning for the following season. "What you give to the land, the land gives you back. In like measure to which she is worked and cared for will the coming harvest be plentiful."

Where, in practice, labor is structured through such logic, productive activity is strongly directed to ensuring the prerequisites for the following production cycle. In that sense, the logic or calculus can be considered a reflection of, and a consciously applied precondition for, the pattern of historically guaranteed autonomous reproduction mentioned earlier. An intriguing aspect of Chacán logic is the balance between the individual and the collective. The logic outlined refers to a world which can be overviewed and to some extent manipulated by the individual farming family, which allows him to adapt the degree to which labor is structured as "good work" to his own individual needs and capacities. At the same time the prerequisites and results refer to the community sphere: "Respect" is acquired within the community and in this community it is converted into "help." Both of these terms, and the interconnections between them, refer, in other words, to social networks in the community. This also applies to "capital": it is supplemented and reallocated through social networks. Finally, essentially the same holds true for "pacha mama." Specifying the precise manner in which the good mother earth must be worked in order to show respect for pacha mama is deeply anchored in community traditions. Even the rotation of individual fields covering various sections of community land was specified directly by the community.

A second intriguing aspect of this calculus or logic is that it is supported and thought legitimate by just about everyone. Although clearly there are differences in the way work is carried out in practice, the tenet that one must "work properly" (i.e., respect pacha mama to produce a good harvest) is upheld by everyone. In contrast to other agricultural zones, such as the Italian Po valley region, the people of Chacán have developed no alternative to the calculus sketched here for legitimizing any other way of working. Naturally the calculus outlined does not mean that what happens in the fields will be uniform. Insofar as a calculus displays itself in a particular style of farming, there are naturally within that framework a number of possible variations, such as Hofstee (1985) made clear. The prerequisites that farmers have at their disposal—i.e., the measure to which they can depend or draw upon social networks to mobilize ayuda, pacha mama, and, where necessary, capital—varies from farmer to farmer. Also, the measure to which a farmer thinks these networks ought to be mobilized will differ: in other words, within each household, the concept of "working properly" will take on its own specific form depending upon such factors as the availability of family labor, the relationship between family labor and the number of mouths there are to feed, and the time perspective held by the family.

Thus, both the practical application of this calculus and the measure to which it can be achieved vary and allow for considerable flexibility in adapting farming to inter-household dynamics. Within this system there emerges an important degree of heterogeneity in the fields. Such a mechanism can also be discerned in Chacán itself. Asked who achieved the highest production levels (who gathered the best harvest), 52% of the farmers replied that is was the rich who had sufficient capital, land and labor (either through their own family or through ayni or wage labor relations). External factors, however, can also occur which make adherence to the calculus impossible. In the following sections of this chapter I will investigate in this connection the extensive and often abrupt commoditization of agricultural practice in Chacán. In an ongoing process of commoditization, the notion of a "good harvest" is replaced by the notion of the current market value of crops. And "working properly" can, in the eyes of the farmer, become primarily a cost measuring exercise where costs must be kept to a minimum.

Although the process of commoditization is relatively advanced in Chacán (even compared to northwest European agricultural systems), Chacán has not yet developed a calculus that corresponds to this high level of commoditization—that is, one where the increasing relevance of market and price relations is understood and legitimized as a guiding principle for agricultural practice and where, as in the E-calculus described for Italy, market and price relations are translated into the organization of labor and the development of production. There is, in Chacán, no such alternative to the prevailing calculus. That is striking. Recent research in the Netherlands, the Italian Mezzogiorno, and Ireland shows how diverse calculi are handled. Some of these respond to a high level of commoditization, while others are basically aimed at maintaining a certain degree of autonomy (see Maso, 1986; Bolhuis and van der Ploeg, 1985, especially Chapter 4; Leeuwis, 1988; Long et al., 1986). Although it can only be speculated upon here, this phenomenon is understandable. A calculus in which institutional and economic preconditions are made the explicit and exclusive starting point for agricultural practice would be an exceedingly fragile construction, for the relevant institutions are very unstable. A change of government can result in a major shift in the credit requirements of the Agrarian Bank. Development projects financed from abroad come and go. The instability of agricultural markets speaks for itself. Adequate technologies through which rapid growth in scale could take place barely exist, and insofar as they are available, scarcely show any advantage over indigenous techniques. And, finally, the ecological dangers of an imaginary E-calculus are here great and can clearly be seen. The "neglect" of pacha mama, or the interrupting of rotation systems to allow the most profitable crops to dominate, can perhaps offer short-term financial reward but are in the longer term disastrous.

Perhaps here lies an important part of the drama of the Andes. Confronted with an unmistakable undermining of their own specific rationality (as outlined in Figure 3.5) these farmers lack a changing conceptual scheme (a new calculus) within which the material changes occurring (such as rapid commoditization) can be interpreted as rational. What remains is a feeling of inadequacy, which one might call, along with Levi Strauss, a certain "tristesse." They know how to work the land "properly" but feel themselves forced to "hasten." "Now it is impossible, the world has gone mad," is the lament of an evening at the edge of a field; "it doesn't work anymore, the world is crazy." "You still work hard, but proud is something you can't be anymore." The tension between the existing calculus and the increasingly adverse circumstances results in feelings of powerlessness and dismay, without the real causes being clearly identified. Notions of the past are coloring what they now miss: "Ay, there used to be pure gold planted in these fields!" And the growing incapacity to produce a "good" harvest is reified by characterizing the farmers in question as "drunkards . . . and the drunkards are increasing" (where a "drunkard," or *borracho*, is a man incapable of doing any good work).

The pattern of relatively autonomous, historically guaranteed reproduction, described simply but graphically by the farmers in Chacán as "*trabajar por cuenta propria,*" ("working for one's own account," for oneself), not only implies specific dynamics but specific vulnerabilities as well. Thus, as Bernstein (1977) argues, "The crucial moment in the penetration of [traditional agriculture] by capital is the breaking of its cycle of reproduction." A "squeeze" can arise not only on the level of

Table 3.2. "Can Those Who Work with Credit Become Rich?

	No	Maybe	Yes
Had not borrowed in previous season	31%	19%	50%
Had borrowed in previous season	53%	12%	35%
Difference	+22	-7	-15

exchange relationships but also according to Bernstein, "by rural development schemes which encourage or impose more expensive means of production (improved seeds, tools, more extensive use of fertilizers, pesticides, etc.) with no assurance that there will be increased returns to labor commensurate with the costs incurred." Finally, there is "the precariousness" of small-scale peasant production. Indeed production is so organized (through the use of different ecological levels, diversification and also by the application of livestock as reserve funds) that failed harvests are not able so easily to bring the whole reproduction of the enterprise and farm family into danger. However, an increasing shortfall in autonomous reproduction is naturally not ruled out, and this can encourage borrowing, which brings us to the next category, los medios.

Los Medios

This concept, which defines a particular category of farmers, is closely associated with that of acquiring credit. When asked "Who works with credit, the rich, poor or middle farmers?" 52% replied the medios, 24% said the poor, and no one thought that the rich would work with credit (24% did not respond). In answer to the question "Can those who work with credit become rich?" 42% thought not. The striking thing is that it is the farmers who borrow money who are the ones who think one cannot become rich through credit (see Table 3.2). They had not abandoned the possibility of becoming rich ("the hope of getting a bit ahead"), but their experience of credit had led to skepticism rather than optimism.

With increasing incorporation into the capital market, a quite different pattern of production and reproduction emerges. Reproduction (and thus production too) becomes market- and future-dependent. Peruvian researchers often speak in this connection of commercialization, or *mercantilización* (Ccori, 1982; Tupayachi, 1982; Villasanta, 1982). Bernstein talks of "commoditization" and Pearse (1976) defines the same process as "incorporation." The terms are different, but they all refer to the process by which market and price relationships penetrate the core of production. Thus means and objects of labor increasingly enter the process of production as commodities. In practice this scheme can emerge in many different ways—for example, through paying for labor with a part of the harvest in either cash or kind. However the most common mechanism through which this scheme becomes a reality is by financing the entire production cycle with short-term credit. And it is this that at some decisive point impedes the application of the calculus with which farmers used to structure their labor and production.

Two formal credit institutions were operating in Chacán: the Agrarian Bank, oriented to the middle farmers with a relatively large amount of land, and the Proderm Program, which aimed at the upper layer of the poor and the bottom layer of the middle group (Madueno, 1980). Of the fifty-two farmers, 42% had worked without credit in the previous agricultural season; 41% had worked with credit from the bank; and the rest had worked with credit from Proderm. The situation is not a static one. Of the farmers who had not borrowed money in the 1982/83 season, 77% had had some previous experience with formal credit mechanisms, and of the farmers who worked with credit during our research, 31% intended to work without borrowing the following year (i.e., they would work "for their own account").

Credit mechanisms have thus become an everyday part of life, in the sense that practically all farmers have experienced them. At the same time, the situation obviously fluctuates, in that periods of taking up credit are interchanged with periods of working for one's own account and vice versa. The fact that so many farmers have experience with credit, and that credit plays an increasingly large part in financing production costs, is linked to how its role within the global process of rural development is perceived. Up until the 1970s, agricultural economists generally held the view that the role of formal credit in the "capital formation" of small-scale farm enterprises was rather limited. Next to their own labor and savings, and the use of informal credit circuits (Firth, 1964:30), formal credit would amount to, at most, 20% of the "real resources allocated to capital formation" (AID, 1973:XX, 5). Such observations coincided with or mirrored the northwest European agricultural development experience. There emerged a general awareness of the fact that: (a) extensive substitution of the farmer's own resources by borrowed capital brings about substantial differences in farm development, and (b) an evaluation of such differences would show largely negative consequences for those enterprises that were heavily indebted (see Dijkstra and van Riemsdijk, 1952).

In the 1970s, however, two studies introduced a "new school of thought, which attributes to credit and financial markets an importance in economic development exceeding that usually recognized" (AID, 1973). These studies (Williams and Miller, 1973; McKinnon, 1973) maintained that not only was 20% an underestimation but that credit should be seen as crucial for stimulating other contributions in capital formation, such as personal savings. Be that as it may, since the 1970s small farmer credit programs have become a substantial part of agrarian policy in most of the Third World. In Peru there has been a similar rapid expansion of credit. Haudry (1978:79) states that agricultural credit grew from 3,775 million soles in 1966 to 24,215 million in 1976. Measured in real terms, credit provided by the state-controlled Agrarian Bank doubled in this period—but it should be noted that this was only a fraction of the total amount of credit taken up. Salaverry (1981:14) concludes that it amounted to some 27% of the total. Commercial houses, transporters, and the like, provide another 49% and the remaining 24% originates from non-commercial sources, principally the family. The volume of credit provided by the Agrarian Bank is thus an underestimation of the size of total commercial credit.

In 1966, the amount provided by the bank amounted to only 8.4% of the total value of production in agriculture. By 1976 it had risen to 19.1%, and from 1976 to 1980 further expansion took place to as much as 41% of the total value of production (Ministerio, 1981:20,27). Production for home consumption is included in the total value of production. We are seeing, therefore, very rapid and substantial commoditization which goes much further than the case of Italy (Fabiani, 1979). The use of farmers' own savings is rapidly being replaced by the use of short-term credit. This process is partly related to inflation and is closely linked to the substitution of parts of the agricultural labor process by inputs manufactured elsewhere. The use of industrially produced cattle feed, veterinary products, chemical fertilizers, pesticides and "modern" seed is rising rapidly. The purchase of such non-factor inputs in 1970 was still only 8.9% of the total value of production, but by 1979 it constituted 16.7%. The use of these commodities often leads to the borrowing of credit—and thus to further commoditization.

Credit mechanisms themselves often lead to further commoditization. The provision of short-term credit (and this is the biggest slice of the credit given by the Agrarian Bank) is based on a specific interpretation of the non-factor inputs per hectare to be applied. In practice credit operates in the following fashion: the bank opens an account in the name of the farmer or cooperative but allocates the credit to various commercial agencies. The borrower can then pick up the prescribed inputs at these agencies. After the sale of the harvest, however, he pays both interest and capital back to the bank.

It needs no imagination to see that such a structure gives rise to a strong and often one-sided prescription for agriculture. Credit is especially orientated towards commercial crops. Credit mechanisms stimulate the production of particular kinds of crops. Originally only a fraction of potato production was financed this way, the reason being that 41% of national potato production was consumed by the farming households themselves. Meanwhile, 29% was exchanged through non-market mechanisms (treque), and only 30% of the total was handled by the market (Eguren, 1981:11). If credit mechanisms, for whatever reason, now penetrate this sector, this necessitates increased commercialization. Thus, taking up credit—itself an essential element in the commoditization process—generates further commoditization. In this sense, the observation that "credit facilities are an integral part of the process of commercialization of the rural economy" is correct (World Bank, 1975a:5).

Often there is a high degree of internationalization behind many of these credit facilities. For example, in 1983, 54.9% of all agrarian credit provided in the Department of Cuzco was financed by funds from the EC and the Netherlands (Haudry, 1984: part 2, annexo XIV,1). One of the consequences is that the criteria operative in the international capital market also become operative in the fields of Chacán. Peruvian funds used within the framework of these programs must also satisfy the same criteria. The "universal market" of Eisenstadt (1963) is thus indeed created.

This universal market is not complete, however, for besides the capital markets and credit agencies mentioned, capital may also be mobilized through social mechanisms. Compania is one such way. Capital is mobilized through temporarily cooperating through labor or land with someone able to contribute the capital. Ayni also functions to mobilize capital. As one medium farmer put it,

> "The rich are quite happy to lend, they earn well from it. They give you maybe 100,000 soles to buy fertilizer, and for that you have to go and work for him for ten days. Every month two days, by oneself or with your wife who helps in the kitchen, depending on the arrangement. The days that you work only take care of the interest. At the end of the season you have to pay the 100,000 soles back with what you earn from the harvest. And if that doesn't go well, ay, then you have to work for him another ten days the following season and still pay the money back."

Under this form of ayni, reciprocity has largely disappeared. It has become an asymmetrical relationship which carries an element of exploitation. Nevertheless, there remain striking differences between formal credit lending and capital mobilization through informal mechanisms. Compania entails the spreading of risks between both parties. With ayni there is, in principle, a more flexible time span than is the case with formal arrangements—a difference appreciated by one farmer when he said,

> "In order to get a new loan you always have to have the last one paid off, the bank is very strict about that. If you haven't settled then they send someone round with a lawyer and they take your livestock in payment. That's why many people have a deep mistrust of the bank. With Proderm it's a bit easier but even there you're well advised to think twice."

Land, labor and capital can be linked to the different markets or mobilized through various social mechanisms. Table 3.3 provides an overview of these different mechanisms. The clear trend in the communities in and around Anta Pampa of replacing socially regulated patterns of mobilization in favor of an increasing market dependency was already indicated in the foregoing discussion. I will now explore comparatively the effects of commoditization on the farm labor process. I will do so through a discussion of soil fertility, the reproduction of seed potatoes, and finally, by analyzing the process of production itself.

Farm Labor: The Production of Soil Fertility

The farmers of Chacán farm at different ecological levels, from the pampa to *muy arriba* (i.e., the high altitudes). The pampa is the lowest and most level land and is suitable for irrigation. From time immemorial these pampa lands belonged to the communities, although a long process of land encroachment led to the growth of the great haciendas. One peasant depicted the hacienda situation thus:

> "The *hacendados* worked the land badly. Here in the community at that time we were 600 families with at most 1,100 hectares of land for our use. Thus we went to the hacienda to work but there wasn't really much work and we were obliged to go to Cuzco or the Valle Sagrado to find work."

Table 3.3. A Summary of Mechanisms for the Mobilization of Factors of Production and Non Factor Inputs

Production factors and non factor inputs	"Traditional" or non-market mechanisms	"Modern" forms and mechanisms
Land	inheritance, marriage; communal decisions; compania; use of communal grazing lands; collective or individual use of occupied land	purchase/sale of land; renting and hiring (i.e. incorporation into land market)
Capital	compania; savings; family capital; loans from friends; informal credit from shop keepers; earnings from work elsewhere; earnings from craftwork; etc.	incorporation into capital markets (short- and medium-term loans from Agrarian Bank and PRODERM development program)
Labor	ayni; compania; faenas	incorporation into labor market
Traction	ayni (oxen for labor following rules governing reciprocity); use of communal tractor	incorporation into market for machinery services (hire of tractor) or hire of oxen
Farm implements	ayni and craft-made implements	hire or purchase of needed implements
Fertilizer	manure produced on the farm; guano obtained through exchange; interchange with other communities; communal planning and control of rotation schemes	purchase of fertilizer (often combined with incorporation into capital market)
Seed	own production and selection; exchange with compadres from other ecological zones (papa de regalo)	purchase
Knowledge	craftsmanship and art de la localité gained through experience and embedded in community norms about good farming	external prescriptions and control by market agencies, Agrarian Bank, rural extensionists and development programs

Then at the beginning of the 1970s a large cooperative was established by officers of the land reform. It was a failure in all respects (Casaverde, 1979; Egoavil, 1978; Matos Mar and Mejia, 1980). Finally the lands were "taken" by the peasant communities (Rocca et al., 1980; Cencicap-Anta, 1980).

> "The cooperative worked just like the hacienda and that's why we occupied it and took it over. We have the titles and have studied them well: that land belongs to the communities. Finally we divided up the land taken. The rich farmer and the poor all got land. We are better off for that. We are now 1,000 families with 2,200 hectares of land. We have more grazing land, more animals. Men go less to Cuzco and Valle Sagrado for work. More children now go to school."

In the pampa lie the *maizales*, plots of land on which maize and potatoes are grown in turn. Higher up lie the plots of land which cannot be irrigated and where cultivation is dependent on rainfall. These are called *temporales*. They are plots which necessitate a complex system of rotation. The farmers of Chacán are generally of the opinion that with careful working and the application of a good rotation system, these plots can bring a similar yield to those in the pampa. At very high altitudes the yields naturally fall.

The rotation system is important for the production of soil fertility. At present the most frequently used system of rotation is to plant potatoes in the first year; barley, wheat or maize in the second; and in the third mostly beans, after which, where possible the land is left fallow for a year before beginning the cycle again with potatoes.

There are many variations possible on this general pattern. Some farmers lengthen the fallow period; others eliminate it in order, for example, to extend the growing of barley and potatoes. However, growing more potatoes in this way has detrimental consequences for soil fertility. Chacán farmers say that when the period for growing potatoes is lengthened and the fallow period is decreased, the land becomes "colder" and "tired." Rotation and the frequency of potato growing influence the number of pests and disease that occur in the different plantings. The percentage of land lying fallow and the percentage under potato cultivation fluctuates considerably. An average farm with 8.4 topos has 2.7 topos in use for potatoes (SD = 2.2) and 1.5 topos lying fallow (SD = 1.4). The percentage given to potatoes can vary from 9 to 72%! And the percentage for land lying fallow can vary from 0 to 33%. Of the fifty-two farmers in the sample, sixteen had no land resting.

One of the most frequent topics of conversation in Chacán is the falling potato yields:

> "Pucha! Fifteen years ago or so, the potato harvest still gave good returns. We used to be able to collect a lorry load from one topo, and that meant a lot in those days. Then you could still enjoy yourself. . . . After the harvest everyone bought his four crates of beer to get through the nights happily."

It would seem obvious then to relate falling potato yields to increasing pressure on land, a relationship which would proceed via the elimination of the fallow period from the rotation system. Comparison with earlier studies (Sabogal Wiesse, 1966) shows that the fallow period has indeed been noticeably reduced. However, this reasoning is not sufficient. To begin with, since the invasion of cooperative land there in fact has been less rather than more pressure on land. Furthermore, increasing pressure on land does not mean per se reducing fallow periods. Technically, it should also be possible to intensify and stick to the rotation scheme. Also, the reduction of fallow periods does not need to result in "tired land" and poorer yields. With better fertilization and rotation, one should be able to maintain or even improve the level of soil fertility.

Naturally, fertilizing plays a large part in producing soil fertility, in "respecting pacha mama." Again one meets great variation. The application of chemical fertilizer for potatoes varies from 0 to almost 600 kg per topo. There are also noticeable differences regarding other crops which indirectly influence the results of potato cultivation. In addition to chemical fertilizer, different types and amounts of dung are used. Some swear by sheep dung; others prefer cow dung which is sometimes bought by the wagon load from neighboring communities. The use of chemical fertilizer dates from the mid-1960s, when a fertilizer factory was built nearby in Cachimayo. A development program directed by the University of North Carolina introduced a package for potato growing that included fertilizer and credit. Until then, island guano and urea were primarily used. For many farmers buying fertilizer and credit are closely bound together, both historically and practically. Expenditure on fertilizer can amount to a third of total costs. To fertilize one topo reasonably costs the same as fifty-five days' wages (Madueno, 1980).

On what does such a difference in the degree of fertilizing depend? In Figure 3.6 the total sample of fifty-two farmers is divided into four categories. The rich are those who possess the means to produce well without needing to call on credit institutions. The second and third categories are composed of farmers who use credit from the Agrarian Bank and Proderm. The fourth group are the pobres, who grow only or primarily for their own consumption and use no chemical fertilizer.

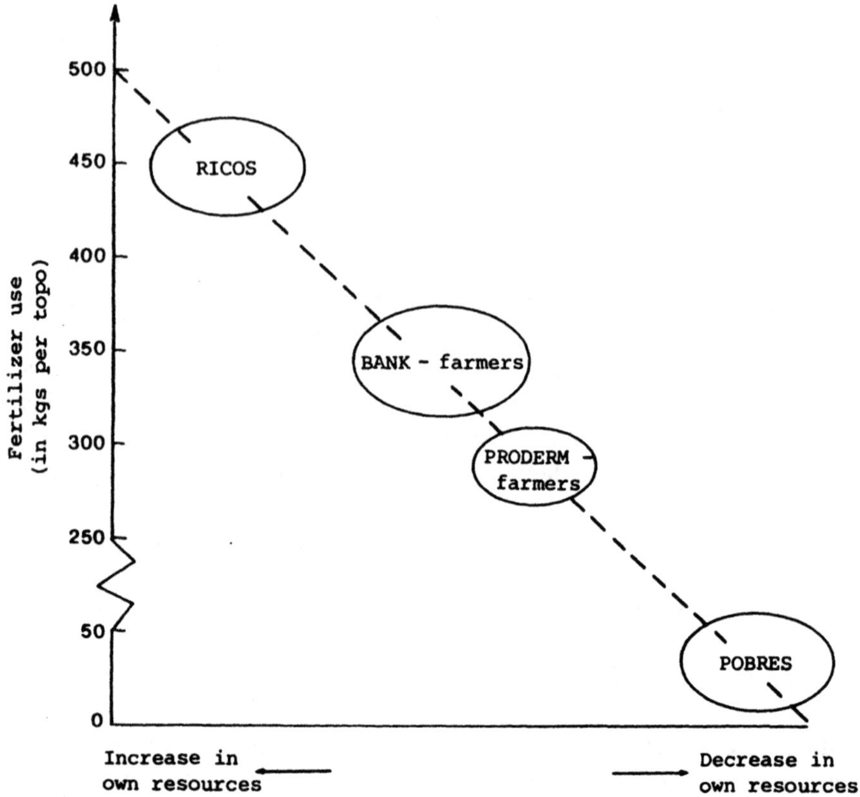

Figure 3.6 Relationship between fertilizer use, amount, and origins of working capital, Chacán (n=52)

The amount of fertilizer applied bears a strong relationship to the category to which the farmer belongs. The ricos use the most chemical fertilizer, at 420 kg per topo. The variance within the group is relatively high (SD = 98). The amount applied by farmers who borrow from the bank is significantly lower, the average being 310 kg per topo. Here there was also a fair degree of variance (SD = 64). A closer analysis shows that this variance is related to the availability of private means with which to supplement the amount borrowed.

Entering into loans is more than simply substituting private for market-related resources. Farmers who take credit apply substantially less fertilizer per unit of land than the rich farmers. Farmers who take part in the Proderm Program use an average of 295 kg per topo. The variance is minimal, partly because of the supervision of Proderm officials and partly because of the specific form credit arrangements take. Farmers do not receive the money itself but a specified amount of fertilizer. Finally, the poor, who have barely any private resources and do not wish or are unable to take up credit, use no fertilizer or at most only one or two bags.

Yet why do farmers who borrow money use less fertilizer per unit of land than the rich? To explain this one must take several factors into account:

1. The bank deals with schemes in which the production costs and amounts loaned are specified, but the farmer can borrow the maximum amount and use it for purposes other than those intended. (This is not ruled out by Proderm but is more difficult when credit is in kind.) Some farmers may borrow more money than that formally laid down for the scheme, particularly if they have the necessary security (land, livestock, or a lorry). However, farmers usually borrow less than the bank will in principle give them.
2. During interviews and in the sample questionnaires farmers were asked whether the amounts borrowed were deemed sufficient. The majority (83%) said no but added that they had not wished to ask for more.

   > "If I had enough money of my own then I would naturally buy more fertilizer; logical, then I would have a bigger harvest."

   Time and again it appeared that in the case of borrowed money farmers felt obliged to think or calculate in ways other than would be the case if they had sufficient private resources:

   > "Look, what is asked from the bank depends ultimately on the person in question, on the view the farmer takes of the business. How many sacks of fertilizer you need for one topo, every farmer must decide for himself. But the truth of the matter is that for some farmers, their hearts sink when they are confronted with the high costs and all those problems with the bank. They think they'll never get their heads above water again and that makes you think in a different way. The farmer who uses 15 bags, he will gather a good harvest. And the one who uses eight, he thinks to himself that things might go amiss."

   This particular farmer borrowed half a million soles. And why not more? His reply was:

   > "Thank God that I did not borrow more. You have to consider the possibility that things can go wrong and take the appropriate measures."

   Another farmer with credit said:

"When it's a question of taking credit you must think things over carefully. We think and observe each year, and so draw conclusions about the best way to work."

3. The pattern for the rich can be described as follows. First, they purchase fertilizer when the harvest is sold and are thus able to buy much earlier than those who have to wait for a new loan. Sometimes the difference amounts to half a year, and with the high rates of inflation that can make a significant difference in the price. In the second place, the farmer with credit has to sell his harvest immediately in order to repay the loans as soon as possible. This implies that, more often than not, he has to be satisfied with lower prices than the rich farmer who can afford to wait for a more attractive market. Third, if the rich are confronted with a failed harvest, they can sell cattle in order to buy fertilizer and ensure a harvest, insofar as it is possible, for the next season. Thus, even with misfortune, the pattern of historically guaranteed autonomous reproduction can be continued. A farmer with bank loans must also sell land or cattle after a bad harvest just to pay his debts, leaving him again without the resources to finance the following harvest. His reproduction remains, in other words, market dependent. The rich can eat into their own capital, but for the farmer with loans, the enterprise itself is brought to a standstill.
4. Finally, credit has its own price: an interest rate of 45% from the agricultural bank and 20% with Proderm in 1982.

In brief, incorporation into credit markets introduces such changes in prices and costs and, above all, in the parameters within which these must be calculated (time span, risks) that a lower application of fertilizer per unit of land, and thus a certain extensification of production, is the result.

This extensification reaches beyond a simple drop in non-factor input use. Farmers who borrow from the Agrarian Bank also put less labor into potato production per topo than the ricos (22.4 as against 28 man-days). Furthermore, it appears that taking credit is linked to a change in the cropping system: the percentage of land devoted to potatoes increases (see Table 3.4). in practice, a certain combination of scale enlargement and relative extensification occurs.

There is an intriguing argument circulating in Chacán which throws light on and legitimizes the farmers' lower use of fertilizer: "It no longer has any force . . . the fertilizer has no power anymore (*ya no tiene fuerza*) . . . if you use too much then you throw it for nothing

Table 3.4. Effects on Taking up Credit on the Proportion of Potatoes in the Cropping System, Chacán.

	Total area	% Potatoes	% Fallow
Ricos	10.0 topos	26	16
BANK-farmers	11.1 topos	36	20
PRODERM-farmers	4.9 topos	44	5

on your field . . . it used to be much stronger." Such references to a lack of strength seem at first sight to be an irrational rejection of technological possibilities. The argument seems no more than an appeal to magico-religious belief. But this argument, held by some 65% of the farmers interviewed, nevertheless has a grain of truth in it. Taken by itself fertilizer is the same now as it was in the past. But related to changing farm practice, the argument contains an interesting reference. Those who subscribe to it—we kept systematic accounts of this—cultivate on average more extensively than those who say that fertilizer is what it always was: 23 man days per topo versus 28. Fertilizer will indeed "lose" its productive potential as the cultivation pattern becomes more extensive. A high percentage of land devoted to potatoes, a lower labor input, and a lower application of fertilizer per unit of land, taken together, form a coherent pattern. Manipulation of one factor, for example, amount of fertilizer, will (holding all other factors constant) bring about only minor or even counterproductive results. In this respect it is symbolic that the shortcomings of a new (more extensive) style of cultivation are related to the innovations which marked the destruction of the earlier, more intensive style of cultivation. Hence the validity of the lament "ya no tiene fuerza."

The relationship sketched in Figure 3.6 can be illustrated through several biographies: "When my father was still working" said Pedro,

> "we used to strew *'guano de isla'* and urea on the fields. We harvested more than now, 35–40 sacks (the sacks he refers to here are sacks which hold about 80–85 kilos). Later, but that was when we still worked with our own money and when fertilizer was much cheaper, we strewed about 8 sacks on each topo. That was according to our own criterion of course, and we harvested about 32 sacks a topo."

In brief this is how Pedro in his simple way depicted his former position of being a rico. Then he went on to describe a period of being medio:

> "After some setbacks I had to go and borrow. I have been up to my neck with Proderm. They gave us 6 sacks of fertilizer per topo, that wasn't so very much, a couple of sacks less than we had been using ourselves. But that's what they said, that we had to work two topos with that, which gave us a poor harvest, only twenty-four sacks, while normally we were getting thirty. And the second time the harvest failed because of the frost and *la rancha* [a potato blight]."

Payment of interest and capital were thus impossible. Even after selling his cattle, Pedro still had an outstanding debt of 42,000 soles; a relatively small amount, but for a poor farmer this is equivalent to 100 days working as a wage laborer. "How am I ever in the world going to pay it off?" he asked.

> "The harvest failed for several reasons, but it isn't my fault. We worked hard . . . and the worst thing is they won't give me a new loan, so I can no longer plant to earn and pay off the last debt. They won't lend me more because I still owe them money. We're really ruined. So what can I do now? I shall sow a few bags of potatoes for our own use, and for that I will use dung. Fertilizer is out of the question. I think that I can still harvest maybe ten bags of potatoes."

So, Pedro's circumstances have been reduced from being a rico, to functioning as a "credit borrowing farmer," to the condition of being pobre.

Of course moving in the opposite direction is also possible. It is one of the explicit aims of Proderm to loan money to subsistence farmers (the pobres) to allow them to produce "more and better." Isidoro had something to say about this:

> "The money that I can now borrow for two topos of potatoes is a real incentive to work. . . . That's common sense. Finally I have the chance to become a good farmer. You cannot otherwise buy fertilizer or pesticides or whatever. That is now possible and that's why I feel inspired to produce more."

In short, the perception of credit mechanisms depends on previous experience and style of farming and on whether or not one has available sufficient personal resources. However, one might question how far the borrowing of credit for these smaller farmers provides a real basis for enterprise development. As we have already seen, credit introduces changes in crop rotation plans which in the long term appears to be untenable, not only because of soil fertility problems but also because reduced diversity greatly increases market-induced risks. Falling potato prices increasingly threaten the continuity of the enterprise.

From the point of view of los ricos credit is a step backwards, whereas from the condition in which the poor find themselves, it is a step forward, though of course not free of risks since it creates new ones. Both the data collected by ourselves and the project documentation of Proderm indicate a high "drop out" rate. After two or three years of borrowing credit the farmer ceases to do so and returns to the status of "pobre" (see also Haudry, 1984:28). Ignoring the subjective views of the farmers, one can state that credit is necessary insofar as it compensates for a reduction in personal resources. However, that compensation is not able to neutralize the degradation (i.e., extensification of production) of agriculture. On the contrary, incorporation into capital markets becomes one of the structuring principles of this degradation. Should the provision of credit dry up, then the general slide from rich to poor (i.e., the pauperization of the rural population) would simply be speeded up. Credit provides a half-way house, that of the medios, or of the "poor with credit." To expect that credit would also be functional for agricultural development is illusory. As we have seen, the provision of credit introduces relationships into the organization of labor and production which preclude intensification.

Farm Labor: The Reproduction of Seed Potatoes

Looked at superficially, the reproduction of potatoes seem a simple enough task. From the total harvest a part is set aside annually to provide seed for the following harvest. The smallest potatoes are usually chosen for this, preferably those from the best and strongest plants. However, behind this deceptively simple appearance lie rather complex processes of selection, propagation, adaptation and improvement. In their turn these processes presuppose a spatial organization of the enterprise as well as the (non-monetary) exchange of seed within an extended social network. Finally, such a complex process of selection, propagation, adaptation and improvement is unthinkable without the farmer having a basic knowledge of taxonomy. With this knowledge—a crucial part of their art de la localité—farmers coordinate the process of seed reproduction in a way that harmonizes with other parts of the labor process, such as the production of soil fertility and the method of potato cultivation.

Most farmers in Chacán use about thirty varieties of potato. Some of them are associated with the different ecological levels available to the farmer. Some plots (mostly the lower-lying fields) are planted uniformly with one or at the most two varieties. In contrast, other plots show an impressive variety. Three to seven varieties per plot is the common pattern. Brush et al. (1981) noted likewise that "varietal heterogeneity of native potato fields is one of their most important attributes." This heterogeneity has been deliberately created with specific goals in mind, not only, as is commonly maintained, to minimize the biological and economic risks, but also to maintain the genetic stock which makes renewal or innovation possible. The small, often minuscule *chacritas* which a substantial number of farmers use are of vital importance for this work. A reservoir of not directly used genotypical stock is maintained in them for possible further development. Brush and his colleagues give a fascinating picture of such chacritas, which they define as mixed fields (Brush et al., 1981:81, fig. 2). They maintain that such "laboratories" have a crucial function: "The crop evolution of the cultivated potato is closely linked to the mixture of species and genotypes which promote hybridization and crossing between ploidy levels and among clones" (1981:80).

Apart from the varieties that farmers actually use, they generally know dozens more. They also know who grows which varieties, where and in what manner, under what conditions and with what results. As far as we were able to ascertain, the farmers of Chacán were able to provide a detailed picture for an area covering a radius of 15 to 20 kilometers. They are able, when necessary, through the exchange of information, to obtain a sample of any other variety. In this connection they speak of *papa de regalo* (gift potatoes). Such gifts are first sown in the chacritas, subsequently planted out in the various fields, and then multiplied as needed in order to serve as seed potatoes for a substantial part of commercial production. This takes a cycle of several years, at least four, and often more. This specific organization of space, time and social networks guarantees, to quote Brush, "a) the maintenance of numerous genotypes over space and time, b) the wide distribution of particular genotypes, and c) the generation or amplification of new genotypes" (1981:73). Crucial to this is "a regular system of nomenclature, organized in a taxonomic manner" (1981:85). The selection of seed and also the mutual exchange that takes place within what are often widely dispersed networks, assume a capacity to recognize and put a name to the different varieties. Moreover there has to be a common language for the dissemination of this taxonomy between larger groups.

After the pioneering work of Conklin (1955) on Hanunóo agriculture, numerous anthropologists and also biologists have mapped out similar "folk taxonomies" (see Brokensha et al., 1985, for a recent summary). However, even for researchers familiar with this literature, it is still a fascinating experience to listen to a farmer who has with him a basket of seed potatoes. For us—researchers mostly trained in agronomy from an agricultural university—it appeared nothing more than yet another basket of amorphous potatoes, as amorphous as the peasantry must have been for Marx when he described them as a "sack of potatoes." But for the farmers this same container was a basket full of diversity. Endless varieties were picked out and reasoned over, sometimes the precise names for them being disputed. And with the name-giving (which is strongly associated with morphological characteristics) a great breadth of knowledge was activated—of where and at what height and under what conditions each variety could best be cultivated and with what results in terms of the harvest, taste and processing; who had had positive and who negative experiences with the variety and the reasons for this. In brief, this taxonomy, as a body of communally shared knowledge, is the pivotal point for the social reproduction of seed potatoes.

Alongside this specific taxonomy with which farmers handle their genetic stock there is also a basic knowledge of the particular fields. The farmer accumulates this knowledge through the reproduction of soil fertility. What Mendras (1970:47) wrote of the French peasantry is equally valid in the Peruvian highlands: "The traditional peasant knew all the minutest details of his fields: the composition and depth of the arable layer which often varied from place to place; its rocks, humidity, exposure, relief and so on. The result of long years of apprenticeship, work and observation, this knowledge . . . was the basis of his skill as a farmer. . . . He felt as if he had "made" his field and knew it as the creator knows his creation, since this soil was the product of his constant care: ploughing, fertilizing, rotating crops." Taken together, this detailed knowledge of fields and seed variety (of phenotype and genotype) forms an essential dimension of the art de la localité, specific knowledge developed through labor. It is in this way that the high level of adaptation of potato varieties to very different ecological conditions is achieved. Knowledge of the different phenotypical conditions on the one hand, and the genotypical variety on the other, is an essential precondition for this. Perhaps the term "adaptation" in this connection is somewhat misleading. It seems to suggest that the process is finite, that once adaptation is reached then a stationary state is entered upon. And that is indeed an argument which often lies behind the setting up of rural development projects which hinge on the introduction of "improved" seed. As Oasa wrote on research on the internal discussions of one of the leading research institutes in this field (IRRI in the Philippines), "Yields, it was felt, were stagnant because traditional agriculture had reached its limits" (1981:202). However, such an as-

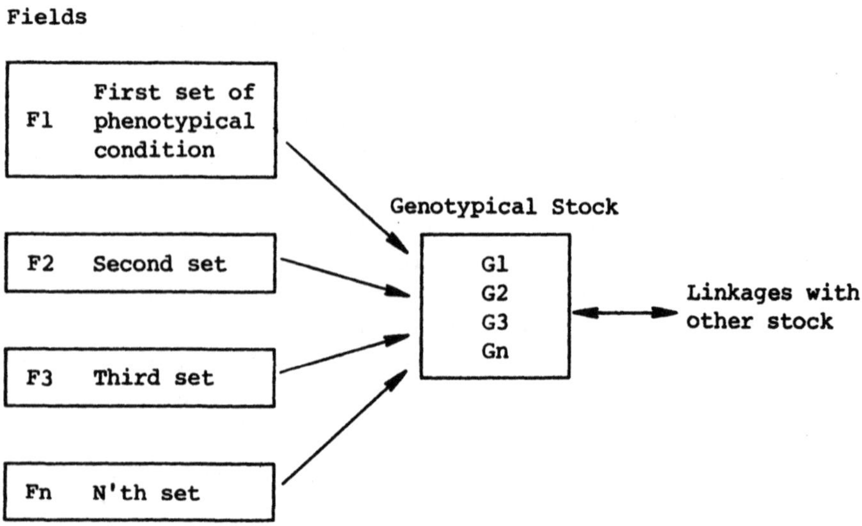

Figure 3.7 A schematic overview of the dynamics of indigenous seed-potato reproduction in the Peruvian highlands

sumption—which seen historically leans to a large extent on the "poor but efficient" thesis of Schultz (1964)—is in my opinion unjustified.

Propagation, adaptation and improvement of seed material forms a unity within the system I have outlined. Let me now enlarge somewhat on the aspect of improvement, or what is technically known as "upgrading." This concept forms the main thread of the schema presented Figure 3.7.

A farmer knows and manipulates a large number of phenotypical conditions.[3] At the same time he is able to draw upon a thorough knowledge of genotypical stock. For each field (for each set of specific phenotypical conditions) he selects a particular variety. Thus adaptation arises within the framework of the art de la localité (the specific and communally coordinated knowledge of fields and varieties). Trial and error and the insights thus gained imply that adaptation is continually being refined. However, phenotypical conditions should not be seen as static. They are the object of farm labor: drainage can be improved by improving the drainage system; leaching and the loss of nutrients it causes can best be prevented by the gradual terracing of the fields; through the building up of a herd more dung can be obtained and the structure and fertility of the soil are thus improved. In short, phenotypical conditions can be altered step by step and thus improved (though naturally the opposite is also possible). And it is precisely this change that prompts a renewed adaptation of the genotype. From the exchange of genetic stock within social networks, new cultivars are selected as well as varieties which match with the improved phenotypical conditions (which are mostly those genotypes known for their higher productive capacities). And in turn, improved phenotypical conditions can be used to enlarge and improve genetic stock.

Although this is only a rough sketch, I hope that three things are clear. To begin with, the dynamics of double adaptation (the mutually coordinated production of soil fertility and seed selection) bring about continuous improvement. In the second place, this improvement falls to a very large extent within the "domain" that can be understood, controlled, and manipulated by means of farm labor. This is not only a question of affording a degree of autonomy to this process, but also of how the specific forms of continuous improvement are geared to the possibilities, perspectives and limitations of individual farming families. Each phenotypical change takes place within this framework. If certain changes imply more labor than is available in the household, then they can be temporarily left undone. In other words, there is a considerable degree of flexibility. It is also important to stress the crucial role played by the art de la localité in this ongoing process of improvement. The continuous confrontation of a broad spectrum of genotypes with diverse phenotypical conditions offers the farmer some insight into promising phenotypical changes.

Asked for the reasons behind a somewhat deviant use of a particular plot of land, one of our respondents said:

> "Look, I have noticed that this variety likes good quality ground, so I allow it to become quite dry, then I harrow again and make sure there is enough dung with plenty of straw in it dug into the ground, preferably to quite a depth. . . . But you have to take care because this potato doesn't like too much water once it has rooted well, and so I have taken this hill, and made enough channels to ensure that the rain drains off quickly. Yes, if the gods favor me then this will be a wonderful field."

It is tempting to quote more such observations, but the core is always the same: through the interaction between plant and field, carefully observed, interpreted, manipulated and evaluated, the farmer develops a specific knowledge which can justifiably be described as "art de la localité." This knowledge, in turn, serves as a guideline for the continous improvement of labor, objects of labor, and means. In other words we are dealing here with a highly dynamic system. It is intriguing that such a dynamic system is more often than not substantially undervalued by agricultural scientists today. As the celebrated biologist Prakken (1965:149) remarked, evolution in plant breeding has been taking place for thousands of years, an "evolution which happens under two influences: through the selective working of the new milieu (as modified by new techniques of cultivation and harvesting) and through the more or less conscious selection by some cultivators, who will have removed the notably bad types and reserved the especially good ones for planting out." This double dynamic springing from the simultaneous improvement of phenotypical conditions and genotypical material can also be discerned today. On the basis of field research, Brush et al. (1981:80) are quite explicit about this: "Selection may be observed in fields, in the terminology for fields, in the technology applied to different fields, and in the farmer's objective in planting different fields." They conclude that "undoubtedly man's role as a selective agent is felt on all levels" (1981:73). However, such observations and conclusions have become marginalized in modern agricultural science, which views "traditional" agricultural systems, almost by definition, as stagnant. A dynamic process is ipso facto an exclusive function of scientific innovations. Such a stance is nothing new. Referring to the disciples of the New Husbandry tradition (of the eighteenth century) Slicher van Bath (1960:263) writes "they were so greatly convinced of their own excellence that they painted the 'old' in violent colours. They considered themselves too much as the bringers of light in the darkness of ignorance and backwardness to be able to see the situation as it really was."

Of course in the "situation as it really is," endogenous potential for development (such as that sketched for seed potato improvement) can be blocked by politico-economic processes which generate chronic impoverishment. However to conclude on the basis of this that agriculture is intrinsically backward or "stagnant" is not only unjust but closes off all kinds of development possibilities. Clearly the pattern of seed potato improvement sketched here has undergone rapid changes in recent years. The reproduction of seed has become externalized under the control of specialized enterprises and limited to certain areas. Seed potatoes have therefore become a commodity. Closely linked with this commoditization process is the arrival of "modern varieties" that are more often than not developed in experimental stations.

This process is clearly observable in Chacán. In the lower lying irrigated ground, only one variety, mostly an introduced variety, is sown. In the higher zones farmers still grow more varieties per plot, mostly varieties which they have propagated themselves (or have obtained as "gift" potatoes). Fields worked with the credit and technical assistance of the Proderm Program are the exception. According to Proderm records, in these fields only one "modern" variety is grown. This accords with the overall aim of the Proderm Program which is "within three or four years to bring about a total renovation of potato plants" (Haudry, 1984:I:64).

This typical spacial distribution repeats itself at the national level. Within the national panorama, the Mantaro Valley predominates as the production area for seed potatoes. A whole new stratum of *semilleristas* (seed producers) has developed here specializing in the cultivation of seed potatoes. Next, there are areas of the coast which have developed as production areas for potatoes for consumption. The necessary seed is entirely purchased—sometimes from the Mantaro Valley, sometimes on the international market (Chile and the Netherlands are two important suppliers). In the Mantaro Valley the propagation of potatoes, at least those destined for sale, is concentrated in the hands of the large-scale farmers, who produce them in the lower areas of the valley. The commoditization of potato seed is a recent appearance. Specialized enterprises for their production have only existed for the past seven or eight years (Benavides, 1981:41; Franco and Horton, 1981:54). Typically enough it was not so much the farmers who seized this "market opportunity" but the traders and transporters; their specific market knowledge and the relationships built up by them in the potato market became very useful for setting up businesses for the propagation of potato seed. According to existing studies, semilleristas have at their disposal an average of 7.6 hectares, of which 4.3 hectares is on average sown with seed potatoes. This is many times greater than the average for other farm enterprises. Such enterprises are not only heavily geared to the production of commodities (the high proportion of land devoted to seed potatoes is a striking example) but are themselves highly commoditized. For instance, the land needed is mostly rented on a yearly basis. Of the total resources used in the cultivation of potatoes (i.e., labor, inputs, traction, etc.), 75% is mobilized through commodity relations. Personal resources reproduced within the enterprise itself thus make up only 25% of the total, whereas the figure is 31% for other farmers in the lower areas and 70 to 75% respectively for those from the higher and the very high altitude zones of the Mantaro area. This division seems also to apply to seed: the semilleristas buy 60% of their seed, as compared with the other categories of growers who buy 40, 10 and 5% respectively.

The earlier ecological specialization described by Mayer (1981) and discussed by Skar (1981) was kept in equilibrium through reciprocal exchange and a specific spatial ordering of individual farm enterprises over the several ecological "levels." This pattern is now giving way to a whole new ecological specialization linked to the degree of commoditization, which increases as one comes down the mountains to the economically more commercialized lower areas. It is here that commoditization is most advanced, as the example by Horton et al. (1980:25) in relation to seed potatoes shows.

It is worth noting that some of the semilleristas have become even more embedded in commodity relations by delegating a part of the seed propagation to farmers in the higher zones. Contracts are concluded in which the stock and working methods are closely stipulated (Franco et al., 1980:61). Some contract out phyto-sanitarian control which entails the diagnosis and control of plant diseases. As Franco et al. (1981:61) commented, "They contract qualified entomologists who prescribe the time and frequency of chemical treatment."

As mentioned, improving seed potatoes has become a major focus of scientific research in the past ten to fifteen years. The CIP (Centro International de Papa) in Lima has become one of the cornerstones of an international network (CGIAR) for the scientific production of new varieties and strains. The CIP is a nursery which has given birth to many "modern" potato varieties. The pattern is conceivably this: the CIP designs and tests new varieties and brings them onto the market. Selected growers, such as the semilleristas from the Mantaro Valley, propagate the seed, and it is subsequently sold and transported to Anta Pampa. The Proderm Program then furnishes these "modern" varieties as credit in kind to the farmers of Chacán. In this way spatial links arise which in some ways are far more complex, and in other ways simpler, partly because the ingenious methods of regulating the balance between several ecological levels can be eliminated. Be that as it may, it is clear that the regulating principles and driving force of the new spatial and socioeconomic relationships differ markedly from those which earlier integrated production and reproduction in potato cultivation.

The scientification of seed potato reproduction combines in complex ways with the process of commoditization. Scientification assumes a certain commoditization, but it also intensifies commoditization. The externalization of seed improvement to specialized propagating stations (such as the CIP) implies that seed, at least improved seed, becomes a commodity. Hence next to *papa de regalo* ("gift potato") is the term *papa mejorada* ("improved potato"). But the commoditization of seed is only the beginning of many complications. Scientific plant breeding begins with defining a so-called "ideal plant type" (see Oasa, 1981), i.e., a plant characterized by a certain conversion of energy measured in grams/growth per day or characterized by certain yield levels. To this ideal a new genotype is then created. From this genotype the phenotypical conditions are derived under which the ideal becomes feasible. Such conditions cover the amount and composition of the nutrients in the soil (i.e., fertilizing); the transport of nutrients which is closely related to the working of the soil; the availability of water during the growth cycle (i.e., kind of irrigation and drainage); and the elimination of growth-hindering factors which is closely tied to combatting weeds and plant disease. These phenotypical requirements are then tested in experimental plots where, by definition all phenotypical conditions are controllable and, according to common belief, manipulable. If one subsequently wishes to apply the "new genotype" on the farm, these phenotypical conditions will have to be copied as exactly as possible (see Figure 3.8). In order to employ the "improved" seed successfully, the assumed phenotypical conditions must serve as the blueprint to be realized on the farm. Hence, the farmer, whose own specific phenotypical conditions are his starting point, is now confronted with a blueprint of immediate and interdependent demands relating to necessary phenotypical changes. This leads directly to the question of whether such a blueprint of numerous interdependent changes is realistic.

A second, more implicit consequence of this way of working is the need for an abrupt acceleration of the commoditization process. The scientific design of a new genotype and its phenotypical specifications rests on unambiguous definitions. Thus, the application of chemical fertilizer is seen as less equivocal, more standardizable and thus more easily controlled than the application of green manures or a change of cultivation technique (see van Noordwijk, 1985). That is to say, the specification of phenotypical conditions is mostly done in terms of standardized, industrially produced technology. So, the application of "improved" potato seed requires a major acceleration of the use of fertilizer, herbicides and pesticides because the "improved" varieties are constructed on the assumption of their synergic application. The financing of this entails the need for credit: thus, in addition to using the market for most inputs, it becomes necessary to enter the market for capital. Other phenotypical specifications may imply a heavier exploitation of the land and therefore the use of the market for machinery. High yield levels may also outstrip the limits for mobilizing labor under ayni and other social mechanisms, thus entailing integration into the labor market for the harvesting.

In short, the use of "improved" varieties not only implies at enterprise level a commoditization of one of the most important labor objects (namely, potato seed), but it can also compel a much more far-reaching commoditization process. A part of the necessary labor and means become commodities in the production process. This can, as I explained earlier, clash with the specific calculus used by farmers for organizing their production and labor.

Chains break at the weakest link. The implementing of a blueprint in the field is quite another matter to its realization within the narrow confines of scientific institutes. Moreover, the purchase and use of each new commodity in the field is not a neutral operation: it implies entering

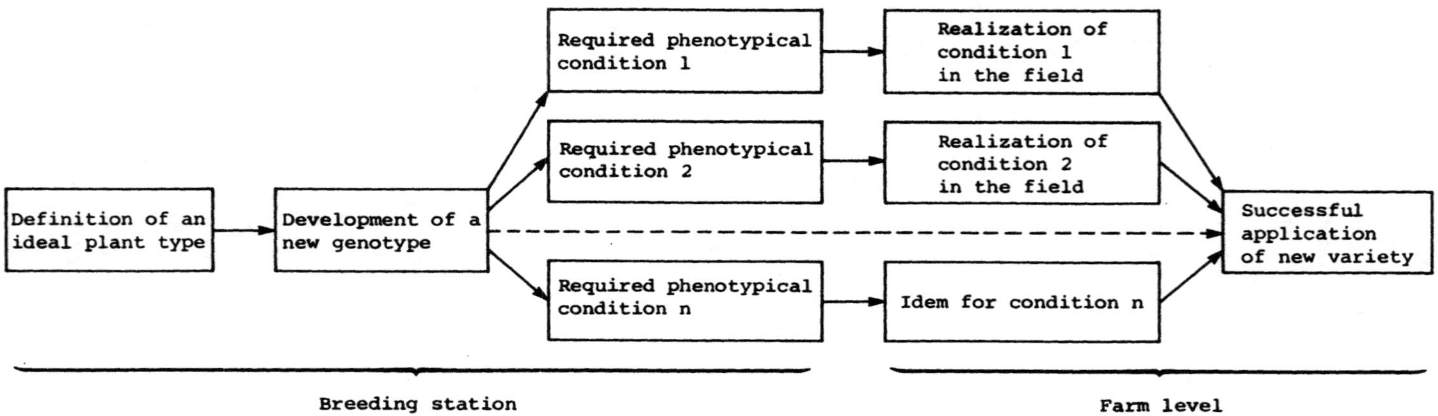

Figure 3.8 A schematic overview of the dynamics of plant breeding and subsequent "innovation" at farm level

into new and often antagonistic relationships with commercial houses, banks, intermediaries, and so on (Pearse, 1977). From Hardeman (1984) it appears that most farmers are able to achieve only a fraction of the prescribed packet, thus realizing only a fraction of all phenotypical demands. The overwhelming majority (about 80%) of farmers implement only a part. One consequence of this, in the case of potatoes, is rapid genetic deterioration, since application under sub-optimal conditions brings about degradation, often within a period of three to four years. In this way, the mistrust of new technology is reproduced. The farmers also exclaim, *ya no tiene fuerza* (it no longer has force) to the "improved" potato seed introduced. But, in the meantime, they may have lost a considerable part of their own stock of genetic material, which means that new, "improved" varieties have to be found anyway.

Farm Labor: The Process of Potato Production

Potatoes are planted in *temporales* following a period of fallow. The ground then has to be broken up. This is called *barbecho*. The task is sometimes done by tractor which can be hired for seven thousand soles per hour. It can take from one to one and a half hours to open up one topo of land, depending on the terrain and the depth and number of furrows desired. Other cultivators continue to use ox ploughs for this task. When this has been accomplished the soil is broken down further (*yondear*), then again ploughed, harrowed and levelled, using one, but usually more than one team of oxen. Some farmers use as many as six teams. During this time the fields are buzzing with people and ox ploughs trekking after each other in ever-repeating patterns.

The third phase is the actual sowing or *sembrio*. Here also many people are to be seen. At least one and usually two pairs of oxen are used for the "opening" and "closing" of the ground. If the task is to be really well done then an activity called *golpear* will follow the "opening," whereby, with simple tools, the remaining clods are broken down to form very fine soil. The women then drop the seed potatoes into this finer soil. After them the men follow applying a dose of fertilizer—a task that must be done with care in order to avoid "burning" the young seed by dropping it too close, or reducing its effectiveness in the early growth phase by dropping it too far away. Finally come the second pair of oxen to "close" the ground, and frequently a third team follows to compact the earth. This is done with what the farmers call "an aeroplane." Although a majority of farmers use two pairs of oxen per topo for working the land (a few are limited to one pair and quite a large number [30%] use three), the majority use only one team of oxen for sowing. A farmer lacking ox teams can usually mobilize them through the ayni system, and they can be hired (for 1,500 soles a day plus food and drink for the ox drivers). In terms of labor, the difference between farmers is even greater. The labor needed to sow one topo in a single day varies from 4 to 11 people (the average being around eight: 7.8 with an SD of 1.7). Part of this labor is mobilized through ayni, part will be family labor, and some will be hired through the market. Opinions over the advantages and disadvantages of ayni as against the hiring of daily wage laborers (*jornaleros*) vary considerably:

> "I prefer ayni because the people come early, about seven in the morning. It is the custom with wage laborers to begin much later. Also the costs of ayni are much lower. True, you have to take care of food and drink, but then later you also eat and drink at someone else's expense."
>
> "You never know with wage laborers if they will turn up. You can rest assured that those working with you in ayni will come. And what is more people work steadily on through the day and work much harder than wage laborers do. Naturally everybody knows they are being watched. Everybody knows precisely how he would like to see others working his own fields."
>
> "Ayni doesn't suit me. You always have to give a day's work back. So you are bound hand and foot and it is a pure waste of time having to sweat away in somebody else's fields."
>
> "If you consider it carefully, ayni is as expensive as wage labor. See what it costs on food and drink for all those people! And then you might well say that they work for nothing. But that's a mistake. You pay them back with a days' work. Ayni is actually much more expensive. The problem here is that some people are afraid of money, and those people look down on you if you don't want to work in ayni with them."
>
> "It is becoming very difficult to get together enough people in ayni. There are fewer people all the time who are willing to do it. That takes all the pleasure and usefulness out of it."

One of the things which greatly complicates the development of ayni is the rapid rise in this part of Peru (particularly in the rural areas), of Protestant groups such as the Jehovah's Witnesses, derogatorily spoken of by some as "sects." Followers of these groups must not drink alcohol, which makes it impossible for them to participate in ayni. Some observers who know the area well suggest that many farmers join these "sects" in order to have an excuse for backing out, in a socially acceptable way, of this network of obligations.

Be this as it may, an interesting link can be demonstrated between the use of ayni, wage labor, or a mixed form of both and the input of labor into sowing. Farmers who mobilize the necessary labor solely through ayni use 8.3 labor units per topo; those who only draw the extra labor needed via the labor market use 6.5 units per topo. Interestingly, those who use a mixture of both use a slightly higher labor input (8.4) than those who use only ayni. Ayni, as indicated, has natural boundaries since one must have enough family labor to repay the obligations incurred. Thus, using ayni and labor through the market, makes a higher input of labor possible than using ayni alone. It is quite another story when hired labor is used to substitute entirely for ayni. A complete dependence on the labor market leads to a marked drop in the labor input in sowing. The same pertains to the lampa.

The first lampa is a task carried out soon after the young shoots appear. Fertilizer is once more applied, the plants are mounded up, and the furrows between the plants are stirred somewhat in order to give the roots more air. Those who hurry this task are said to do it *a la ligera* (lightly). Others, who spend more time on the task, believe that in this growth phase proper care must be taken of the earth and the plants. The second lampa follows a number of weeks later. No more fertilizer is applied, but otherwise the activities are similar. The labor taken on to complete a lampa in one day varies from two to ten men per topo per day. On average, 10.9 labor units per topo are used for both lampas together. Here also, ayni and wage labor relations, or a combination of both, can be used for mobilizing the necessary labor. Those who work only in ayni take on 6 men/topo/day per lampa; for those who use both it is 6.1 men/topo/day; and for those who use only labor from the market 4.9 men/topo/day.

Subsequently, weeding will be undertaken once or twice again. Some farmers perform this task hastily while others make a thorough job of it. In this case, the various weedings require up to 15 man/days per topo, the average being 5 (SD=3.2).

After this the land is sprayed. Some farmers make do with one treatment; others spray as many as five times. The effectiveness of the treatment depends on having working capital and the material available at just the right moment, as well as on a correct recognition of the pests.

Harvest is another time when the fields are alive with people. During this time of year the farmer mostly sleeps in his fields in order to guard against theft. The crop is harvested, transported and then sorted with the help of between five and twelve men. A part of the harvest is first selected as seed for the following year, mostly from second or third quality potatoes. The biggest potatoes, the first quality, are sold, and the rest are kept for family consumption. Sometimes those who have helped with the harvest are paid with produce.

In summary, the method of cultivation has six phases (barbecho, sembrio, first lampa, second lampa, weeding, and the application of chemicals). A noticeable variance can be seen in each phase. Together these differences, ranging from very intensive to extensive forms, make for considerable variety in potato cultivation (see also Franco et al., 1981).

Harvest Estimates and Per Hectare Yields

Yields per hectare fluctuate significantly. Apart from the differences in soil fertility already described, rainfall, disease, early frosts and many other factors have a decided and capricious effect on yields. We measured yields per topo during our research, asking specifically about external circumstances beyond the farmer's control. The data given are for normal undisturbed harvests. The answers ranged quite widely, from 1,700 kg per topo to 4,500. Naturally a check on such answers is needed to establish their reliability. This was done in numerous ways. The amount of seed used was always registered. According to key informants, this is the most reliable datum a farmer can give. From this it is possible to calculate the yield. They were also asked how many men worked for how many days on one topo during the harvest. In this way the harvest per man/day could be calculated. They were then asked for their harvest figures from the previous year. The division of the harvest into seed potatoes, personal consumption, and that which was sold was also registered to provide further control on the internal consistency and reliability of harvest estimates "in good circumstances." Finally, an entirely different control was possible, namely, the harvest figures for other crops (grain, sorghum, maize and beans). We accept that a "good farmer"—i.e., one that structures his labor to reap the highest possible harvest—does not only work well with potatoes.[4] He will work in a similar way with other crops, and there ought therefore to be some similarity between the yields. And this does seem to be the case.

All this does not imply that the estimates given (and where necessary corrected) are reliable in an absolute sense. As we well know, farmers tend to overestimate their hectarage. When they talk about one topo the actual cultivable area amounts to no more than about 0.85 of a topo. This goes for practically all farmers. In a relative sense then, the harvest estimates given by the farmers, if they are to be judged reasonably, are all about 15% to 20% too low.

Towards an Explanation of Heterogeneity in Farm Practices

As noted above, yields in potato production vary widely. This is not due to chance. To a large extent it is dependent on the various ways in which farm labor is structured. It depends on the production of soil fertility, the reproduction of genetic stock, and the methods of cultivation during the actual process of production. In order to attempt a more quantitative analysis, a scale was constructed for each of these dimensions. For the production of soil fertility this was simple. For each of the elements already mentioned (amount of land lying fallow, amount under potatoes, quantity of dung, quantity of fertilizer), each farmer was allocated a score of one if he scored better than the average for the whole data sample and zero if the figure was lower than the group average. If these four scores are then totaled, they give a simple but useful total score for soil fertility.[5]

Intensity of cultivation can also simply and adequately be quantified in a scale. For preparing the land, for sowing, for both lampas together and for weeding one point was accredited if the labor input for the particular task was higher than the average. A reasonable distribution emerged: 18% of the farmers scored zero on all tasks and 21% worked more intensively than the average over the complete range of tasks.

A simple quantification of the reproduction of seed is much more difficult, not only in terms of measuring techniques but also for conceptual reasons. To begin with, there is no undisputed yardstick which runs from "bad" to "good." Good potato seedlings are as good as they are used. Thus the use of such a variable in a general model can cause considerable problems of a multicolinearity type.

The following analysis is therefore based on only two elements: "intensity of cultivation" and "production of soil fertility." If both of these aspects of farm labor occur simultaneously, then we can speak of farm labor structured as craftsmanship in every respect. Table 3.5 gives the correlation coefficients of the links between the production of soil fertility (1), intensity of cultivation (2), and yield per hectare for potatoes (4). The interaction factor (1×2), i.e., "farm labor structured as craftsmanship" (3), is also added.

One can see that the interaction factor has the highest correlation with harvest yield (r=.48). However, production of soil fertility and intensity of cultivation also correlate significantly with harvest yield. These results lead to the following question: do production of soil fertility and manner of cultivation, independently of each other, exercise an additive effect on hectare yields, or is the concept of craftsmanship (operationalized as the interaction of the two) necessary to explain

Table 3.5. Correlation Coefficients for the Relationship Between Production of Soil Fertility, Manner of Cultivation and Yields per Hectare, Chacán (n=52)

variable	(2)	(3)	(4)
1. Production of soil fertility	.16	.33	.33
2. Intensity of cultivation		.22	.41
3. Farm labor structured as craftsmanship (3)=(2)x(1)			.48
4. Hectare yields			

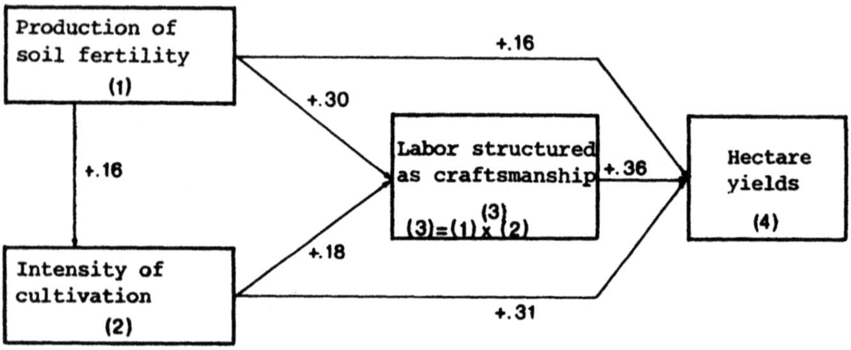

Figure 3.9 Path diagram showing the effects of farm labor on yields, Chacán (n=52)

yields. The calculations show that a greater part of the variance can be explained when using an interactional rather than an additive model.

Figure 3.9 expresses this interactional model in the form of a path diagram. It shows that the effect of labor structured as craftsmanship is stronger than the direct effect of soil fertility and manner of cultivation taken separately. To a certain extent this result is self-evident, although not trivial. Where a high level of soil fertility is produced, along with an intensive form of cultivation, the yields are high. So "good farming" leads to "good results." Recent literature on agrarian development mostly stresses the introduction and diffusion of innovation: the transfer of "high value input packages." The link between farm labor structured as craftsmanship and harvest yields highlighted above puts into perspective this one-sided emphasis on the adoption of chemical fertilizer, new varieties, and the like since it is primarily where labor is structured as craftsmanship that the introduction of innovations like fertilizer begins to bear fruit. A more detailed analysis

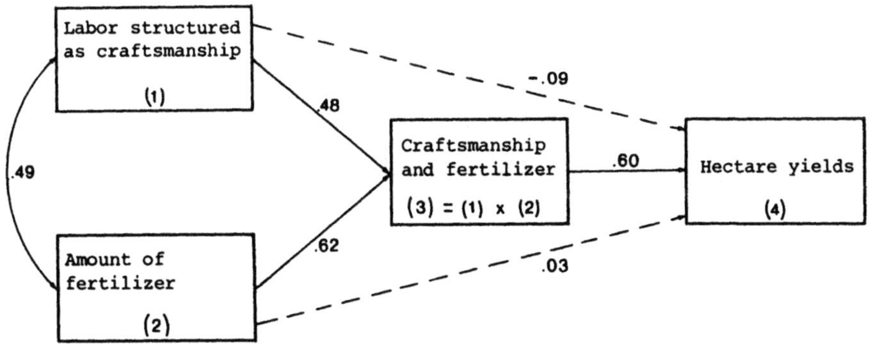

Figure 3.10 Path diagram showing the effects of farm labor and fertilizer on yields, Chacán (n=52)

confirms this line of interpretation. If, for instance, we isolate the use of chemical fertilizer from the whole of farm labor, then the following relationship is revealed: Figure 3.10 shows in the form of a path diagram the interaction between labor structured as craftsmanship and the application of fertilizer. Fertilizer on its own has no significant direct effect on yields (p42=.03) nor does it have any significant effect when it is combined with "modern seed varieties." Hence, packages of modern inputs do not achieve much if divorced from the quality of labor.

On what, then, does the structuring of farm labor depend? An answer has been partly provided earlier. Incorporation into capital markets (in this case mostly as credit from the Agrarian Bank or Proderm) has a negative effect on the level of fertilizer used and therefore on soil fertility. Incorporation into the labor market (i.e., the substitution of ayni by market mechanisms), certainly if it is "complete," has a negative effect on labor input and thus on the intensity of cultivation. A partial incorporation into the labor market has, on the contrary, a positive effect. It would also seem that taking credit leads to an increasing preference for wage labor as against labor exchange (ayni). The connection between these aspects of incorporation and their interaction on labor structured as craftsmanship (the central term for explaining yield per hectare) is more closely analyzed in Figure 3.11. Incorporation into the market for capital exercises a direct and negative effect on labor structured as craftsmanship (p41=−.21). The indirect effect is −.29. Thus the total effect on yields per hectare of incorporation into capital markets is overall highly negative. In other words, where the different elements of the labor process (such as labor force, seed and fertilizer) increasingly take the form of commodities, farm labor process becomes restructured. One result of this is falling yields.

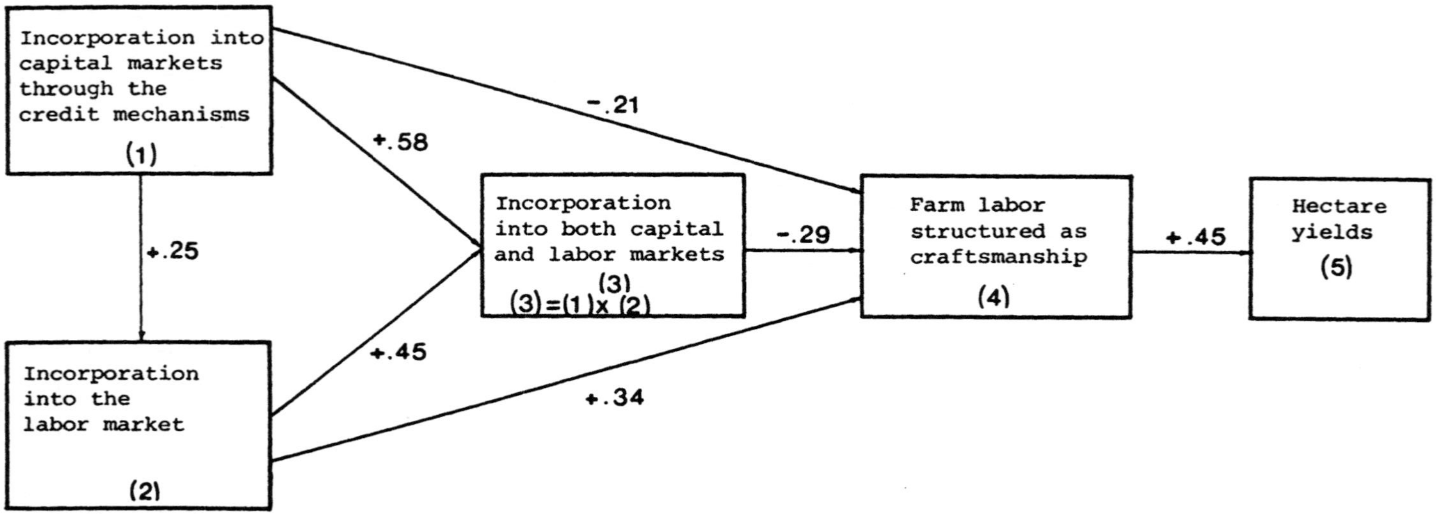

Figure 3.11 Path diagram showing the effects of incorporation into capital and labor markets on the structuring of labor (n=52)

In his comment on small farmer credit in the AID spring review, Mellor (1973:85) notes that "increased cropping intensity probably offers more direct benefits in the long run to the small farmer than the high yielding varieties and yet I find the attention much more towards high yielding varieties and the problems of credit for those than for increasing intensity." Indeed, through many credit and integrated rural development programs, a simple but solid train of thought runs: with new technological packages, productivity, production and income can be raised.[6] Since working capital is missing for the initiation of this springboard to growth, temporary credit is required. It is apparently as simple as saying "good morning." Thus, "Success in Small Farm Credit Programmes has generally been linked in the minds of men to the only readily available quantitative indicator, repayment rates" (AID, 1973: XX, 14).

However, credit is not just a "simple supplement" for working capital. It is an element in a historical process in which the gradually eroding personal resources of farmers give way to production factors and non-factor inputs obtained via the market. New and more complex commoditization patterns arise which penetrate deeply into the labor process. At the same time new production relations become dominant, which lead to a decidedly different application of the goods and services in question. Mellor suggests that "the capital requirements per acre of moving into intensive production may be so large that a farmer must think in terms of increasing his permanent working capital and may need at least insurance of a continuing line of credit and perhaps an increasing line of credit" (1973:89). The mechanism of credit as a relation of production, however, introduces the opposite of this: a short time horizon, high-risk insecurity, and an increasing rigidity as far as buying and selling in different markets are concerned, so that higher costs and lower benefits occur, requiring further extension of credit. In short, mechanisms of credit prompt extensification.

The mold of integrated rural development, in which most actual programs are shaped (including Proderm) is double-edged: the introduction of "modern," potentially more productive inputs (and the implementation of necessary infrastructure) is combined with "integration" into markets. "Rural development is concerned with the modernization and monetization of rural society, and with its transition from traditional isolation to integration with the national economy" (World Bank, 1975b:3). Hence, incorporation into the capital market fulfills a strategic function: through credit, farmers are in a position to buy the "package" of modern inputs. The preceding analysis offers the grounds for judging the effect of such programs. Modern inputs provide, in principle, the possibility of substantial increases in production, providing they are used within a framework of labor which remains structured as craftsmanship. However, it is precisely this specific structuration of the farm labor process, which is made difficult, though not totally impossible, by the effects generated by credit mechanisms. Thus, the possibility of intensification offered by improved technology is at the same time closed off. In this way a pattern of underdevelopment is reproduced which nevertheless does not eliminate the possibility of analyzing matters differently, as demonstrated above. Most development institutes interpret the failure of "real" development after implementation of particular programs as evidence of (a) the poor response of farmers (or environmental difficiencies), and (b) the need to carry out more such programs!

Notes

1. This is how specific "circuits of value exchange" (Barth, 1967) are created. Breaking through these specific circuits is quite often a specific feature of entrepreneurship; circuits which until a certain moment were separated are then combined (Long, 1977). This is often seen as a neglect of social conventions. Moerman (1968:144) concluded that "those who use the market more efficiently than their neighbours are the villagers who, for these and other reasons, are criticized as being calculating, aggressive and selfish. Studies of village economies in India and Latin America support the observation that successful village entrepreneurs frequently fail to maintain the common peasant values of equanimity, generosity, loyalty to kinsmen and conspicuous piety. . . . To put it bluntly, it is not uncommon for villagers who are ambitious, enterprising or successful to be 'sons of bitches' in the eyes of their fellows." In Chacán, wholesale traders amongst the villagers are perceived in a similar way. It applies also to the ricos who no longer wish to participate in ayni and companía exchange.

Indeed, should all ayni and companía arrangements disappear, the poor would especially be in an awkward situation, for they would then have to pay for the oxen, tools and sometimes money that they now mobilize through these mechanisms. They would then be obliged to sell some of their produce, used presently entirely for home consumption, or they would have to work elsewhere to invest in their plots. This too implies a further closing of opportunities for them.

But the ricos too would suffer serious problems. With the disappearance of ayni they would have to contract more wage labor and would have to pay the full market price for this instead of the "shadow-price" now existing in Chacán. Thus costs would rise and flexibility decrease. This in turn would urge more extensive schemes of cultivation.

2. Nemchinov (see Shanin, 1980) followed the same analytical approach: he conceptualized differential (intracommunity) levels of commoditization into a pattern of social stratification. Notwithstanding this similarity, I maintain that

in this case (the Chacán community) one cannot speak of class differences. They are relative differences between farmers within the same class. Beyond that, these differences are highly variable over time.

3. Strictly speaking one should use here the term "environmental conditions" and consider the interaction of environment and genotype as phenotype: E * G = F. For practical reasons, however, I choose to talk here of phenotypical conditions, i.e., those environmental conditions that are of direct relevance for the emergence and reproduction of a specific phenotype.

4. This assumption was checked and validated in the campanía study included in Bolhuis and van der Ploeg, 1985.

5. The validity of this score was then checked through its correlation with the seed/harvest ratio (ref. Slichter van Bath, 1960). There turned out to be a strong association (see Bolhuis and van der Ploeg, 1985: 311, table 6.14).

6. Unless the total output raised provokes a decline in prices; low price elasticity of demand causes then a negative income-effect for rural producers (see Scobie and Posada, 1977).

4

Peasant Struggles, Unions, and Cooperatives

This last case study focuses on cooperative farming in Alto Piura, an agricultural area lying in the foothills of the Andes in northern Peru. It is an area with a long militant history of peasant struggle (the beginning of which is well documented in Castro Pozo, 1973; and Albujar, 1969). Its roots are to be found in a pithy local saying, "*Tierra sin brazos y brazos sin tierra*" ("land without hands and hands without land"), which captures the essence of the particular pattern of agrarian development that brought extensive agriculture and high unemployment into the area. The two phenomena, at least in the eyes of relevant local groups, are closely related.

This chapter introduces two related dimensions: cooperative agriculture, and social struggle led by peasant unions whose aim was to intensify agriculture and at the same time raise the level of rural employment. Enormous potential is entailed in this struggle. As I shall argue later, production could be raised, in the short term, by at least 50%. That would be twice the actual agrarian growth of the last twenty years (Alvarez, 1981). One of the prerequisites for this, however, is the attainment of the political and economic autonomy sought by the farmer's movement. That is to say that the external economic and institutional relationships which now condition the production process, must be substantially reduced.

Luchadores del Dos de Enero

"Combatants of the Second of January," *Luchadores* for short, is the name of the production cooperative at the center of the following case study. At the beginning of the 1980s it was a cooperative of 400 members, had between 1,000 and 1,500 hectares of land under cultivation (depending on the rainfall) and had considerable livestock and pastureland. Rice, maize, sorghum, cotton, bananas, citrus fruit and occasionally sunflowers were grown. It was once the domain of the Rospigliosi, a classic family of landowners (*gamonales*), but after a massive land invasion, carried out just after the New Year's celebrations of 1973 (hence the name), it fell, nominally speaking, into the hands of the workers. Nominally, because as shown by the long history of strife in the years to follow, effective control of production does not come per se with the setting up of a cooperative. The struggle carried out by Luchadores was militant and on a massive scale. I will discuss in some detail the content and form of the struggle since the current literature on peasant movements pays little attention to these new forms of strife, forms which are not so much geared to the redistribution of wealth as to the expansion of social wealth—and thus from the outset assume direct intervention in the realm of production. In some senses Luchadores is unique. That does not in any way reduce its relevance for socioeconomic research. With reference to Mondragon, a unique experience with production cooperatives in Spain, Thomas and Logan (1982) argue that it is precisely because such experiences are unique (in that there are no historical precedents) that they throw a whole new light on what until now has been regarded as more or less a rounded-off field of study. In what sense is Luchadores unique, and what is the theoretical and practical meaning of this unique experience?

As a unique microcosmos Luchadores presents an unusual combination of general patterns and specific responses. General in the sense that in Luchadores, as elsewhere, the consequences of a high level of incorporation and institutionalization can be seen to take their toll in the stagnation of agrarian development and the progressive cutting back of manpower, "tierra sin brazos y brazos sin terra." Atypical and thus unique is that, thanks to the unusual union structure and its accumulated experience, workers have:

1. successfully developed a reasonably well-defined plan for intensifying production and raising employment;
2. over time, and with some fluctuation, come to grips with the relations of production (in short, with the cooperative as an enterprise); and
3. been able, with some degree of success, to achieve their plan to intensify and increase employment opportunities, which has gone some way towards stemming the stagnation, extensification, and reduced employment that is generally the case in the area.

It is this plan and its partial realization that makes Luchadores unique—not because it would not be a repeatable experience somewhere else, but because it shows the way that could be followed in many other places.[1]

The History of Luchadores

For a clear understanding of the history of Luchadores, a periodization is necessary because marked fluctuations in power relations have occurred over time between rural bourgeoisie, the farming class and the state. Important economic changes have also taken place. Other factors, such as the social composition of Luchadores, have remained more or less constant. The union was also a permanent factor during the research period. However, the interaction between these changing relationships and constant factors is such that several aspects of development are influenced in ways that are also constantly changing. It is useful, therefore, from the beginning, to start by looking at these very different periods during which development has taken place. They are the following:

1. The hacienda period (up to 1968).
2. The period of the *parceleros*: in 1968 the great hacienda was divided up into smaller parcels or areas of land and sold to the "rich from the city." This maneuver served both to avoid the consequences of the land reform and also to liquidate the union of the time. It lasted until January 1973, when the people of what is now Luchadores invaded all these areas and demanded the formation of a cooperative.
3. The cooperative was established in 1973. It was managed by a Special Committee, which was set up and run by government officers.
4. From 1974 to 1976 Luchadores functioned for the first time as a self-administered cooperative.
5. Then began a two-year period characterized by increasingly penetrating military intervention. The enterprise during this period was administered by a Junta Interventora, whose aim was to eliminate farmer control over production in order to make the enterprise "economically healthy."
6. At the end of 1977 a massive strike broke out aimed at getting rid of the Junta Interventora and returning control to the farmers. From that year on, Luchadores again functioned as a cooperative. During this period new problems arose which made the union decide to go further than just defend employment, wages and better working conditions; the union decided to place itself at the head of the cooperative.

7. Thus in 1982 began a new period, a period in which the union, under the slogan, "Let us work and fight," tried to alter the course of the enterprise. The year 1982 was also a time of severe drought which clearly affected the union's project and led to all sorts of drastic adjustments.

The periodization outlined will be used to describe several aspects of development in Luchadores which at first sight appear to be purely technical, such as the area planted, employment, yields per hectare, profits and losses, etc. This less customary way of writing history allows one to show not only the factors and problems related to these technical indices but also, and more importantly, that employment, production, yields, etc., are not enacted outside the class struggles of the countryside but rather follow, as compass needles, the power relations and development of that social struggle.

The Development of Employment

The graph in Figure 4.1 shows employment over time in Luchadores. We shall comment on its fluctuation and relate it to the periodization already set out.

1) The Hacienda Period. The period up to 1968, the time of the "patron," is characterized by a level of employment which is relatively low when viewed in the light of later developments. Furthermore, between 1960 and 1968 employment followed a downward trend. At the beginning of 1960 there were still 400 permanent workers, whereas in 1968 there were 250. Although these figures may not be completely accurate, they do reflect the normal trend of employment figures in capitalist organized farming. With production geared to profit and the use of mechanization to substitute for labor, employment took a gradual and sometimes a brusque turn downward. Theoretically speaking, the reduction can be compensated for by new investment which creates other employment opportunities, but in a situation where agriculture is subordinated to other economic sectors such investment seldom appears profitable. The margin of profit is generally higher and less risky in trade, industry, and speculation circuits. The words of Rospigliosi, one-time patron of the hacienda, illustrate the point: "Ha," he said, "this whole hacienda here is nothing more than a stable for my horses." Rospigliosi visited his "stable" by plane once every three months. His words also illustrates how small and subordinate agriculture was when considering the economic activities of this entrepreneur as a whole. The profits were creamed off and invested elsewhere. A low and falling level of employment was and still is the consequence. A

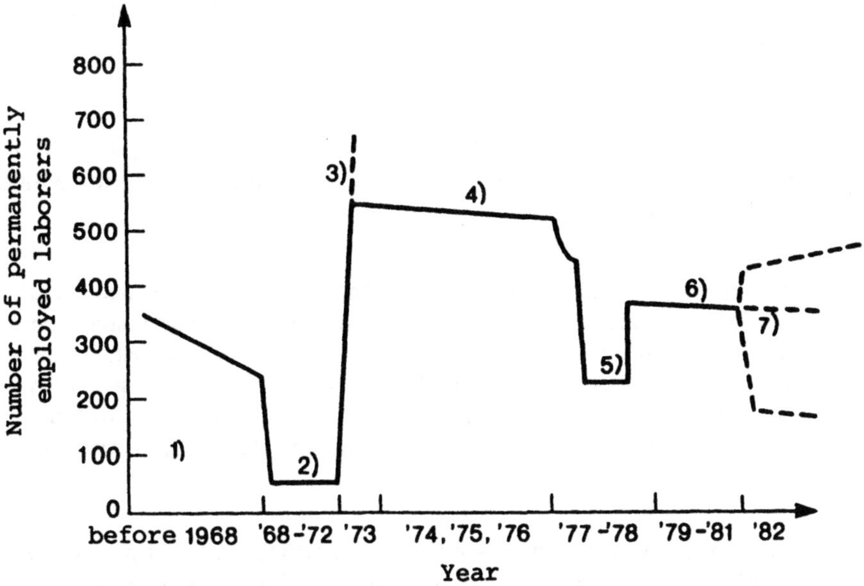

Note: The note numbers in the graph refer to the numbered sections in the text.

Figure 4.1 The development of stable employment in Luchadores

second consequence was that labor was reduced to its cheapest form: that of "temporary" labor with its lack of security and benefits. Even though a worker might be employed on a permanent basis, his legal position could be virtually reduced to that of "temporary" laborer. It was in the context of this widespread practice that the union arose. As Leonides Palacios, the union's secretary general, said,

> "Don't be taken in: the fact that we once had 400 permanent workers here was the result of our efforts. Before that, before we had a union, the hacienda kept on at most only 100 permanent workers. They were people who fell in with the wishes of the patron and who also occupied key positions, the irrigators, the ox drivers, the livestock men and of course the officials. Thus one arrived at about 100 paid faithfuls. For the rest there was no security. Then the idea of a union slowly matured, which has cost us a lot of strife. You see, the Rospigliosi, the owners of the business, sat in Lima and could reckon on every political support there . . . and so recognition of the union was held up for a long time, but we fought and we got our union. We demanded the status of permanent worker for all those who worked full time so we no longer had to walk the streets on account of some caprice of the boss, or the market, or the weather. We also demanded the correct payment for overtime, and

> for a bonus so that you can give your children a little something at Christmas. To cut a long story short, we demanded that the law be adhered to."

The situation that developed as union efforts made progress led to problems at enterprise level. The first losses occurred—something which is still talked about:

> "This led to the boss declaring the business bankrupt. They said there was a loss of 8 million in 1968. Ha, now we are a cooperative and even with a loss of 80 million we are still in business and ticking over. In our opinion it was a bogus argument, the losses were a pretext, the bosses have always been able to obtain enough money. The fact was that the union was not in their interests and so they were looking for excuses to eliminate it."

And indeed bankruptcy appeared to be nothing but a prelude to a second phase: the carving up of the hacienda into smaller plots.

2) The Parcelero Period. On October 18, 1968, the Rospigliosi family divided the hacienda into twenty-eight smaller plots, which were principally bought by the rich of Piura, the provincial capital lying some 80 kilometers from the hacienda. Owners of commercial firms, agronomists and bankers bought plots of between 80 and 120 hectares each. The Rospigliosi family got out of farming. "With the liquidation of the large enterprise," said one of the members of Luchadores,

> "200 workers were dismissed. Most of us suddenly found ourselves without work. The owners of the new plots maintained only a very small number of workers in permanent employment, forty-five in total. The rest of the work was carried out by people who were brought in from far away. Humble people. And you know the whole splitting up of the farm took place long before the land reform was announced. Yes, it was a means of breaking the union."

Thus, on the land of the earlier hacienda, "middle-sized" properties developed, administered in an efficient way by the upper middle class of Piura. The level of mechanization was heavily increased, infrastructure improved (above all boreholes and pumps), the most profitable monoculture, rice, was opted for, and the permanent workforce was frozen at an extremely low level of around forty-five workers.

3) The Invasion. After substantial preparations, the land of all twenty-eight middle-sized properties was invaded, or "taken" according to the literal Spanish expression (see also Luna Vargas, 1973).

"They had not destroyed the union. That remained alive through links with the forty-five permanent workers and above all as a dream, as a memory, and as a demand by all those who had been thrown out of work. We saw the land which needed more labor, we saw how people were brought from afar to work here for a pittance, we saw how our children sometimes fainted from hunger at school. So a group arose who talked further and got to work studying all the measures needed. We had the help of comrade Andres Luna, the support of comrade Dr. Ruffo Carcamo, of the schoolmaster from the village and of all the others who identified themselves with us. We all decided to unite and 'take' the enterprise. We occupied all the gates, the women to the fore . . . then the police couldn't shoot, and our women wanted to be out in front, because they were even more left than us men; they saw the hunger in the house much more than we did. That was the 2nd January 1973. Everyone helped, workers, non-workers, with more than 2,000 we threw the white-heads from Piura out; we chased them high into the mountains—everything in the struggle for the recovery of the right to work."

Two key elements are to be noted here for an understanding of the history of Luchadores:

1. The social base of the union was broadly defined. Neither the objectives nor the social base of the union were to be limited to the permanent workers of the enterprise but, in the last resort, included the whole population of the district.
2. From the start the union began the struggle with a programmatic focus: the plots would be taken with "the right to work for all" in mind. What is articulated in this demand clashes unequivocally with the dynamics of bureaucratically controlled land reforms and cooperatives. With this kind of program at the outset the basis was laid for permanent contradiction.

Why this last demand? Let Leonides Palacios, present secretary general of the union who was involved in the takeover, tell us:

"With the land alone you could do nothing, certainly if you took it over in accordance with the land reform regulations then in operation. Officials came to tell us that there was only room, at the most, for 250 people and maybe at the outside 300. So the union once more took the reins in hand, because say what you like, the union stood iron firm during and after the takeover, and said that they in no way accepted that, that everyone in need must qualify as a member and be able to work. And that was about 700 at that time."

Intensive preparation preceded the land reform in Peru (DL 17716 was decreed in October 1969 and in the following years supplemented with various rules). Calculation of the so-called "economic holding" played a key role in this preparation. "Economic holding" is a concept that indicates the number of hectares (taking into account ecological and economic conditions) that can be worked by one man (or family) in order to give him a reasonable income. These calculations were put together by the Iowa-Mission of the USA (Figueroa, 1975). By applying this rule to a greater area, manpower levels (*cabida* in Spanish) for a cooperative, for example, can be calculated. In calculating an economic holding the following elements are crucial:

1. whether to keep the level of agricultural development constant or to change it (for example to intensify);
2. the level of technology (in principal also changeable);
3. the current or changing nature of price relations;
4. town-countryside relations (on what terms should agricultural wealth provide for other economic sectors, how much, and in what way?) (see also Quijano, 1973:423).

The Iowa-Mission excelled by holding the first three factors constant and by raising in an immediate sense "the exploitation of the countryside by the town" with the introduction of a "savings quota"[2] of 25% (Convenio, 1970a and 1970b). Applied to the total agricultural area of Peru, the economic holding so calculated resulted in productive work for 149,538 families (van der Ploeg, 1977:236–240; and 1982:218–219). That means plus or minus 10% of the economically active population of the Peruvian countryside. Projecting national and international price relations on the post-reform situation in order to deduce future manpower levels is not only a vehicle of underdevelopment; it is also an instrument for the continuing marginalization of large parts of the agricultural population from the production process. It is how the misery of land reforms is born. Manpower levels are frozen at levels generally lower than previously. Practically speaking, this means that intensification of agricultural production (and thus agrarian growth) is already doomed in advance; there is insufficient manpower available.

In Luchadores the misery of the reforms rivalled the misery of the rural poor because when *the economic holding* rule was applied to the land they had newly occupied (calculated for the Peruvian coast at 6.7 hectares per man) this led to a *cabida* (an employment level) of 250 men. The union demanded work for all, and this meant for 700 men. Why?

> "For us it was as clear as day. We saw clearly that the enterprise had enough potential, enough resources. There were possibilities for creating more work. We knew all those fields you see, like the back of our hands, and we knew what could and couldn't be done . . . the fruit of the algorroba which is now wasted, and the pastureland which is not used, an excellent centre for the fattening of cattle could be made, there are outstanding plots for planting fruit, and then there are all the simple but very important things like the removal of field boundaries through the better use of water, and the improvement of production itself, because make no mistake, where for the sake of argument 100 can be produced, with better cared for production you can obtain 200 per hectare. But for that, everything must be better worked. For that you need more labor. For us it's as simple as that."

And Augusto Cruz, who was very much in the forefront during the occupation, made the following argument:

> "Then we knew very well that by fully planting all the land here there would be work for everyone. You must make the land bloom to have work for all. . . . All that talk of 250. In the hacienda, for instance, they never planted a between harvest (*campaña chica*), a second crop after the rice harvest. That in itself increases work opportunities enormously and it also means that work is then more evenly distributed throughout the year. But our arguments, the language of the fields, fell on deaf ears."

In summary, by intensifying and diversifying production, as well as by enlarging the production base (the removing of boundaries) there could be work for everyone. "And since then," Cruz added, "we have demonstrated clearly enough in practice that is indeed the case." However, what was such an evident project in the eyes of Luchadores not only clashed head on with the objectives of the land reform at the time but also with the economic policies of today's regimes.

In general terms, one can argue that intensification, diversification and expansion (a) demands substantial investment in the countryside, and (b) implies that a greater share of the wealth generated should remain in the countryside. Neither the one nor the other has coincided with economic policy either then or since (Fitzgerald, 1981). Does that make the project a utopia? Before going more closely into the theoretical arguments on this question, it seems to me desirable to follow first the actual history. Luchadores managed to push their ideas through. The ministry, however, with its formal rules, still tried to chip away at the initial list of 700. Thus the "handbooks" mentioned that only family heads could be members of the cooperative that was to be formed—a somewhat problematical rule for young men who did not know how to turn their engagements into marriage quickly enough. Be that as it may, when the cooperative began there were 614 members, more than double the number possible according to the economic holding formula.

4) The Fourth Phase. This is the period in which the cooperative as such gets going, although in the first year (1973) it was still under the control of a Special Committee composed of government officials. They started work with the 614 member-workers. The area planted rose substantially, and for the first time a "between harvest" was sown. By introducing new varieties, in rice cultivation among others, yields rose. But new problems began to surface. The Agrarian Bank turned out to be one of the decisive factors, expressed by one informant:

> "To the degree that the bank makes money available, you can plant more and work better; and to the degree that the bank turns off the credit tap you are sentenced to less and poorer quality work."

Although I will not elaborate here on this specific problem, it must be noted that the cooperative is highly dependent on the bank. This is so because before and during the land reform a gigantic decapitalization occurred which the farmers could do nothing to prevent, and because cost/benefit relations were such that internal saving was not possible.

Such costs, however, are viewed in various ways. During those years numerous studies were carried out by the bank and other state apparatuses. In one of them it is concluded that "the cooperatives (and with them Luchadores) absorb more labor than is justified by the carrying capacity of these enterprises, so that an excessive increase in costs is generated, which is an obstacle to good economic results" (SINAMOS, 1975).

So the years came and went, until in 1976 latent conflict burst into the open. The bank refused further credit. Suddenly payment of wages was impossible. However, the people of Luchadores decided to go on working.

> "Then it was hunger in the belly and scouring the river banks in the evenings for those forgotten beans . . . because to let the business get ruined . . . that never . . . we thought then."

For twenty-one weeks people worked without pay. Some of the problems that resulted could be compensated for by the women. They were engaged in a whole number of economic activities that generally come under the name of "informal economy" which help to tide families over a temporary loss of income; cultivating of food crops, engaging in retail trade, running small *chicherias* (beer shops), and selling local dishes in food stalls.

Next, the cooperative decided to share a part of the harvest (mainly rice) among the members. That, however, led to direct intervention. The state closed the mill. The "twenty-one weeks" constituted an event that was later often referred to. But it should be emphasized that this episode did nothing to diminish the will to build the cooperative, even though it bit deep.

However, to work for so long without wages finally forced some families to consider alternatives. Need forced a number of members "to seek their living elsewhere, even it was only to haul cargo in the town." One member reflected,

> "I don't know, perhaps it was only the *machos*, the plucky ones who cleared off. But it could also be the other way round, that the plucky stayed and the skivers went. Who can say."

The cause of all this was the bank's unwillingness to provide further credit, or to be more precise, the bank's demand that Luchadores should effect a rotation of personnel—fourteen days work, fourteen days off for each person. This would have meant a halving of the effective wage bill and also employment. At the same time the scheme allowed the bank to cut credit by 50%. The demand for such a rotation would be repeated more than once. But Luchadores rejected this demand, as they did later, since it implied the surrendering of one of their principal gains, work for 600 people. It was this that made the men decide to work on, all of them, without pay. But the gradual desertion taught them that this response was less than adequate. "You only burn yourself once," said Norberto Cruz.

> "There was plenty of work at that time, only the money was missing. The bank would not part with another cent. They wanted a rotation from us, but no one wanted it. Must you let your stomach also rotate? So we continued working, also so as not to let the farm deteriorate, what else could you do when the fields were full of crops? We also wanted to put the bank under pressure. . . . Ha, they still owe us for those twenty-one weeks. However, we say here that you only burn yourself once. No one wanted that again, to work another twenty-one weeks for nothing."

Plenty of work but no money describes the conflict in a nutshell. When the members of Luchadores say there is plenty of work they are referring to intensive agriculture. For them it is a norm which in innumerable ways permeates their thought and speech. "You must take good care of your crops," "make the land bloom," are expressions which all point to a felt need to practice an intensive kind of agriculture. The bank, however, calculates the amount of credit and thus the weekly wage bill in terms of extensive farming.[3] In the eyes of Luchadores this results in "no money to work."

5) The Junta Interventora. In January 1977 the state intervened in all eleven cooperatives of Alto Piura, which from then on were obliged to work under the control of a Junta Interventora. This Junta represented the most direct effort of the state to reorganize internal relationships in the cooperatives in such a way that they conformed better to extensive, large-scale agriculture. The core of the Junta's action program was the imposing and implementing of rotation of personnel, which amounted in effect to a 50% reduction in employment (see Figure 4.1). At the same time the area under cultivation was reduced, even though there was enough water. The council chosen by the farmers was dismissed and the daily administration came into the hands of the Junta. A fourth line of action was a systematic reduction of the labor employed per hectare per crop, referred to as "economizing on labor." A fifth measure was to increase mechanization, for which a call was made on external contractors in the hands of former landowners.

Let us look more closely at one of these measures—the economizing of labor. In banana cultivation, as for that matter in cultivating other crops, production rises by the degree to which one puts in more and better work. In a sensitive crop like bananas, where the "care" with which you work is very important, this holds a fortiori. From the specific *planillas* (specified wage bills) one can deduce that up until then, on average, 140 man/days per hectare per year had been allocated to bananas. The Junta reorganized the work so that a bare 117 man days per hectare per year were worked. Because of this cost reduction, according to Luchadores, the land "was a deal worse worked. . . . There wasn't enough manpower to do the work properly." The consequence was deteriorating plants, sharply falling production and damage that would last for years. Another consequence of imposed rotation was the continued desertion of members. As Augusto Cruz said:

> "we were darned well getting thinner. We got more and more fed up until the moment came when the union, in a peaceful way, was obliged to make its point of view clear again, that rotation had to come to an end and that the social rights that had been denied us for more than fifteen months should be honored. But the Junta would have nothing to do with it. They never once listened. We began increasingly to see that all of this was a way of making people so fed up that they would migrate, so that the land would be left empty, and then the great landowners, the

gamonales would be able to return again. . . . After taking experienced advice a strike was called for the 8th August 1978 with no limits set. Phew, that was quite something, that went directly against the military, against their whole system . . . it was not easy but after a real struggle, which went hard, they finally buckled. That was on the 8th September.

There were other conclusions to be drawn from this strike, as Palacios, explained:

"Within three days of striking, the military were here. They demanded that our leaders and advisers, like Ruffo, should come to Piura to 'negotiate.' But the union told them categorically that the problems were solved in the fields, not in the city. After thirteen days it became really tense; some began to say that they no longer believed in saints you could not see. . . . Finally a senior military delegation came here to us. We had created a lot of pressure, there were other cooperatives on strike. To begin with they paid us all the dues which we had not received for fifteen months. They paid here in the fields. . . . So you see, suddenly there was money, after years of repeating that there wasn't any. But we had other demands. . . . After an impressive march on Piura they again gave way. The Junta was removed, and rotation scrapped. So we resecured the right to work. You can't imagine it, how joyfully we went to work, with our own people and more than 500 temporary workers. There was work enough, and the damage had to be repaired . . . and then there is something else. The Junta was supposed to make the business economically sound, but imagine, they left the business with more debts than when they started."

6) A Cooperative Again. At the beginning of 1979, Luchadores began to work once more as a cooperative. The permanent workforce recovered to a level of 420 workers—higher than during the period of intervention but not as high as the 470 there had been previously. The difference can be explained by the desertions. During 1979 Luchadores achieved in practice what the Junta Interventora had always hoped to achieve. As Oyala put it:

"We happily got cracking, even with more than 500 temporary workers . . . in a manner that was more disciplined than ever. The strike had been a hard school. Indeed we worked very hard in 1979 and made a profit of 79 million."

Again the years came and went. The years 1979–1981 saw a slight drop in permanent employment, from 420 to 400. This was the result of a legal problem for which Luchadores could never find an adequate answer. When a member retired, he had no rights other than his pension,[4] something which caused increasing criticism:

"You work your whole life, your sweat remains on the land you love, but the lot you are left with is the rubbish heap. What we must move towards is that one of our children automatically becomes a member."

In this period, new problems began to manifest themselves, which were largely related to internal control, the commitment with which one worked, the role of the engineer, and the degree of mechanization desirable. These questions are treated in more detail later.

7) The Drought of 1982. Then began 1982, the year in which the union placed itself at the head of the cooperative, among other things, as Tonga explained, in order

"to push the enterprise forward in a way that would enable it to create more work for others who badly needed it. . . . The land was worked with a lot of energy and enjoyment, the seed-beds were full to overflowing . . . and then, aay, we were struck again with the cursed drought."

During the four previous years drought had manifested itself in different ways but at the beginning of 1982 it was acute (Vela Suarez, 1982). The level of employment projected for 1982, around 550 permanent workers, was thus impossible. The bank again demanded rotation and in the meantime halved its credit. But instead of dropping to a level of 200 men, they shared the work between all 400 and the union, now at the head of the enterprise, managed to maintain permanent status for all of them.

An analysis of employment in Luchadores cannot be ended without discussing the temporary workers and small farmers (here called *chacreros*). Only by including them can total employment and its development be assessed. The relevant developments are treated briefly here, as the central elements will crop up again elsewhere. To begin with, at the boundaries of the cooperative an important development was taking place. The number of chacreros was continually increasing, going from less than 50 in 1968 to around 400 in 1982. Bit by bit, cooperative land which was not yet in use was "taken," partly in an organized way by the surrounding communities and partly as a spontaneous movement. As far as land itself was concerned there was still no problem, but there was and is a problem regarding water. Water bears an important relationship to employment in Luchadores. On other cooperatives even land being used was frequently invaded, but this did not happen to Luchadores—maybe because of the high level of employment but maybe also because of the relationship between Luchadores as a cooperative and the landless:

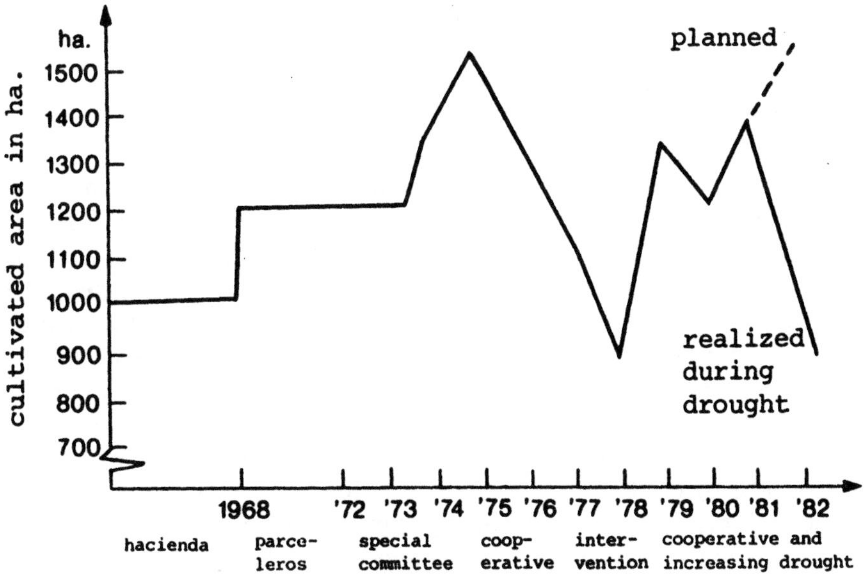

Figure 4.2 Cultivated area in hectares over time

> "Never in their living days have we put the police on to them, as was common in the days of the patron. They are poor farmers like ourselves, in a manner of speaking they could be your uncle or your brother. . . . No, that would be a step backwards, to be knocking our heads together."

The fluctuation in the number of temporary workers over time has been extremely erratic. In the period of the parceleros, 1968–1973, there appears a clear complementarity—fewer permanent workers and more temporary workers. With the cooperative the case is mostly reversed; many of the full-time "temporary" workers were given "permanent" status. During the Junta Interventora permanent employment was laid aside and the number of temporary workers dwindled almost to nil. The reduction of total employment was then consistently implemented. In the later years of the cooperative, from 1979 on, development was not so much a question of complementarity, but of another phenomenon. The number of permanent workers remained stable, and the number of temporary workers fluctuated according to the vagaries of nature and the bank.

The Development of Area Under Cultivation

The development of the area sown annually is summarized in Figure 4.2. Three phenomena become apparent:

- the huge increase in cultivated area with the transfer from hacienda and parceleros to cooperative;
- the *grosso modo* negative tendency ever since; and finally
- erratic yearly fluctuations.

The water supply from year to year and in the long term is an uncertain factor. "And that is not only the fault of Holy Joseph," explained Tonga, during a procession in which San Jose was asked for water, "there used to be more water." All Luchadores would agree. The volume of the river that flows to and through the fields of Luchadores did not drop only because of decreasing precipitation in the sierra. An important factor was that chacreros and comuneros higher up the river had taken over much of the pastureland and had sown rice, yuca, and maize—often three harvests a year. This drew off a large volume of water. The increase in the number of chacreros meant that the distribution of available water gradually moved from a ratio of 60/40 to 50/50, to the disadvantage of the cooperative. This would have been unthinkable when the patron still formed the epicentre and raison d'être of the whole system of water distribution. No one dared point a finger at "the water of the patron." With the land reform, however, this system, already in a state of breakdown, quickly disintegrated. It is true that the state played an important part, yet control over water rights shifted to the cooperatives (at least in part). Endless and often bloody "conflicts between brothers" were the sad result. Luchadores refused, for the reasons mentioned, to take such a role. But there is much more to it than this. At various times the cooperative has invested substantially (part in capital, part in labor) in cleaning out wells and in buying and repairing pumps and motors. The water holding layers (with depths ranging from 3 to 35 meters) form a substantial natural resource which at least can tide farmers over short periods of drought (after droughts of four or five years even these water supplies dry up). A complete use of this resource, however, is prevented by numerous factors.

First, there is the oligopolistic nature of the commercial firms in Piura, the only ones that can do certain repairs and sell new material. Repairs to pumps, etc., entail very high costs and endless delays. Often the most valuable parts of equipment are pilfered and replaced with almost obsolete parts. Last but not least, machinery can be returned "repaired" but still not function. Luchadores added an error to this. Their own workshop for maintenance and repairs had from time to time been neglected in the previous years in their desire to prevent unproductive expenditure. A second factor, which rules out the increasing use of subterranean water, is the very high and rising price of diesel oil. Unlike many other countries Peru does not subsidize the use of diesel for agricultural purposes. Finally, if acute drought occurs, the bank recalculates the credit given in terms of the reduced water supply. In this way drought is not only a natural catastrophe but becomes a definitive economic one as well. The moment that working capital for pumps, diesel and repairs is at its most urgent, there is suddenly not a cent available. This might be logical in terms of risk avoidance as defined by the bank, but for agricultural production subjected to the vagaries of nature it is a catastrophe.

There were two feasible solutions which had been talked of for the past eighty years. A partial solution would be to build reservoirs. The definitive solution would be to tap off part of the water of the river Huancabamba and divert it to the rivers which flow to Alto Piura. At the moment, all this water is lost in the immense Amazon tributaries. There were in fact plans for a diversion. However, this was not to Alto Piura, which lies close by, but to the unpopulated pampas of Olmos, many hundreds of kilometers away. It is there that the Belaunde regime thought it could more easily achieve its economic and political objectives for agriculture (a "competitive agricultural sector" and "the stimulation of middle-sized farms"). In the thickly populated Alto Piura region, cooperatives and communities would undoubtedly present their own demands. Tonga was indeed right when he said that Holy Joseph was not the only one to blame for the drought.

On the July 21 there were several road blockades in Alto Piura. Thousands of cooperative members and chacreros were involved. The central demands were to divert the Huancabamba river to Alto Piura and to extend short-term credit to fight the drought. The second half of 1982 saw a period of escalating strife. Initially it was even amusing. The national press wrote that Belaunde was moved to tears by his visit to Alto Piura. Angry voices contended that he wept from the tear gas which the riot police used to keep the 10,000 demonstrators at bay. The demonstrators had marched on foot (across 80 kilometers of desert) from Alto Piura to the regional capital Piura. Then, at the end of 1982, fell the first dead. The year 1982 was dramatic in other respects as well. The drought, which for several years had been biting deeper, had become acute. The rivers had a minimum supply, and the bank therefore soon halved the amount of credit they had already promised and again tried to impose rotation on the cooperatives. Nine of the ten cooperatives in Alto Piura finally capitulated.

Luchadores continued to stand firm. Increasing pressure from the base and a short strike made it clear to the administration that rotation would not be contemplated. With men and might the wells and pumps were patched up. Fast growing food plants were sown to fatten livestock, and 15 hectares were given over to market-gardening in an attempt to both rationalize the use of scarce water and to provide a cheaper food supply in the area where hunger was beginning to bite. Part of the livestock was "half-illegally sold off," along with other mechanisms, in order to ensure the payment of wages. Part of the rice and sorghum harvests were saved, and the "surplus" labor was used for the large-scale recovery of the banana plants and the irrigation system. A number of typical incidents took place. Engineers from the bank came to inspect and encountered workers repairing fences around the sorghum fields. They complained about such unproductive work for which payment could not be permitted. Things almost exploded when one of the big firms arrived—first with a bailiff and later with a detachment of riot police—to reclaim ploughs and other machinery. The obligatory repayment for 1982 had not been made "because the bank didn't lend us anything," said the people of Luchadores. It is a typical occurrence. Bank and commercial companies mutually control an important part of the rolling stock of the cooperatives. The risks are practically one-sidedly off-loaded onto the cooperatives. The year 1983 indeed showed that Satan has two faces: the drought was suddenly broken, but villages and crops were carried away by the heavy downpours. However, Luchadores again set to work, almost in despair. A thousand day workers in addition to the regulars were set on because in such conditions it was impossible to use tractors. In spite of, or maybe because of the extremely unfavorable conditions, their efforts succeeded and the cooperative achieved a resounding profit for the first time.

Development of Yields

Hectare yields in Luchadores (given in Figure 4.3) show some notable developments, which deserve comment. Average hectare yields are a little higher, though not much, than in the hacienda period. In order to understand the significance of this, two other developments need to be taken into account: first, the situation in other cooperatives, and second, the decapitalization at the transfer from hacienda to cooperative. In practically all cooperatives, at both national and regional levels, hectare yields dropped during the land reform. Ever since there has been talk of progressive extensification.[5] This process can be verified in neighboring cooperatives. In Carrasco the cotton harvest dropped from 20 to 18 *carga* per hectare; in Alvaro Castillo and in Morropon and Franco it was even more. This makes the slight rise in Luchadores quite remarkable, particularly as Luchadores was so decapitalized during the formation period of the cooperative. Lamented Augusto Cruz,

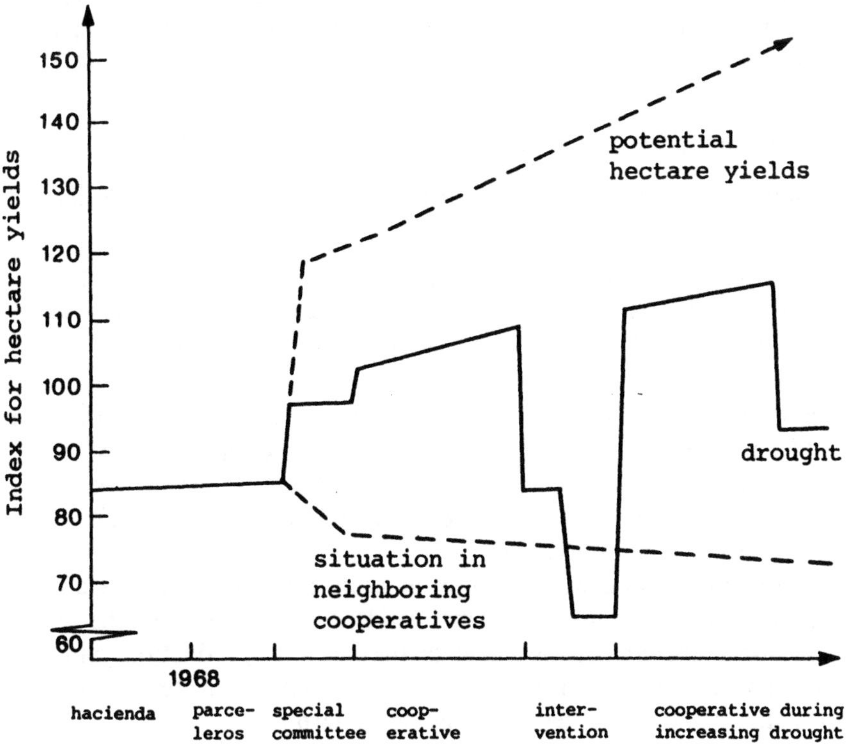

Figure 4.3 The development of hectare yields in Luchadores

> "Earlier in the hacienda, ay, there were machines everywhere, everything there was, from bulldozers to airplanes. While now we must wait endlessly . . . and by that time the plants have already suffered heavy damage."

Don Jaramillo added,

> "There used to be beautiful harvests. But it was also much easier, because the bosses had it easy, they had everything ready at hand. They had tractors, motors, credit, everything. . . . Everything was their own. That was very different from the way things are today. The cooperative hasn't even its own tractor; we have to hire everything and that's why the rest move ahead with their millions."

To these factors, which highlight how much more difficult for them circumstances are (their slight increase thus all the more an achievement), one can add still others, such as the deteriorating exchange relations, so often mentioned. They are keenly aware of the consequences of all this as several quotations indicate:

> "Look, if we had our own machines, and could carry out the necessary activities on time, and I'm not yet even talking about better leveling so that you could irrigate more carefully, then naturally production would sharply rise."
>
> "With other prices and costs, you could really help your plants."
>
> "Because with the present price of fertilizer one is all too likely to fertilize less."
>
> "Thus they have impoverished our land and deprived us of power. That shows that the gamonales are most decidedly not gone and slavery has not disappeared. Those from above are still sitting there cutting us out."

In short, taking changing circumstances into account, extensification would have been the logical thing to do. One sees it with most cooperatives in their dropping hectare yields. That this has not been the case in Luchadores shows that there are indeed forces at work which actively promote intensification.

Although one can characterize production in Luchadores as relatively intensive—i.e., in relation to global tendencies elsewhere in the countryside—one should not lose sight of the fact that production could be even more intensively carried out. Figure 4.3 also indicates the hectare yields that could be achieved in the short term.

Development of Total Production

Total production, in principle, is nothing more than the multiple of area sown by yields per hectare. Figure 4.4 gives the development of total production over time in Luchadores. Between the first and the second periods (i.e., from the hacienda period to that of the parceleros), a substantial increase occurred. A second mill was installed, as the capacity of the old mill was insufficient to cope with the greatly increased rice supply.

There is again a slight rise in total production under the Special Committee. That was in 1974, when the cooperative functioned for the first time as such. Total output reached was higher than it had ever been. In the year following (1975), production dropped somewhat. Generally speaking, sharp rises and falls in total production are likely to remain a characteristic, primarily because of the irregular water supply. However, the graph in Figure 4.4 also indicates another, equally important, factor behind the erratic progress of production: power relations between farmers and the state apparatus. For example, in 1977 and 1978, there was sufficient water, yet total production in Luchadores reached the lowest point in its history. Those were the years in which

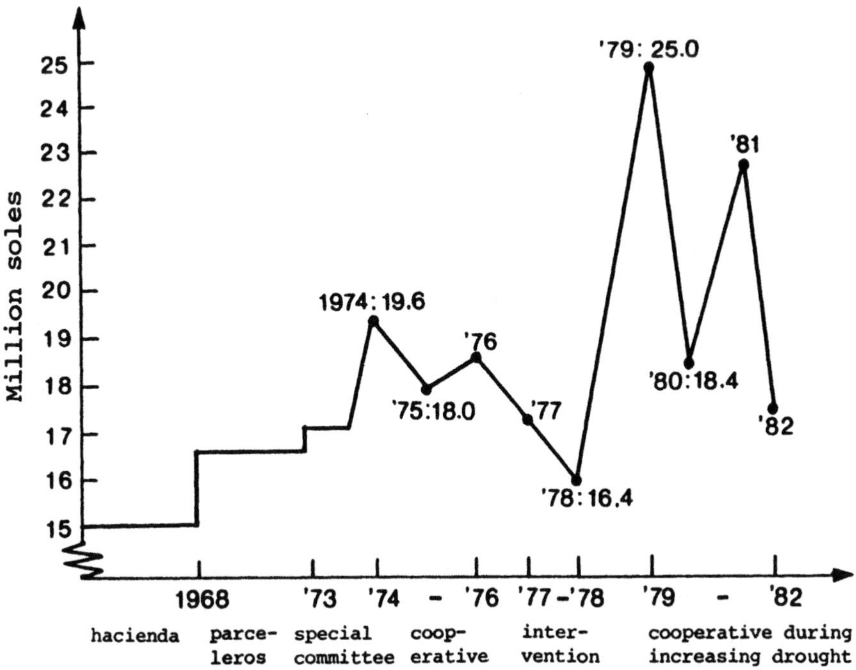

Figure 4.4 Development of total production in Luchadores (in soles of 1973)

control was taken from Luchadores and placed in the hands of the Junta Interventora.

When the direct producers again recovered some control of the production process after the strike of 1978, production rose once more in a significant way: 1979 is de facto the year known for the highest production. Thus, Figure 4.4 highlights an essential relationship: that control by the direct producers of the production process is a prerequisite for progressive agrarian development. Maybe this relationship appears all too evident, but it has certainly not lost its relevance, especially at a time when progressive incorporation and institutionalization is undermining this control.

Profit and Loss: A Historical Analysis

From an analysis of economic results through time (see Figure 4.5) four phenomena emerge:

- the cooperative is characterized by a chronic state of insolvency;
- the losses are in marked contrast to the situation current during hacienda days;

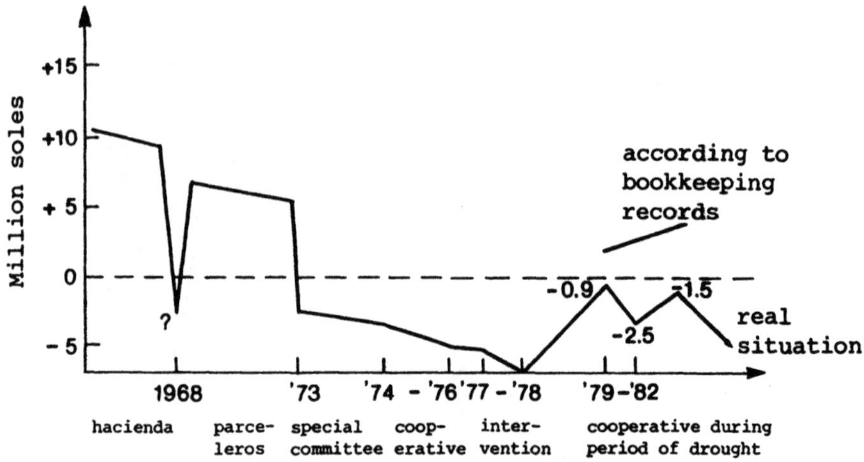

Figure 4.5 Losses and "profits" per year (in soles of 1973)

- the Junta Interventora also did not manage, even with the imposition of personnel rotation, to overcome this tendency;
- and finally, in the years 1979, 1980 and 1981, we have a curious phenomenon to contend with: real losses continue, but in the accounts, "profits" appear for the first time.

To begin with, to what are these continual losses attributable? Of course deterioration in exchange relationships makes it more difficult to obtain positive business results. But there is more to it than this. There are, after all, cooperatives that do make a profit. One might possibly hypothesize that this is precisely because they adapt themselves to prevailing price relationships (through among other things, extensifying production and reducing labor).[6] In Luchadores such an adaptation was ruled out, both because of the high employment level and because the social base continues to demand that "the land be properly worked." The price of maintaining such an intensive manner of production, which is inconsistent with the "economic context," appears to be the occurrence of persistent losses, losses which in their turn lead once more to strong decapitalization. Luchadores has not managed to keep the machine workshop in working order, let alone renew the machinery. A second consequence of incessant loss is progressive indebtedness. In the short term this often takes the following form: the costs for year t are paid off with the loans for year t+1, where inputs are financed with a bridging loan from the buyers that then again must be redeemed with the loan of year t+2. This leads to the question, up to what point is the cooperative doomed to take a loss in defense of its (relatively) intensive form of agricultural practice? Profit and loss then might not be the most adequate yardstick for evaluating the enterprise's economic efficiency—though it is clear that the institutional environment uses such a yardstick, and sooner or later the enterprise can go bankrupt from continual loss.

Another point that must be taken up here is the "double line" in the years 1979 to 1981: profits on the one side, losses on the other. While the cooperative actually continued to take a loss, a profit appeared in the annual report in the form of profit sharing. This was 30,000 soles for each member in 1979, 40,000 in 1980, and 50,000 in 1981. The misrepresentation of loss to profit rests on current accounting techniques. The specific position of the engineer-administrator of the time was crucial here. Not only did he teach the "office" and the accountant how to manipulate the figures, but he had an ulterior motive in doing so. As is practically everywhere the case, the engineer-administrator of a cooperative is awarded a hefty proportion of any profits. Alongside this is a second and maybe more significant "interest." The engineer occupied an important position in the cooperative, initially with the support of the administration. He presented himself as the person "who would from then on help the cooperative." At this ideological juncture the showing of "profits" naturally takes on strategic significance. The "profits" showed de facto that the enterprise was thriving—thanks to the engineer. It was how things were then perceived. In the introduction of the annual report of 1980 the administration wrote:

> These then brothers are the general themes; more detailed information will be furnished by our administrator, engineer Chiroque, who is directly and primarily responsible for our enterprise and whom we thank for all the help he has given us to enable us to carry out all our duties as well as possible. Again we thank our engineer for a good result in these difficult times. (Cooperativa, 1981:3)

The negative consequences of these (temporary) internal relationships and the corresponding "double bookkeeping" cannot be left unstated. In the first place, they obscure the real situation of the enterprise and the destructive effects of agrarian policy: profits as opium for the people. At the same time they encourage the development of the idea that an intensive form of production rests on an individual instead of on collective labor.

> "Let's face it, an enterprise such as ours needs a good agronomist, although he is not indispensable. Make no mistake, we are not impressed

by the engineer who comes playing the tough guy, who suggests that production is thriving only thanks to him. There is healthy production in proportion to things being properly done. That must each one do, as befits also the engineer. But then for that he needs to darn well get out of his jeep." (Sergeant Vilchez)

In 1981, the specific position of the engineer came under scrutiny. He had again spoken of profits. The bank, however, had sequestered 300 head of cattle. The two did not seem to match, even though the harvests in 1981 were good.

"It appeared a lie: in 1981 we produced well, good quality and a lot. And nevertheless we made a loss."

By chance, the rising discontent concerning the engineer-administrator coincided with a change in the law which specified that from then on each cooperative must appoint a manager whose role would go far beyond that allowed the former administrative officers and engineers. Thus confronted with a further expansion of the role of engineer-administrator, already considered problematic, Luchadores decided, not without fierce internal conflict, to elect one of their own members to the manager's job. They appointed Cuevita, a young man and a farm laborer like the others. He had followed an agricultural training program at the cooperative's expense. But, of course, this did not solve the problem of how to clarify the day-to-day running of the cooperative as an enterprise. This problem is made more complex by the fact that an important part of it is played out in Piura—often directly between the bank and commercial establishments with hardly a minimum of participation by cooperative members and administrators.

Trabajemos y Luchemos ("Let Us Work and Fight"): Towards a Progressive Intensification of Production

In the previous section we typified agricultural practice in Luchadores as relatively intensive. It is intensive in comparison to neighboring cooperatives, and also intensive in comparison to what was practiced in the earlier hacienda. The social struggle to enlarge and defend work opportunities was acknowledged to be the driving force behind this relatively intensive manner of producing. However, we describe it as "relatively" intensive because hectare yields are still far from what could, in practice, be achieved. This is the theme of this section: the possibility of raising employment and intensifying production a step further. The first link between employment and production in cooperative agricultural enterprises is, of course, formed through the goals to which production is geared. Santiago Rocca, in a recent study (1982b) of Peruvian sugar cooperatives, gives a clear theoretical overview of this:

> Theoretically, and assuming an identical allotment of nonhuman resources, "twin" enterprises facing the same market will employ different levels of labor if they have different objectives and goals. The enterprise that maximizes profits will not contract units of labor force unless the marginal product is equal to its price. . . . The enterprise that maximizes net income per worker—net income is understood as the residue left after paying the costs of non-labor factors—will employ workers up to the point where the marginal productivity of the last member does not increase net product or income per man further. The enterprise that maximizes the level of employment subject to a minimum level of attained income will hire labor up to the point where the costs of non-labor inputs or factors of production plus the total income accepted by its workers are at least equal to the value of production itself.

He illustrates this argument as follows:

> To allow a quick and simple comparison let L denote the number of workers; W the wage or income per worker; X the production function; P the competitive price of X; K the quantities of land, capital, and depreciation allowances; Pk the average weighted price of those payments and the subscripts a, c, and e refer to the net income per worker, profit and employment.

Figure 4.6 shows that the profit maximizer enterprise will produce at Xc and employ Lc workers, the net income per worker maximizer, Xa and La, and the employment maximizer Xe and Le. Assumed are (1) labor homogeneity, (2) perfect competitive factor markets, and (3) perfect competitive product markets. It is also assumed that social capital remains constant through time, that is, enterprises neither grow nor contract. The equilibrium of the employment maximizer enterprise has been drawn considering We as the minimum income per man accepted by the workforce.

Arrus (1974) uses linear programming to show how different goals, such as those put together above by Rocca, indeed lead to different production levels. Implicit in this whole discussion is the assumption of a constant function of production. The position taken up on this function shifts with a change in goal. Alongside this, a second dimension can be added to the problem: the "shift" of the production function itself as a consequence of explicitly orienting praxis towards it, or as

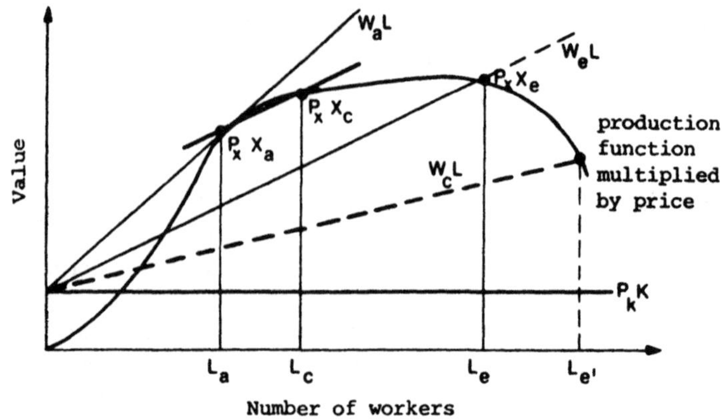

Figure 4.6

they say in Luchadores, "through more and better work." One of the necessary conditions for this is a lowering of the degree of incorporation and the often constrictive external prescriptions and sanctions over the way work must be done. This point will be explained in more detail in the following pages. However, even a consciously implemented structuring of labor that results de facto in intensive production levels is not something that is created from one day to the next—certainly not given the complex division of labor in a large organization such as Luchadores. In Luchadores, "social strife" means lowering the degree of dependence on external markets and institutions and structuring labor in such a way that an optimal use of productive potential becomes a reality. As Palacios explained:

> "If I say that we still have a long fight on our hands to be able to work more and work better, then I mean that as well as having a battle on our hands within the cooperative against sloppy example and working mindlessly. . . . I also mean that we must continue our fight with the bank and the commercial establishments for better prices. We need to fight for both, without the one the other makes no sense."

If we put the possibility of raising technical efficiency (through the progressive increase of collective craftsmanship) into the schema of Rocca, it is clear that:

1. hypothetically speaking, a substantially greater rise in employment is possible (L), from Le to Le3, without needing to lower wage levels, and
2. the struggle for employment shifts from being primarily a redistributive problem to a question that in essence arises out of how

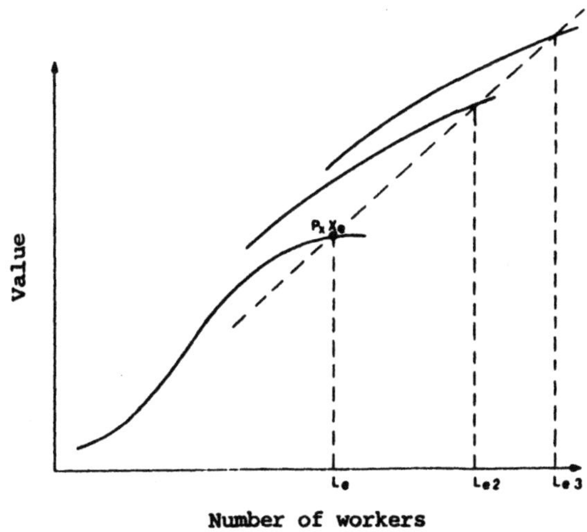

Figure 4.7

> to organize production and how to manage its development: it is not the distribution but the production of social wealth, represented in Figure 4.7 as a "shift" of the "function of production," that becomes the center of attention, debate and struggle.

Again, Leonides Palacios, secretary general of the union, formulated the problem of progressive intensification ("to work more and work better") in the following terms,

> "For us it is beyond dispute that you can work better with more labor. That we know from our own experience. In the rice fields, for example, you see it clear as day. Earlier when ox drivers still chased their teams through the mud, ah yes, then the land was outstandingly well worked. First a group with a deep biting plough went in, followed by others with the so-called 'flying machine' which leveled the ground superbly. Then came groups of workers who made the *pozas* [miniature dikes for precise regulation of water]. In this way you got ground that was really well-worked and you had no more trouble from weeds and obtained a high level of production. In the time of the parceleros we saw the same. Some applied the economy to their fields; others, on the contrary, worked the land as well as it was possible. Among those parceleros was a gringo, Soloman Whiskey or something, he may have been called something else, but that's what we called him. Ay, but that Whiskey knew how to set men to work, he let them 'chew and digest' the land until it was truly ripe. [*Machacar* is the local term for the first irrigation that makes the land soft for the plough]. Then it was ploughed and then 'crossed' [an expression which refers to a second ploughing directly at right angles to

the first], then came the harrow and then the 'equalizer' [for fine levelling] so that you got really well-prepared ground.

And although Whiskey practiced direct sowing [here concerned with rice, where generally transplanting is the method applied] he obtained a very healthy production, as high as 60 *fanegas*, simply because of such good ground work. You see, high production, simply because more work was put in.

In the cooperative you see similar examples. In 'the bath' [a section of Luchadores' land], 60 hectares were sown directly, but it went wrong because we were short of labor, the ground was not well worked, and everything is now overgrown. And that could have been avoided if more labor had been set on. You must give the land her due. If you work her badly then you get a persistently bad harvest. You can in a manner of speaking set ten men on a hectare and it will give you 45 fanegas. But you can also tackle the job with 15 men, and with five extra men you can work the land well and the production repays that. Then you harvest, through better work, 55–60 fanegas. More work gives higher production—if that wasn't so what would we then be on about. And more production is more income."

"More and better work" leads to a good harvest, which gives a higher collective income. The terms are different but the structure of the argument is virtually identical to that of the I-calculus which we laid out in totally different circumstances. In the same way, giving the land her due is seen as a prerequisite for working in such a way that a good harvest is achieved (impegno). Such a coincidence ought not to be surprising: it reflects the degree to which intensification, which depends on craftsmanship, embodies the ratio of farm labor.

In the course of 1981, the union decided to place itself at the head of the cooperative. One of the principles that union intervention made normative can be generally described in their slogan "Let us work and fight." Palacios had this to say:

"That principle was one of the most important decisions of the general meeting. And for us that combination of work and struggle is very logical. Without the inspiration of work and production you can't fight. We must also be able to see the results of our fight in more work and production. If that wasn't so, why would we still go on with the struggle? Yes, with all the comrades here, all these 'coconuts,' that principle has really caught on. It is widely discussed and there has been a marvelous upsurge of energy to be seen in the fields. . . .

"Working well in itself is a form of fighting, because it means that we achieve what the patron never managed and what the state, the new patron will never manage: to make the land bloom, so that there is work and food. That is the way it is. And to be able to work well the fight must be extended."

The way mapped out by the union "to work and fight" was not the result of just talking and then underwriting together an abstract principle. On the contrary, there were fierce discussions. But through such conflicts the union was able to develop its arguments further:

"Also here there were many who said that 'working well' was absolutely not to the advantage of the *campesinado* [farming class], that it served only the interests of the state, the new patron. The more we produce the more of our wealth the state will claim, the more exploit us. So why? We recognize that up till now it is largely the government that has controlled production, but suppose that we produce more in order to be able to demand more of our rights and have a better standard of living? Now, demands must be founded on something, you can't just arrive with nonsense. You must base your demands on production and returns. We can't go on producing at a loss, because how can you ask for a greater share of the profits if there is no profit! You have to have arguments; if there is nothing how can you claim more? And you know there is something else going on behind all this. The workers themselves lose their pride and their preparedness for action if there is no production and thus nothing to demand. If we have water, money, and the fields are well fertilized, yet we, through shortchanging ourselves as workers, lose . . . well, then is it clear that you are not in a position to argue."

From the gradual intensification achieved in Luchadores in the past ten years it was possible to make some future predictions. A certain phasing could be distinguished as feasible in the union's plan. In the first place, short-term hectare yields for the various crops could be substantially increased: working from data to be presented later, production could be raised by 30% and employment by 20%. Next, a between harvest could again be implemented, which would raise the intensity of land use another 30%. This system was practiced earlier in the cooperative but the decrease in membership led to its abandonment. If the water supply could be improved and membership could again be raised, this "second campaña" could be reinstated. Finally, with a permanently improved water supply it should be possible to substantially extend the area under cultivation as well as introduce more intensive crops and diversify production.

In the following analysis of the degree to which intensification is possible in practice, I limit myself to the first phase: the short term raising of hectare yields. The first "step" assumes:

Table 4.1. Present and Potential Hectare Yields

	Present level	Potential level	
Cotton	16 cargas/ha	20 cargas/ha	25%
Rice (by transplanting)	50 fanegas/ha	60 fanegas/ha	30%
Bananas	7,000 units/ha/month	25,000 units/ha/month	250%
Sorghum	5,000 kg/ha	7,000 kg/ha	40%

1. The availability of more resources with which to work.
2. Increasing control over and improvement of water distribution.
3. An effective raising of employment.
4. And, as they say in Luchadores, "a better will to work," by which they mean that the organization, implementation and evaluation of (collective) work must be explicitly geared to higher productive results.

The interrelations between these four elements speak for themselves: of course, without the necessary resources and better water control, increasing employment is senseless. And vice versa, if better water control and availability of more resources is not accompanied by an increase in employment, there can be no talk of intensification. The given style would simply be reproduced in more favorable circumstances so that profits would rise. Production, however, would remain on a similar level. Improved water control and greater availability of resources with which to work do not transform themselves automatically into intensification, let alone into the raising of employment. Social forces must consciously intervene in the production process before improved conditions actually result in the raising of employment and the intensification of production.

The question of hectare yields was discussed in detail with section heads and workers. Table 4.1 summarizes these discussions: it gives an estimate of the present yields (under normal conditions) and of attainable yields. The problems most mentioned as standing in the way of achieving potential yields turned understandably around water management, and especially around the scarcity (!) of labor and the shortage of working capital. Organization and motivation of "labor" were also problems which frequently came to the fore. Let us take rice cultivation as an example.

Elauterio, a young worker, is section head of "the bath," where rice is usually planted. In contrast to some, mostly the older heads of sections, he only became section head after the cooperative was in effect.

> "On the grounds of our experience we know that we can harvest 70 even 80 fanegas of unpeeled rice on the better pieces of land. However, because of careless and rough handling of the seed beds, and frequently through errors in transplanting, through not weeding and fertilizing on time, production turns out to be substantially lower. Usually we have to cope with one or more of these problems at the same time. . . . Sometimes we have to draw the water off the rice, because you know the water problem is not only a question of how much, but of coordination and of how accurate the irrigating is, and that can all count in production. If you must then still fertilize you burn a proportion of the plants. . . .
>
> "Sometimes also there are problems because there are not sufficient workers. All too often the bank does not give enough money on time, and so the best time to weed slips through our fingers. If you plant a lot of rice in the cooperative and you then have to weed with the couple of members that we have here, of course you are never finished. Then you begin in April and in May and the end is never in sight. At such times we need temporary workers, hundreds, but if the bank sits tight on the cents, then there is not much you can do except watch the crop wither. Look if it takes you one and a half months to weed instead of one then it means you are fifteen days behind schedule and those fifteen days wreak their havoc through the whole cycle. You fertilize fifteen days late, so develops a total delay, plants degenerate, they don't grow as they should. . . . If you have the means to work properly then plants get cared for, and at the most appropriate time. As farmers we must attune our work to the rhythm of the plants."

Rice cultivation is characterized by a virtually endless series of alternatives in the matter of applied technology, work methods, and costing, with each combination known for its specific consequences for production and employment (Angladette, 1966; Grillenzoni, 1974; Spijkers, 1983). One of the differences relevant to Luchadores is between direct sowing and transplanting by hand. Intensity can be noticeably different for each sowing system, but on the whole transplanting needs more labor and results in higher hectare yields. In Luchadores transplanting is usually opted for, although the ideal cannot always be realized. Elauterio said:

> "Theoretically everyone can and should transplant, at least when more labor is available in the enterprise. But in practice there are two prohibitive factors, the bank and water. We have already spoken about the bank. With regards to water, the seed beds must be made ready before the rains in the sierra begin, because once the water begins to flow down,

every minute counts. The transplanting must be done then, and therefore the seedlings must be ready. If the rains are late then you set the first wells full by pumping up water. And there the story begins again—diesel oil, the upkeep of wells and pumps, it all costs money. Expensive diesel and no credit. We do what we can with our own energy and sacrifice, but you know what I mean when I say they are a fine bunch of youngsters, those who run Peru . . . because we get upset if we have to use direct sowing again, in order to avoid such problems."

The problems sketched by Elauterio can be illustrated and deepened with the use of recent research. Figure 4.8 compares "transplanting" with direct sowing. Within transplanting, three grades of intensification are distinguished. The data are based on empirical research of the Ministry of Agriculture (MAA, 1978). The most intensive pattern is encountered among the small, unincorporated farms. Next are data concerning the middle-sized farms, where alongside family labor, wage labor plays an important role. Finally, there is the cooperative sector, where alongside transplanting (extensively done), direct planting also frequently occurs. The research cited is representative in the sense that the most typical small- and middle-sized farms and cooperatives were chosen.

Of course, this does not mean to say that the sequence given in Figure 4.8, which goes from intensive to extensive, coincides per se with the given organizational division. In principle a cooperative could realize yields similar to those of the *minifundia* section. Luchadores illustrates that possibility. It must be further emphasized that the image presented in Figure 4.8 is comparative in nature, thus not to be interpreted as the rendering of a historical process, for growth of harvest yields are possible even at the lowest levels of intensity. That will primarily be a "bought" growth, that is, a growth dependent on high-yielding varieties, new fertilizer combinations, and other techniques which result in an improved cost/benefit ratio. The most intensive production level, which rests on a high labor input, will experience a qualitatively different growth dynamic: growth will be produced in and through labor. Nevertheless, in terms of impact and outcomes, the sequence presented can be used to analyze decisions concerning rice cultivation and the conditions under which those decisions must be made.

First, it shows that with an increase in labor input per hectare, production also increases, and vice versa. Next, it shows that the costs of inputs and machinery rise by the degree to which labor input is lowered. Thus the question is one of a labor-substituting technology aimed primarily at cost reduction. Finally, it shows that transfer to

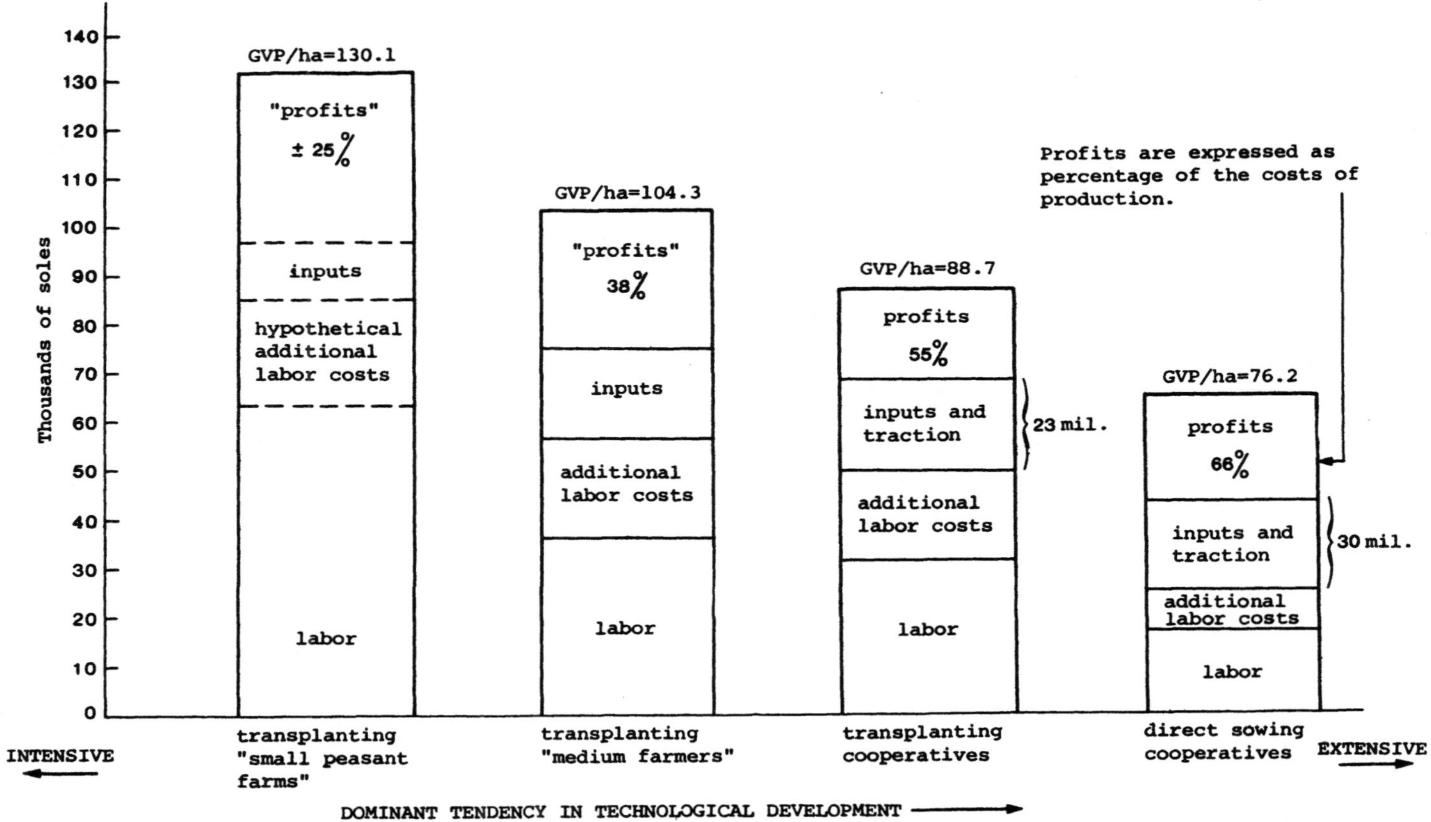

Figure 4.8 Intensity levels in rice production (MAF 1977) (1977 prices)

direct sowing is the logical completion of this process: costs per hectare are further reduced. However, this cost reduction is not neutral; it brings about a reduction in production. We can speak here of regressive substitution.

In practice such a substitution takes many forms: herbicides may have to be used because there is insufficient labor for weeding. This usually requires the use of aircraft, which can also be used to spread fertilizer, and even the sowing can be achieved this way. The same happens when the land is less adequately prepared or with direct sowing, as there is certainly more weed growth with direct sowing and the weeds get a headstart on the rice. That leads almost inevitably to the application of herbicides. With more weeds plant diseases develop and quickly propagate (which almost never occurs with transplanting) and this again leads to intervention. In summary, substitution of labor represents a vicious cycle with one step forcing the next: technology creates its own problems and subsequently requires its own solutions, which again create new problems (van der Ploeg, 1983). The same cycle is related to another significant element. Costs are not only lowered, but changes occur regarding structure: dependency on external markets and institutions increases. If one now starts from a scale of alternatives concerning rice cultivation, as illustrated in Figure 4.8, one might ask in what way the relevant aspects of relations "between town and countryside" condition and direct the way of farming.

To begin with, one must again point to the freezing or even reduction of permanent employment in large agrarian enterprises such as those created by the land reform. The available labor (the number of permanent workers) is usually only sufficient for extensive land use. The sequence given in Figure 4.8, localizes the effect of credit limits and risk avoidance—two characteristics of the agricultural bank's policy. Practically from the beginning credit is limited to the most extensive schemes, and only after "inspection" will the bank grant extended credit. If farmers want to intensify further then it is at their own risk. The risk is smallest with an extensive form of cultivation. Costs are low, the cost/benefit ratio relatively high.

The continuous deterioration of exchange relations (stagnating harvest prices, rising costs) will also work in favor of extensification, certainly if, through high incorporation, prices and costs penetrate to the heart of production.

It is not possible here to give a detailed description of banana, sorghum and cotton cultivation, but the conclusion remains the same, even though the practical problems of each crop are always different: yields can be significantly raised. The problems which prevent this are clearly recognized and have been defined in a way which points to their solution. The common denominator of the solutions presented can be described as a reduction of the level of incorporation and institutionalization; in short, the increasing of autonomy. Although in the quotations cited the emphasis lies primarily upon the capital market and agricultural bank, it should not be overlooked that, in principle, similar problems prevail in other markets and institutions—such as in the market for machine services and the "pools" that operate there; in the cattle market and with the traders; in the markets for seed, fertilizer and other inputs and technology; and with the commercial companies.

However, in Luchadores people are acutely aware that merely changing the relations between an enterprise and the external markets and institutions is not in itself sufficient. The social contradictions within an enterprise as a whole are also of great importance. Internal relationships and factors may transform and strengthen the working of external factors (i.e., make them a reality within the enterprise), or they may hinder just such a transformation.

"One has to spend" versus "One has to apply economics" are two criteria which frequently clash in the field. "Applying the economy" is the norm reflecting bank policy. As far as the cooperative is concerned, it refers to the need to lower costs (to adjust the way one produces to the scarce availability of "means") and pursue an extensive form of land use. The opposite norm, generally symbolized in the expression "It is right to spend," leans towards an intensive form of agriculture. The situation in Luchadores is such that the two norms frequently clash. They are the basis of arguments used by different groups to criticize each other and even fight.

Such internal strife occurs on a number of levels. Production is organized in sections, each with a chief. He decides—within certain limits—how the work will be done. There are not only differences of opinion between chiefs, but these are constantly discussed and debated. In turn the chiefs are in direct contact with the executive and the managers. The latter form the communication channel with the bank. So it is understandable that between chiefs and executives (plus manager) debates and conflicts also often occur. Finally, there are the workers. If in the fields they don't agree with the decision of the chiefs, then it can get rough. The frequent general meetings are another platform for airing views.

On all these levels the criteria stated play an important role in the daily "game." Let us take a closer look at some typical episodes. An older chief, Don Jaramillo, said,

> "For me it's much better to aim for a somewhat lower production and make the costs lower, for the profits are the same or maybe higher. So I

think you should control costs, you can't step outside the economy. If a certain activity is normally done with six men to the hectare, and I can do it with four men, then I have saved two days wages. That's why I say the chief should apply the economy."

One sees that the argument turns largely on profit. The relation of that opinion with the past is clear. Lazaro said,

"If I was the boss here, then I know what I would do. I would plough under all the rice and sow maize. You can't produce just to keep all the workers on, certainly not if you are losing money by it."

By no means, however, does everyone see "application of the economy" as a convincing norm. That depends on whether the notion of "better work" has a place as an argument in the formal schema to which "application of the economy" refers. "More work," in a quantitative sense—or increased labor input—is an indispensable precondition for "better work" (in a qualitative sense), as we shall demonstrate later. If one separates "more work" from the notion of "better work" then it quickly seems absurd.

The question is also about the applicability of similar reasoning to, say, a cooperative in which wage costs from the beginning are permanent costs, or to a still strongly artisanal agricultural production process where indeed more work is often the precondition for better work. However, production cooperatives function in a structure and culture which together can be described as a capitalist environment. The repeating of norms which reflect a specific capitalist ratio is also then to some extent unavoidable. In the diffusion of such norms, the administrator-engineers play an important role:

"His concepts [those of the last engineer] on how to produce did not seem right to us. He said, same as all the other engineers we've had here, that working properly is a luxury, something that we shouldn't allow." (Norberto Cruz)

But the same problem partly repeats itself for the present manager chosen by the union, Cuevita. Whoever walks with him through the fields hears a hundred times a day from workers and section heads, "Hey Cuevita old chap, in God's name send me more men, send me the caterpillar and not only with the harrow but with the plough, damn it, it's time to get this land here well in hand and see that she recuperates strength." "Cuevita, see to it that we get cassava stalks, and send me ten Christians along with them, this is no way to work." The general opinion is that section heads have to demand with "their fists on the table," because they are also under pressure. They know that they are judged by the production of their section:

> "Yes, if production is beautiful then you're a good chief, but if it's overgrown, miserable and badly cared for then you get short shrift from these fellows here. Then it's impossible to tell a team of workers anything, quite rightly you hear that you understand nothing, the fields say it all. That's why a section head has to make demands, otherwise they tell you your brains are growing tired."

Thus the social base in Luchadores exerts pressure for intensification in all ways—which forces executives and manager into continual conflicts with the bank and other institutions to find the "means" wherever possible.

Let us now look at the opposing norm, "that one must spend":

> "When it's a matter of the best way of working, you are always left with the same situation, that with low expenditure you end up with low production and with high expenditure you get higher production. Every system has its own justifications. But for me it's clear that we must aim here for the highest production and so we need to spend. . . . In that way you get high employment and that is what most interests us. Creating work for the community is one of our most important tasks." (Elauterio)

Work is not just considered a cost. Creating employment is an "important task." But deviation from the current capitalist perception of production and labor goes further:

> "With sufficient men you can work properly, and if you work better you're surer of production; the risks are smaller. With transplanting it is rare to get plagues, but with direct planting the danger is great. It's just the same with weeding. With well-cared-for plants it's impossible to lose."

It is important to point out that those who opt for spending do not reject the notion of "watching the cents." The concept is put into practice, however, in an entirely different way. It is sought not in a reduction of labor input, but primarily by taking a stand against various sorts of "pillage":

> "Last year we had a fine rice harvest, but because our mill has been closed by the state, it had to be transported to Piura to the Vegas mill. Can you imagine it, instead of opening our own mill or going to Pabur (a neighboring cooperative), they give it to a gamonal. . . . The sorghum

was beautiful but the misery is that the threshing machines of the private owners—just for threshing—appropriated the biggest part of the wealth. Look lets be careful of their 'economy'. Why is there no effort devoted to doing up old threshing machines, or why no deal closed with a friendly cooperative." (Tonder, former president)

Discussions about such norms are not the only ones which are chewed over and circulated in Luchadores. Equally important is the debate over how one should work: with *voluntad* (with a will), or by skiving? The content of both concepts is no mystery: "the skivers, that's common sense, they are the ones who don't work properly" (Palacios).

In describing how men should work, the word "voluntad" (literally, "to consciously want something") constantly crops up.

"Work must be done with voluntad, that is it must be done with dedication and carefully. Otherwise you lose the plants in the fields." (Don Jaramillo)

"To work properly, say with voluntad, you need experience, you need to understand and know everything, and above all feel the responsibility. That applies as much to workers as to chiefs. If experience, knowledge and understanding are missing then work cannot be done with voluntad. But there's something else as well. You must feel affection for the plants; there has to be an interest, an importance given to well-cared-for production. If a withered plant gives you no pain, how can you then speak of voluntad." (Renteria, Velasquez, Octavio)

"Look, it is very simple. If you wander now through Bejucal [a section at that moment covered with sorghum], then all too often you see *focos* [patches of poor growth, clearly visible among sorghum by a difference in color and height]. Well if you see a foco you know that place was not worked with voluntad." (Mamberto Farfan, ox driver and former president of the cooperative)

Thus in Luchadores one sees that working with voluntad means "working properly," an activity orientated to intensive production. Working with voluntad, however, is also important at the individual level: a worker must put his back into it and not skimp or neglect, but also he must not work "roughly." In part this is a question of labor organization:

"A job must not be such that a worker is forced to rush it. With weeding you cannot expect more from a man than 4 pozas per person per day. And then it should be lower if the place is full of weeds; otherwise the poor results are irrevocable, and we can't have that."

For this some form of check or control is necessary:

> "If a well-regulated task is still messed about with, I mean if there is plenty of time for the work but someone hasn't the voluntad to do it properly, then the union is informed and that behavior is criticized and sanctioned." (Velasquez)

Another issue implied in the notion of working with voluntad is the length of the working day. By law this is limited to eight hours, but—as in other cooperatives—a workering day is often shorter than that.

> "Yes, now we have tasks lasting five hours. Because of the food situation in the fields, there is no way you could work for eight hours at a stretch. . . . You can't go beyond one in the afternoon, because if you make the working day longer then God makes you pay."

In comparison with hacienda days, the length and rhythm of the working day has dropped. Then there was the threat of immediate dismissal:

> "You had to work, even if you broke your back, because you were permanently watched and if anything was lacking or not to the liking of the patron you didn't get a cent, or you were promptly out." (Augusto Cruz)

And it is better that things are now as they are. It is better to work on the basis of voluntad and not be driven by threats of immediate loss of earnings, physical punishment or dismissal. And I repeat the general opinion of Luchadores "that if work is done with voluntad then in five hours a good deal of work is accomplished." The problem lies elsewhere—when the hours put in are not done with voluntad but are performed with disinterest, even neglect. The problem exists, and has manifested itself in the course of Luchadores' long history in diverse ways. It is closely linked with people's feelings of well-being. In hacienda days a client system dominated. With sickness, and also at feast times, the patron always lent a little something. Nowadays, in a cooperative almost entirely decapitalized, with chronic financial problems, with a system of "public health" which "only makes you sicker," and with mutual relief funds which time and again fall short because of rampant inflation, such "social assistance" is almost non-existent. The consequences of this lack are clearly felt. As one worker Roque, an irrigator, lamented:

"They say the cooperative is in crisis. There were no problems earlier—the cooperative would always lend you something if you were in trouble—but now there is no money, not a cent. Thus the helping hand when you need it is no longer there. People get bitter, and then voluntad flies to the winds."

Voluntad is also closely linked to how internal relations and other matters are run. The year 1981 played an important role in this respect. It was a year of healthy production and yet the cooperative still took a loss. Suspicions immediately arose that something was amiss somewhere, that corruption existed, or that sections other than one's own were being neglectful. The loss must come from somewhere. If there is no clear awareness of the economic problems and the manner in which state agrarian policy influences events in the cooperative, then mistrust is likely to mount: "There must be a team of skivers here." More serious still, as one worker described it,

"If you once think that, then your own voluntad declines, you ask yourself, why should I go to with such a will, when others benefit by skiving?"

This problem is to be found in every production cooperative. There is no immediate or visible relation between individual input, on the one hand, and the sharing of the benefits of such effort, on the other. In the small-scale farmer sector, a direct relationship between effort and reward does exist (to a certain degree by the nature of things), but there can be no question of such a *direct* relationship in the cooperatives.

There are a number of conditions which demonstrably effect motivation. One is control, which should be two-sided. Not only must workers sanction indifference through their own mechanisms of control—at present via the union structure—but the workers' control of all aspects of the enterprise itself needs to be raised. There is too often a shortage of information and insight—factors which escalate conflicts needlessly. That was at work, for example, in the course of 1982, when the bank suddenly cut back on credit because of the drought. Searching for other funds took time and led to several weeks delay in payment, which led Oyala to confide that,

"At this moment the problem is again rather urgent. There are at present workers who are saying to the others, fellows, why work so hard, you're working for nothing, free, they are not paying you, why wear yourself out. . . . You see, they don't fully understand the calamitous problems that the bank has again brought down on our backs. That's why it rankles

in the fields. When I hear something like that, I am furious. Although my mother taught me otherwise, I thunder it straight out. Damn it what are you doing here then man? Stay at home, or rather go carry firewood for your wife, if you think you're working for nothing. Clear out instead of infecting others."

However, in order to be able to inform others, an overall picture is necessary, but from the beginning one is up against the triangle of bank, commercial institutions, and the administration of the cooperative—a triangle within which many essential decisions are made, but which at the same time forms a labyrinth in which every form of corruption is rampant and, of course, clandestine. A second problem relating to control and having an overview is simply that a complex organization of 400 men does not easily lend itself to operable social control.

In 1982 two suggestions were put forward for overcoming this problem. One of them in fact came close to splitting the cooperative into smaller sections with each working on its own account. For each section a group of around forty to fifty permanent workers was envisaged, an overseeable whole where better social control could be practiced. The second suggestion, which came from the union, was to work in smaller brigades but to keep the enterprise as a whole intact. At the same time to improve the union's system of control, "there should be more mutual criticism," said the union bulletin. It is not appropriate here to go into the technical aspects of both propositions. What is important is that "skiving" and lack of motivation for work were seen as a serious but resolvable problem.

The importance of these internal discussions and conflicts can be seen in relation to the mechanization issue, for if there is no voluntad in the work, a regressive process begins in which mechanization plays a typical role. The engineer Chiroque's view was that "If people won't work then we set machines on." And that's what happened.

Machinery could be used at numerous points in the production process, and the more this was obvious the worse became the quality of work (the more voluntad disappeared, according to the people of Luchadores).

"In fertilizing cotton, a donkey would first be used to open up a furrow between the plants, then we spread the stuff and a second donkey was used to close the furrow. If that was well done then no cotton plants would be damaged by it, that is so much better than when you go through with a machine. It is also much cheaper. But if it's not done carefully then a tractor is more attractive. Then at least everything is done in a uniform way."

The artisanal nature of the process of agrarian production implies that through careful work satisfactory productive results are achieved. If the conscious will for this is absent then industrialization of the production process is inevitable, maybe even preferable, but the possibilities for development contained in craftsmanship are lost.

The Luchadores Plan

As mentioned, the union decided in 1981 to place itself at the head of the cooperative. When this was put into effect at the end of 1981, a plan was formulated for enterprise development which committed itself to the interests and needs of the rural poor. "We decided," said Palacios,

> "that from then on, we in the union would not limit ourselves to a purely defensive position, but would become active in guiding and leading the cooperative. We wanted to go further than just defend wages and conditions. It had to be done in ways other than just placing our men in administrative posts—such posts are likely to disappear before you know it. We needed a clear work plan that translated our principles into lines of practical action and permanent discussion and coordination between the union and cooperative executives."

Practically all the problems discussed above came to play a role in one way or another in the emergence of this decision. One could say that a high level of incorporation and marked institutionalization as well as internal "reflections" had carried the cooperative as an enterprise further from what appeared to be desirable: raising and defending employment and intensifying production. The union plan was a way of redressing the balance. It can be summarized in a few words as a meticulous, step-by-step description of how different crops can be intensified. It is at the same time a detailed interpretation of how operational costs can be lowered. This meticulousness is linked, I believe to one of the aspects of Luchadores which continues to fascinate me—the degree to which not only the workers but also the administrators behaviour remained, in a certain sense, that of small farmers: very precise, with an almost microscopic way of looking at matters combined with stubbornness and risk avoidance. Such a stand can have its disadvantages, but it can also lead to an extremely realistic and therefore powerful disposition. The plan reflects deeply this spirit. Past and present running of affairs were, as it were, put under the microscope, and the possibility of making small shifts which together could have a meaningful impact was carefully scrutinized. Let me illustrate this approach by describing the plan to reduce costs, not only because it shows in practice the possibilities for intensification, but also because the intended reduction was conceptualized as a way to increase their autonomy. The plan discusses the cooperative's dependency on machine pools and believes the answer to lie in rebuilding the workshops (started in 1982), in extending technical in-service training for mechanics, in repairing existing machinery, and in devising a carefully thought-out policy regarding the purchase of new machinery. Likewise, dependency on dealers who monopolize the circulation and fattening of cattle could be reduced by developing their own fattening centre and a production plant for milk. With regard to the commercial companies, one of the first steps to be taken was to work with other cooperatives and seek out independent agronomic advice.

Dependency on the bank was a much more difficult matter. The core of the solution here was thought to be the resuscitation of the banana and citrus groves. Banana production could reasonably quickly supply a regular cash-flow of between 2 and 8 million soles a month. That was not only an important part of the monthly wage bill, but it was money that was completely controlled by the cooperative. The supplementing of short-falls in bank credit or the financing of projects falling outside bank criteria would then be possible. The problem was naturally how to get started. Banks do not lend money for such projects. As the investment is principally one of labor, extra "sacrifice" seemed to be the only approach to the problem. The list could be substantially lengthened, but the illustrations chosen are sufficient to show that the common denominator of the plan was to reduce the degree of incorporation and institutionalization in order to limit costs and make it possible to work and produce more. In economic terms that would lead to higher incomes and relatively lower costs. Employment would rise and agrarian production would be intensified. The driving power behind it all was the desire for progress that would be felt in the fields, for:

> "Only the poor of the fields can achieve that, neither the gamonales nor the new patron, the state, have ever been able to get the fields to bloom."

The feasibility of the union plan was researched in a separate study by Bolhuis (1982). His analysis showed that holding all other variables constant, positive results could be attained through the proposed intensification and specific reductions. From a loss of 80.4 million soles in 1980 they could have made a slight profit of 4 million soles. In 1981 the shift would have been greater, the net surplus could have risen to 30 million soles (see Bolhuis and van der Ploeg, 1985:378–379, for a summary). One could argue that recalculating in retrospect is all too easy and naturally does not form conclusive proof. But as Thomas and Logan rightly assert, cooperative experiences must be seen and understood not so much as a static model but as praxis. That applies also to Luchadores. In 1980 and 1981 no positive results were obtained despite the favorable circumstances. That and the conviction that it could have been possible (as Bolhuis's 1982 study suggests) led to radical changes: the union placed itself at the head of the cooperative. The plan for gradual intensification and for restructuring costs was formulated in terms of greater autonomy from markets and market agencies. New conditions were thus created for getting better results, and in 1983 and 1984, which fall outside the scope of this study, Luchadores succeeded in making a profit.

If we now consider the experience of Luchadores as praxis, as a process where problems and prospects can be consciously anticipated and responded to (the plan is in the last analysis an indication par excellence of this), then it seems to me that the discussion concerning the ultimate viability of the plan hangs on two questions: is progressive intensification possible under the adverse conditions previously outlined? And is a production cooperative an adequate structure for this? Why not? would be my answer to the first question. Why should positive results not be obtainable with intensification? The argument usually used against this is "the law of diminishing returns." Maybe as a didactic principle it is useful, but as starting point for enterprise and sector planning this "law" is entirely inadequate. The assumption on which the construction of a production function rests (i.e., the variation of one or two inputs by holding constant the rest) is entirely strange to agricultural development as a process. Lenin once said—in reply to criticisms on Kautsky's book "Agrarfrage"—"thus the 'law of diminishing returns' does not at all apply to cases in which technology is progressing and methods of production are changing; it has only an extremely relative and restricted application to conditions in which technology remains unchanged. . . . Indeed, the very term 'additional investments of labor'. . . *presupposes* changes in the methods of production, reforms in technique" (Lenin, vol. 5:109, 110).

Recent developments in theoretical agronomy (De Wit, 1981; Rabbinge, 1979:149, and van Heemst et al., 1983) have definitively buried the law of diminishing returns. It is precisely the opposite that is now most frequently discussed, i.e., "the law of increasing returns" (also termed the "Liebig-function"). This phenomenon occurs if all the growth factors are changed at the same time in a well-coordinated way: then increasing returns emerge. The many practical forms and dimensions that Luchadores gives to the concept "working better" is an outstanding example of this principle.

In agro-economic research such as that of Ishikawa (1981), the possibility of increasing returns is advanced as a real option. Surprisingly enough, the law of diminishing returns is still echoing around sociological studies of agricultural development.[7] But that aside, in the first instance the prospect for further intensification is a theoretical one. Specific social relations (at whatever level) can effectively stimulate but also block such possibilities. A certain incompatibility can arise between further intensification and the relations under which it has to be realized. The question then arises as to whether intensification and its interaction with existing economic and institutional relations can be so structured that any incompatibility that arises can be remedied.

That leads to the second question: is the cooperative an adequate organizational structure for progressive intensification? Farm labor in Luchadores is mainly artisan labor. High productivity and economic efficiency are only reached if the many tasks which constitute agrarian production are conscientiously coordinated and expertly carried out. This, according to Nove (1983:88), "calls for flexibility, adaptability and initiative, *at the grass roots*. . . . Commitment of the peasantry to their work is essential." From his critical review of the Russian, Polish and the more positive Hungarian experiences, Nove points out the following conditions for achieving such a commitment. First, individual incentives are essential "to stimulate peasant interest in the outcome of the work they do" (1983:90). That is not a simple matter. If once "the peasant love of the land," so narrowly written off by de Janvry (1981) as a petty bourgeois obsession, is eliminated, then other material incentives often seem counterproductive. Second, Nove touches upon the community level: "Remedies, to be effective, must surely enlarge the decision-making functions of those on the spot, both farm management and the sub-units within the farm" (1983:132, 90). And this leads to a third level, that of the interrelations between the enterprise and the economic-institutional environment. "Operational autonomy of farms," that is to say, "the freedom to chose what to produce and what to sell, the freedom to purchase inputs and a much greater flexibility in internal organization and in organizing the work of peasant members is necessary" (1983:132, 90). Reviewing the Luchadores experience, it can be said that a "commitment at grass roots" is possible but that at the same time a number of strategic limitations operate. The practical possibility of commitment, of voluntad, must be primarily sought in the interweaving of social and economic spheres, in the overlapping of enterprise and community. The enterprise is not isolated from the rest of life for the people of Luchadores.

The problem of commitment cannot therefore be resolved solely within the economic sphere as Fals Borda (1970), and Galjart (1981), among others, all too often assume. The social sphere and the economic sphere are reflections of each other, at least in a cooperative setting such as Luchadores. A skiver might obtain advantages from a wage that remains the same, but the price that he inevitably pays for this is diminished prestige in the social sphere, not only within the enterprise, but in the village, in the bar, maybe in his own family and in his many chance encounters. In moments when he needs help he will be seen as a skiver and treated accordingly. In synthesis, if we define the relations between work input and expected wage as a social relation of exchange, then the definition of benefits and costs encompasses both the economic and the social sphere. And that gives the problem other contours than a purely economic analysis would suggest, certainly if voluntad prevails as an important value.

Thus a first sociological limitation for achieving higher commitment is indicated: "working well" as a value must be shared by a majority in the community, and at the same time, the social networks which bind the community must be solid enough for social control to be effective. Further, and presumably more essential, there needs to be a visible link between input and benefits, not only at an individual level but in general. The fluctuation of voluntad, or commitment, over the years in Luchadores, highlights the importance of such a visible link—and of open administration and management practices both within the internal and external relations of the enterprise.

The argument of whether the cooperative is a valid form can be taken a step further by asking what meaning the development of an intensive form of agriculture, geared to high employment, has for the cooperative. Fanfani has done exciting, though as yet unpublished, research in Italy on an agricultural cooperative in Ravenna. The initial aim was to plan production for the coming years using linear programming. Naturally that required specifying the goal function. It appeared in subsequent meetings that the members wanted to maximize employment, but the leader and technicians wanted primarily to raise profits for investment in future development. Fanfani and his team used linear programming to calculate the effects of these opposing opinions and of several gradations in between. The result showed that with a 10% reduction in profits, employment could be doubled. Similar studies were carried out in Peru (Arrus, 1974). Dropping the technical details and complications of such studies, we want here to stress that they especially demonstrate that the productive structure of an enterprise is not a simple derivative of prevailing economic relations. Within similar relations there are always many solutions possible in interaction with the economic-institutional environment: the "goals" which normatize the setting up and organization of production are decisive. There is a second point: the pressure to raise employment (and thus the need to further intensify) is almost always found among the rank and file of the cooperatives. In an empirical study (Ochoa, 1980) of eighty-seven cooperatives, it appears that also in Peru there is a significant difference between members and leaders on the question of goals. The workers are usually of the opinion "that there are too few members in the cooperative" and "that there are too many" managers.

If we consider cooperative experience as praxis, then the social forces which determine its forms and development are crucially important. Hence, where the fight for employment is experienced as urgent by a large part of the marginalized rural population, a cooperative form of organization can be a valid mechanism for the development of agriculture as well as for the integration of economic and political power. The principle of self-management as a link between the fight for employment and development of the enterprise can result in an intensive style of agriculture geared to a high and rising input of labor.

In the theory of self-management great significance is attributed to aspects of employment and to mechanisms which substitute for the traditional labor market. With respect to Mondragon, Thomas and Logan conclude that "in the contract of association the 'open door' principle—the preparedness to create employment—is very important." Two matters should be stressed here. In industrial production such as that found in Mondragon, expansion of employment is first possible after making (relatively sizeable) investments in new capital goods. Savings or loans are thus necessary. Agrarian production tends to be different, especially where it is artisanal. The intensification perspective creates the possibility of directly raising employment. Of course investments are needed, but they are mostly lower than in industry and are largely a question of investments in labor.

A second matter of essential importance is how and in what way the "boundaries" of the cooperative are to be defined. If these boundaries are closed, i.e., if the numbers are to be limited to members of the original group, further intensification geared to the raising of employment is not very likely. Then, expansion according to extensification, the E-pattern, is the most opportune and evident choice. This means again that the formal cooperative model is not as important for understanding actual developments as are the nature and dynamics of the social forces on which the cooperative depends, as well as the "operational autonomy" (Nove) which the cooperative manages to weave in relation to external economic and institutional forces.

Social Struggle and Autonomy as Prerequisites for Intensification: A Comparative Analysis

In the previous sections, two factors emerged as decisive for intensity of soil use in the cooperative context: social struggle and autonomy. We define here social struggle as the effort to defend and raise employment in the fields, which necessarily assumes an intensification of the production process. The ideological pivot of this social effort is the collective conviction (a collectively carried I-option) that the lands can and should be better worked and that the returns will increase. For this to materialize, ideas and collective convictions need a social base which fights for their realization: that is precisely where the strategic importance of the union in Luchadores is rooted. This union has never limited the scope of its action or its horizons to the permanent workers of the enterprise. It encompasses, rather, "the whole of the rural poor"—a whole which is maybe not always exactly definable but which nevertheless prevails in the fields as reality. The history of the union provides an outstanding illustration. Take, for example, the invasion of January 2, 1973, in which everyone, workers and non-workers, took part. In 1981 the union led a strike of temporary workers who at that time were working on the cotton harvest—thus a strike against their own cooperative. A shock for the administrator-engineer of the day, but for the permanent workers a normal affair:

> "They are poor farmers after all, just like us, perhaps even more wronged than us. . . . If they have longings, then they have a right to fight for them. The blame lies simply with the opposing party. They shouldn't be so stubborn."

The "communal land" founded in 1982 through pressure from the union is a similar illustration of its broad identity.[8] The importance of this broad social base cannot be stressed enough: encompassing a set of interests broader than that defined by the cooperative structure as such meant a breakthrough in the segmentation of the rural people as a political block that had been achieved by the land reform.

Next in importance after social struggle in determining the style of agriculture practiced is the degree of autonomy from markets and institutions (i.e., the inverse of degree of incorporation and institutionalization). Just as the degree of social strife can differ from cooperative to cooperative, so can the degree of autonomy. Diverse causes play a role here:

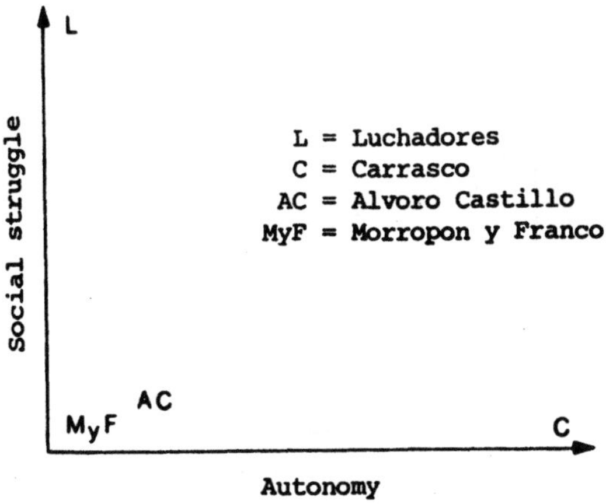

Figure 4.9

1. some cooperatives escaped the decapitalization which took place before and during the reform;
2. some cooperatives have a regular income which cannot be controlled by the bank or state, possibly from the sale of bananas or vegetables, etc., which can supply permanent economic surplus for autonomous use;
3. some cooperatives have such a favorable man/land ratio that their costs are low enough to produce continual profits which allow a high degree of self-financing.

If cooperatives differ in terms of degree of social strife as well as degree of autonomy, then the effects of both factors on the intensity of agricultural practice can be examined through comparative research. This we have done by taking four cooperatives—Luchadores, Carrasco, Morropon and Franco, and Alvaro Castillo. The global position of these four cooperatives in terms of social conflict and autonomy is represented in Figure 4.9. The distribution of the cooperatives over the dimensions discussed follows mainly qualitative arguments, although it is possible to specify "autonomy" quantitatively. I will highlight here the case of Carrasco, which contrasts most markedly with Luchadores (no social conflict and a high degree of autonomy). The autonomy of Carrasco is significantly greater than that of the other cooperatives for the following reasons:

1. Of the cultivated area (643 ha) more than a third (243 ha) is devoted to banana cultivation, which in mid-1982 gave an income of around 2 million soles a week and paid between 60 and 80%

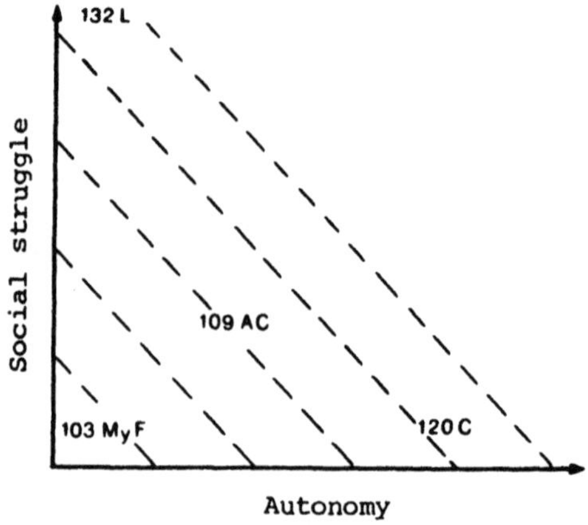

Figure 4.10 Average intensity of farming per cooperative

of the total wage bill. This implies that problems with the bank, although not absent, were nevertheless significantly reduced.

2. The cooperative was formed in 1975, and thus escaped the decapitalization that had occurred in other cooperatives.
3. Finally, the man/land ratio of four and a half hectares of cultivated area per man was the lowest in the area. This made high mechanization possible and kept wage costs relatively low.

These factors ensured that Carrasco produced permanent profits. In 1981 this was 15 million soles, part of which was distributed and part reinvested. However, the relative autonomy enjoyed by Carrasco was not used to intensify production further. Carrasco lacked the social basis to push for such an intensification. There was no talk of social struggle in Carrasco. As far as there was struggle, it was simply about the defending of particular interests, interests which the workers of Carrasco did not associate with but saw as rather separate from the other rural poor.

Intensity of agriculture can be ascertained by two indicators: intensity of the cropping pattern and intensity of the form of production. With regards to cropping pattern, the highest percentage intensively cultivated is in Luchadores, and the highest extensive cultivation is in Carrasco. The greatest and most interesting differences emerge when hectare yields are compared. If we convert the differences in cropping patterns and yields into indices and we subsequently calculate the average per cooperative, then the picture given in Figure 4.10 emerges. Although no further statistical meaning can be attributed to this picture, the interrelations in the graph bear out the hypothesis that social struggle and

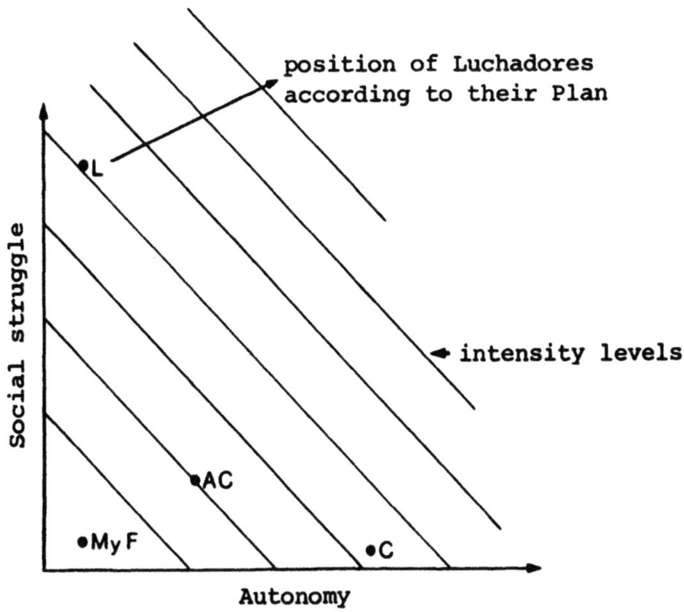

Figure 4.11

autonomy are indeed important forces behind intensity of agriculture in the cooperative sector. What is absent in the graph is the combination of autonomy and strife. And this is where we again encounter the relevance of the Luchadores plan: both elements are central as action lines for practice, action lines which must both result in continuous intensification. Translated in graphic terms, we find the relationships shown in Figure 4.11. One can see here the meaning of the plan as it developed in the course of 1981 and 1982. Putting the plan "to fight and work" into practice as a farming response could mean a break-through in agrarian stagnation. Where the dynamics of agrarian capitalism results in "*brazos sin tierra and tierra sin brazos*" Luchadores is aiming for the opposite: "work for all and fields that bloom."

Notes

1. In the community of Catacaos, in the neighboring valley of Bajo Piura, a similar process of social struggle was observable in the 1960s and 1970s, aimed at intensification through an increase of employment and workers' control over production (see van der Ploeg, 1976 and 1977, and Comunidad, 1974).

2. A savings quota is that part of estimated future GVP that farmers are supposed to save (i.e., allocate to the bank). In practice it refers to the margins that are to be taxed away from the countryside, to stimulate "national devel-

opment." The remaining part of estimated GVP is for the costs of production and wages.

3. To illustrate the hold the bank has on the planning and organization of production, I will describe some elements of loan SAEA-A1-02/81 which Luchadores contracted with the bank for the '82 campaign. In the first place, the funds allocated were administered (as is mostly the case) under an agreement between the Bank and the BID (Banco Interamericano de Desarollo, Agreement 322-SF-PE of 27-6-1972). This agreement grants both the bank and BID the right "to visit and investigate all properties, goods, works and constructions of the cooperative." Second, the cooperative is obliged "to carry out all recommendations of the Ministry of Agriculture" as well as "to supply the bank with all information." Further it is forbidden to interchange machinery, etc., with other cooperatives and communities, since "the cooperative is obliged to use the financed goods and services strictly and exclusively for its own fields." The bank and BID maintain the right to cancel the loan at any moment they consider appropriate.

As to the specific loan, it contains a detailed description of the rotation scheme to be used and the varieties to be sown. For each variety the expected yield is specified. Sixty-five percent of total production costs are to be financed with the loan. The *plan de entregas* specifies how much money will be received each month and how it must be spent. For cotton, for example, 211 million soles may be borrowed, of which 137.5 million must be dedicated to acquiring industrial inputs. This money is to be paid directly to the trading companies. Then there is a *plan de reembolsos*, that is, a plan for the payment of interest and for repayment of the loan. From a certain moment on, the two plans become interrelated: the money will only continue to flow if repayment is effectuated. Loans of different years are also thus interlinked: "the transfer of January (for payment of salaries, as foreseen by loan A1-02/81) will only be effectuated if loan SAEA 1/80 [of the previous year] is completely repaid." On 24 March 1982, loan A1-02/81 was "modified" since "the hydrological situation in the valley is worsening." The total loan was reduced from 550 million to 291.5 million soles. On 31 March 1982 the part referring to cotton cultivation was "definitely eliminated," notwithstanding that the seed and most inputs had already been bought. At the beginning of April the part concerning the cultivation of sorghum was "restructured" to 32 million for payment of salaries in the cooperative and 68 million to the trading companies. However, the payment of the 32 million was then made dependent on (a) "the inspection of the fields," and (b) "the repayment of loan SAEA A1-1/80." At the same time the bank organizes police checks around the cooperative to see whether there are "illegal sellings." Several lorries remain in police custody.

4. The land reform laws governing cooperatives contain no adequate rules for "succession" of an old member by his son or daughter. In practice the same laws exhibit a strong bias against women's membership. Theoretically the assembly of a cooperative could adapt specific rules, but if they do so they will then run counter to the bank, the Ministry of Labor and the Ministry of Agriculture.

5. Alvarez (1981:80) gives the distribution of crops over the organizational subsectors in agriculture. Sugarcane is exclusively cultivated in cooperatives; 68% of cotton is cultivated in cooperatives and the rest by medium and small-scale farmers. Sorghum is cultivated mainly (87.4%) in cooperatives, but only 10% of rice is from cooperatives. For all other crops the percentage for cooperatives is below 10%.

In the 1970s there was an evident extensification in sugarcane production. Yields fell from 169.9 ton/ha at the beginning to 131.4 ton/ha at the end of the decade. Cotton also showed decreasing yields and likewise sorghum. It is the typical crops grown by small-scale farmers (such as beans, coffee and, to a lesser degree, cereals) which showed an increase in yields throughout the decade. A specific analysis of yield differences between the subsectors is to be found in Bolhuis and van der Ploeg, 1985, Chapter 5.

6. One might test this hypothesis using the study of farming accountancy in 1,200 enterprises (MAA, 1978). It becomes clear from the analysis that there is a positive relation between a low labor input and increased "profitability" (see Bolhuis and van der Ploeg, 1985, Chapter 7, note 19).

7. A typical example of the "backwardness" of social scientists is to be found in Warman (1976). Following Geertz's notion of agricultural involution, Warman describes agrarian development in Morelos, Mexico. Between 1930 and 1960 there was a considerable increase in the economically active population in agriculture. The area under cultivation also increased. However, agricultural production saw a more marked expansion. According to Warman: "increases in cultivated area, in capital used and in the labour input explain more or less 50% of the real increase in production" (1976:282); this seems to imply that the technical efficiency of agricultural production rose considerably. Yet Warman offers a different explanation, namely, that the increase in "total factor productivity" results ("if not completely, then to a large degree") from a more lengthy working day and a heavier labor load. He attempts to support this remarkable interpretation by referring to Sahlin's concept of "primitive barbarians" in *Stone Age Economics*!

The crucial point, however, is that agrarian incomes, notwithstanding the considerable rise in production and efficiency, did not rise. This leads Warman to conclude (as so many other researchers of agrarian development have) that "the intensification of agriculture entails a decrease in labour productivity" (1976:303). He asserts this to be a "constant relation": "more intensive production corresponds to an increase in production and a decrease in labor productivity" (1976:299). This is simply nonsense: In the first place, because physical interrelations established within strict limits of time and space (as in experimental stations) cannot be used to understand broad historical processes in which labor as a "factor of production" is itself an active subject creating all kinds of changes—i.e., processes in which the relations between technology, craftsmanship, and political and economic dimensions can and do change. Second, the argumentation itself seems faulty. Incomes did indeed decrease, but was this through the "constant relation" assumed by Warman or through the increased exploitation of farmers' labor? During the period analyzed (1930–

1960) there was an almost continuous and strong squeeze on agriculture in Morelos and the terms of trade got worse (see also Warman, 1976:232–235). Imagine what might have happened if no intensification had taken place among farmers: what would have happened to incomes and to "labor productivity"?

Associated with this rough and illogical reasoning is a neglect of heterogeneity. For some types of farms and for some crops commoditization increased considerably in Morelos. As Warman states: "markets grew, the circulation of money increased. Morelos became 'civilized' through increased contact with the surrounding capitalist world." Social mechanisms for the mobilization of land, labor and working capital disintegrated rapidly and indebtedness became chronic. So "notwithstanding the introduction of fertilizer, hectare yields of maize decreased from 25 to 40%" (1976:228). Even with such data Warman is not capable of distinguishing meaningful patterns in the significant heterogeneity, nor of linking in an adequate way the different changes. He sees only "constant relations."

8. Communal land was a piece of land on which horticultural products were grown to supply the surrounding villages with cheap food during periods of extreme drought (and hunger).

5

Commoditization and the Social Relations of Production

Heterogeneity is neither accidental nor a secondary characteristic of agricultural systems. It is the structural result of the fact that farm labour, as a goal-oriented and conscious activity, takes place under increasingly diverse relations of production. In this respect the relationship between commodity and non-commodity forms and circuits often plays a decisive role. Heterogeneity is not a traditional leftover, nor a simple derivative of earlier, but still surviving structures, as Mellor (1968:260) and Lipton (1968:note 19) assume when they relate substantial variation in agricultural practice to low levels of development. "The less developed an agricultural community, the greater is the inter-farm coefficient of variability of output per acre in the normal year" writes Lipton. Mellor starts from the same hypothesis and relates the phenomenon to the imperfection of the market, arguing that "variation in labour and capital cost may be greater [in traditional agriculture] than in the high-income nation where resources may be more freely mobile. As a result, greater variability in farm organization and operation may occur in traditional as compared to modern agriculture."

Variation and heterogeneity are not limited to peripheral agricultural systems. They are increasingly reproduced in modern, "highly developed" systems. The most significant example of this is perhaps the Northeast Polder of the Netherlands. It is an area of land which was drained in the 1940s from what was formerly the Zuiderzee. In the 1950s the fertile land of the polder was divided into equal plots and handed over to farmers. Farmers from the "old lands" who wanted to migrate to this new land were carefully selected by government officials. Only the "best producers" were considered. A very homogeneous farm structure arose in this way, with farms of equal size, uniform buildings and stalls, and a population of "young and dynamic" farmers.

However, within a decade there were noticeable differences in agricultural practice (Constandse, 1964; Zachariasse, 1974). The average production per hectare amounted to 10,000 kg/ha grain-equivalents, but around this mean considerable variation emerged (SD=4,000 or 40%). Research showed that such differences in intensity were remarkably stable from year to year. Only taken over longer periods did some farms show any variation (Zachariasse, 1979). Figure 5.1 gives the average production of grain-equivalents per hectare for a series of agricultural areas.

In comparative research of the kind undertaken by Hayami and Ruttan (1971) and de Wit and van Heemst (1976), production of grain-equivalents is taken as the unit for judging the level of development in different agricultural systems. Hectare yields per agricultural area is given on the y-axis. The standard deviation is expressed as a percentage of the average level of production and is represented on the x-axis. Results indicate that diversity is to be found in each agricultural system. Marked variation in hectare yields is not limited to peripheral agricultural systems, as is generally suggested in modernization theories, but also occur in systems which are part of "high-income nations." The Northeast Polder, and dairying in the north of the Netherlands and on the Po plain of Italy, are well documented examples of this. However, the obduracy with which such variety is dismissed in the field of policy and theory is as remarkable as the frequency with which such differences crop up, even in highly developed systems. The background of this myopia is undoubtedly the neoclassical postulate that farmers should be considered entrepreneurs for whom, without exception, "profit maximization" is the rule. The notion that different strategies can be followed to develop an enterprise and that consequently different optimums emerge is basically missing. With the help of such neoclassical models one optimum is defined, which is determined by given market and price relations. Empirical diversity is then seen as an expression of varying degrees of successful or unsuccessful entrepreneurship. That is precisely the conclusion of one of the most important studies dedicated to the variety in the Northeast Polder. Differences in intensity were, in the last analysis, seen as an expression of differences in entrepreneurship. And because entrepreneurship is conceived of as a unilinear dimension which progresses from "bad" to "good," only one conclusion was possible—namely, "that it appears reasonable to continue giving maximum support to farmers to strive to become better entrepreneurs" (Zachariasse, 1979:13). More entrepreneurship then, to be achieved through state intervention, is desirable in order to eliminate diversity and the differences in profitability which go along with this.

In such a view heterogeneity is seen as a characteristic of a not-yet-completed reality, as the result of temporary set backs or obstacles that

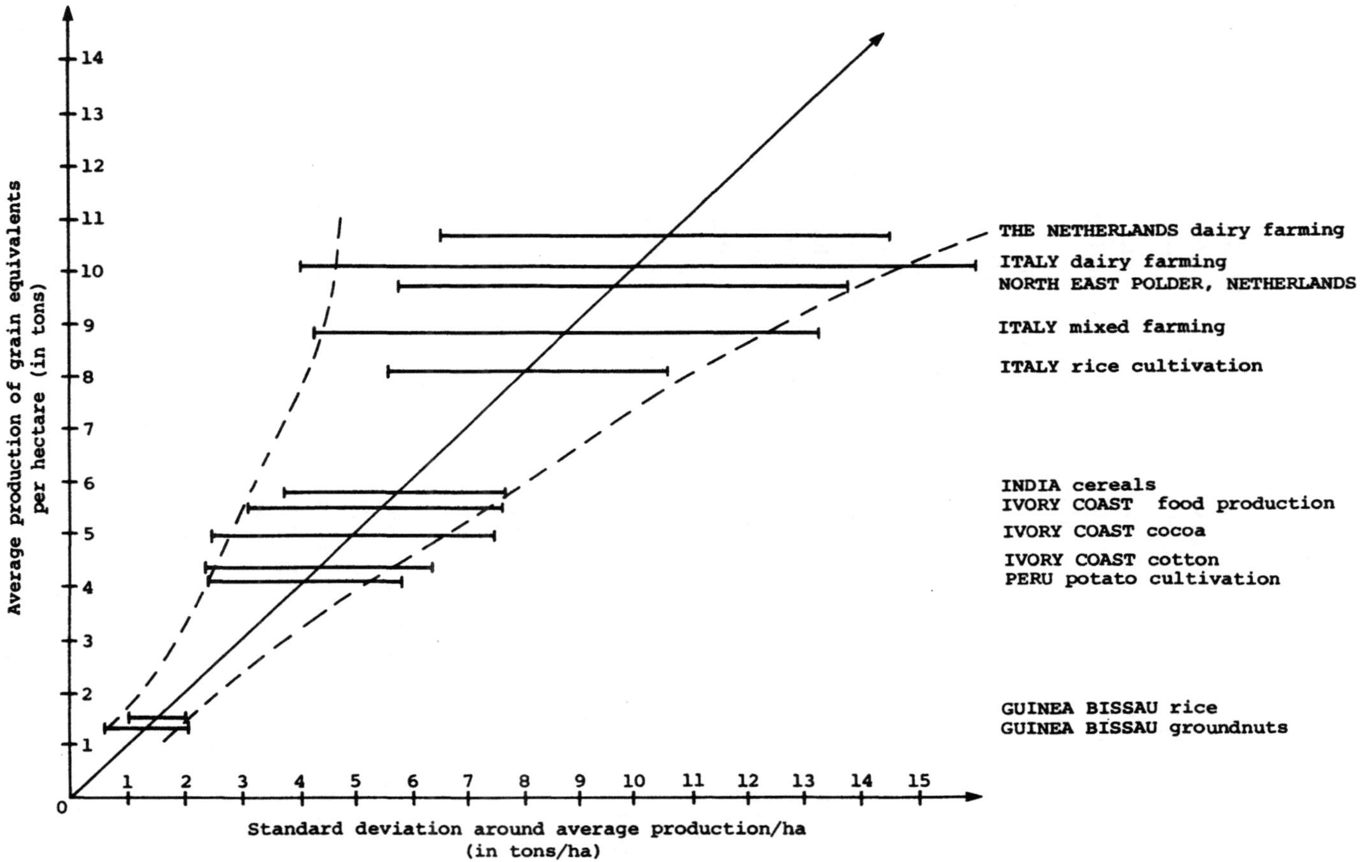

Figure 5.1 Diversity and "stages" of agrarian development

can be eliminated. In the neoclassic model employed in current theories of agrarian and development economics, questions concerning how far agricultural practice is incorporated are irrelevant. Whether market and price relations penetrate the production process, i.e., whether labor objects, means, and labor do or do not appear as commodities has no part in their paradigm because homo economicus (or the "agricultural entrepreneur") ought to calculate and plan as if all markets and market relations were indeed real; as if the indicators which determine farm operation and development were formed in a unilinear and unmanipulable manner by these market relations.

Theoretically speaking, the agricultural systems researched in this book form an integral part of "generalized commodity economies." In Italy, as in Peru, we came across farms which are tied to markets, both on the selling and supply side. There is no question of a "natural economy." But to conclude on the basis of this that there is only one degree of commoditization, namely, a "complete commoditization," as Gibbon and Neocosmos (1985) and recently Bernstein (1986) do, is a serious mistake—in the first place, because micro- and macro-levels are blurred; in the second, because ideal-typical constructions take the place of a theoretical reconstruction of complex and contradictory realities; and third, because the agrarian sector is conceptualized as a uniform category in which diversity can only be a phenomenon of secondary significance.

Value and Commoditization as a Differential Process

In the earlier chapters commoditization was conceptualized as a differential process. It is also an extremely complex process that is multiform and changing, embracing many facets and interrelations. As a historical process, it is characterized not only by an enormous geographical diversity but has displayed through the course of time an often deeply contradictory nature as well as differing drives and incentives. Although at a conceptual level commoditization may perhaps be seen as an unambiguous, "historically completed" process (Bernstein, 1986), empirically it often represents the opposite.

In the case studies, commoditization of the agrarian production process was analyzed in terms of its multiple incorporation into various markets. Incorporation of a farm enterprise into a market—whatever market—implies exchange: goods or services acquired through monitary exchange appear then as commodities in the labor process. The more a farm enterprise is incorporated, i.e., the more the resources used appear as commodities, the more the labor and development process (production and reproduction) has to be organized in *direct* correspondence with current market and price relationships.

Hence, "the logic of markets" (Friedmann, 1980:167) is not an imperative for all farm enterprises which might fall under the general category of simple commodity production. Markets only become a structuring principle when a high level of incorporation has been reached and commodity relations become a reality within the labor process itself, since it is then and only then that labor objects, means, and (a part of) the labor employed appear *de facto* as commodities. As I have shown, commoditization leads to a restructuring of the farm labor process. The cognitive schema (calculi) through which relationships between farm, markets and the labor process are interpreted, structured and reproduced undergo a significant change with increasing incorporation. Thus heterogeneity in styles of agricultural practice and variation in agrarian development patterns also depends, though not exclusively, on the degree of incorporation and consequent commoditization.

The case studies and this conclusion start, analytically, from the premise that one can only speak of commoditization and commodities if there is indeed *exchange* (i.e., actual incorporation into markets). As Marx (1974:III,328) says, "It is commerce which here turns products into commodities." Consequently those goods and services produced and reproduced within the farm enterprise itself, or mobilized through socially regulated exchange, should not to be considered as commodities. They have use value and are the result of concrete labor. That is, they represent a specific value for those who create and use them. This specific social value would be lost if they were commercialized. Chevalier, among others, follows a different line of reasoning on this. He argues that "by commodity is meant not those material values that are actually purchased or sold but, more generally, all those that are *exchangeable* for money and that contain a definite quantity of *value*" (1982:118, italics added). From such a starting point it is easy to overlook the central place that the concrete labor process should have in research on the social relations of production. Such a starting point also dispenses with the need to research the complex unity of production and reproduction within specific forms of the division of labor. Chevalier argues that the concept of commodity implies "exchangeability," and not actual exchange. He concludes by classifying those goods and services which are not exchanged (e.g., part of the harvest, land, labor etc.) as being "subsistence commodities." "They never enter the sphere of circulation, not because they are not exchangeable but rather because their 'abstract' value can be best realized through direct consumption by the producers themselves" (1982:118–9). *Exchange calculations* are thus seen to be crucial even when certain "commodities" are not sold.

On the basis of the case studies such an argument must be disputed. Let us begin again with cows. Whatever the possible sale value is, a typical I-farmer from Emilia Romagna will not sell his "good cow." Of course such a cow is, theoretically speaking, *exchangeable*—the owner will happily even boast about her value, at least in the local bar. But he would not boast of her value in the market, because he would not sell her, except perhaps at the appropriate time. That time would be determined by the farmer himself when he knows for certain that he has a sufficient number of "good" offspring from the cow. In a theoretical exercise of the Chevalier type, the sale of a "good cow," can be presented as an activity in itself, but in reality, i.e., in farming practice, it is an activity which cannot possibly stand in isolation. It has to be coordinated with other activities, must be a logical and consistent part of a general goal-directed strategy. "Exchangeability" would mean that cows of the following generation would not be produced with the help of this good cow but would likewise be purchased via exchange. Precisely because of this, farmers who work with a consistent strategy of autonomous intensification will quickly dismiss any such notion of exchangeability, especially as they are convinced that, like themselves, "other farmers would never sell their best young cattle." In such a case the exchangeability constructed by Chevalier is a farce, and so is his notion of "value." The value of a good cow, in this context, is that it delivers a valuable male or female offspring. Its value arises also from the fact that the farmer knows this good cow intimately and, therefore, also has at his fingertips the basis of an intimate knowledge of the offspring. *This specific value or use rules out any possibility of applying some general notion of exchangeability.* A good cow represents a specific use value precisely *because* she is *not* sold. The specific use value carried by the good cow is that it is one of the means by which the farmer acquires and enlarges his *control over the production process.* If he should reduce this good cow through exchangeability to mere "abstract value," then he would (a) partly be giving up his control over the processes of production and reproduction, and (b) would, in a stroke, undo all the labor previously given to the very creation of such a cow. The "good cow" represents a specific use value. Of course exchange value can be dreamed about but the realization of that value nullifies its specific use value.

In short, the exchangeability and value which Chevalier postulates, are totally unreal categories in farming practice. The same applies to feed and to so many other use value items that are strategic in the farm labor process. Of course feed can be bought anytime and anywhere, but again, self-produced feed represents more than an abstract value. Roughage and concentrates produced on the farm represent again a specific use value, which is real and central to certain styles of farming. Its use value can only be produced, maintained and utilized if it is not bought and sold. The specific use value exists thanks to its non-exchangeability. Self-produced feed is knowledge in material form—precisely because it is a product of real, not abstract, labor. Hence the value of such feed is not abstract but specific. It is a socially defined value. The farmer knows this feed, knows the fields it comes from, under what meteorological conditions it was harvested, knows its quality and grade, and knows how his cows have reacted to it in the past and will react to it in the future. Self-produced feed is also to be seen as independence in material form. The use value of such feed is that the farmer can apply his own insights, unhindered by buyers and sellers and unconditioned by market and price relationships. In short, producing one's own feed (feed as a "subsistence commodity" as Chevalier would presumably call it) just as producing one's own cows, gives the direct producer control over the labor process. It relates to earlier attempts to enlarge this control as well as to the principle of non-exchangeability. Hence, use value emerges as a dominant category.

Chevalier rightly states that "production for household consumption may have nothing to do with the preservation of a 'natural economy.' On the contrary, it may result from the need . . . to obtain the greatest value" (1982:119). Again the essential question is what must be understood by "value." By ignoring the strategic meaning implied in having control of one's own labor process, the question concerning the "practical and positive meaning of usevalue logic" remains unanswered (1982:114).

In the farming styles explored in the previous chapters, use values represented a "practical and positive" meaning precisely because they make it possible to create a certain room to maneuver. By withdrawing at strategic points from the immediate influence of market and price relations, the farmer can organize the labor process more in tune with his own insights and interests. By creating and maintaining a certain self-sufficiency, "you are conditioned by nothing and no one in your work," says an Italian farmer. In short, by bringing the necessary labor objects, means, and labor to rest as far as possible on historically guaranteed reproduction farmers create social relations of production which maximize their own control over the labor process and permit the direct producer as great a share as possible of the wealth produced.

The concept of "farming freedom" has, according to Slicher van Bath (1948), a double meaning. It is "freedom from" and "freedom to"; freedom from ties that bind (among which commodity relations are explicitly understood) and freedom to choose those forms of labor organization and production which optimally fall in with one's own perspectives and interests. This twofold notion of farming freedom is crucial for understanding not only the relations between farm and markets—relations which are constantly changing—but the farmer's conscious role in the organization of these particular interrelations.

The relations between farms and markets often appear to be internally contradictory, but seen from the perspective of the dual "farming freedom" mentioned, they contain an unmistakable, socially defined logic. As van den Akker (1967:139), an old farmer from Friesland, wrote, "the whole farm was organized in such a way that one could help oneself as much as possible and yet spend as little as possible." After a detailed description of the mechanisms for achieving this, he concludes that "in the old farms self-sufficiency was practiced to such perfection that the cleverest statesman [or I might add, academic] could well take a lesson from it" (1947:140). At the same time, however, as van den Akker tells us, these farmers were "like wasps round the honey pot" if there was a new opportunity to make money, "for there was always a hunger for cash."

Certain commodity relations were entered into, but others were, as far as possible, placed beyond bounds. Thus exchange and exchangeability are not universal, as the insights of the old Frisian farmer illustrates, nor everywhere and always applicable in simple commodity production. They are carefully and consciously regulated, for they are related to "farming freedom," to control over the labor process, and to the distribution of wealth.

Chevalier reduces the practical question of why certain goods and services are "subsistence commodities" and others "actual commodities" to a question of savings, and thereby to a mainly conjuntural phenomenon. Because land, labor and certain products are not mobilized by the market (via actual exchange) the farm household achieves "measurable savings in the family budget" (1982:120). Research such as Gudeman's (1978) attempts to give quantitative support to such a thesis. Again, however, such deductive reasoning does not stand up to a careful analysis of empirical constellations. If we look again at the situations highlighted in the earlier chapters, measurable savings do not appear to be crucial. Feed offered in the market can be so cheap that the production of roughage and concentrates on the farm itself produces the opposite of "measurable savings." But the farmers in question continue with the practice. By deriving decisions from a long-term perspective short-term fluctuations on the market appear as irrelevant or may even appear as threatening and dangerous. The same can be said of farmers in Ireland. Leeuwis (1988), in a beautiful case study, describes how a great number of farmers rejected subsidized loans to build new stalls, behaving in a manner which again opposes the notion of "measurable savings." Instead they took upon themselves the extra costs in order to be able to build the stall according to their own insights and means. Likewise, a good cow, whether it stands in a stall in Peru, Italy, Friesland or Ireland, is not sold just at the moment when "measurable savings" could be at their highest.

A certain subjectivity cannot be omitted from the "exchange calculations" farmers make: the time scale used, whether they value their independence, and many other considerations are crucial to the final outcome of exchange calculations, and vital in any decision concerning whether to sell or not. This can easily be shown to be the case even in the examples from Pachitea, on which Chevalier bases his reasoning.

I am not referring here to voluntaristic or atomistic interpretations. On the contrary, it is through the significance of subjectivity in exchange calculations that social relations of production are expressed, and it is through this subjectivity that farmers manage to create room for maneuver with respect to market and price relations.

Markets are arenas. Of course farmers are not the only actors operating in such arenas, let alone the only ones who are interested in the relationship between agriculture and various economic circuits. The confrontation between capital and peasantry revolves to a large extent around this theme, as Bernstein rightly argues (1977). An increasing incorporation into markets and a consequent commoditization and institutionalization of the labor process are practically always the result, even if not the direct intention, of intervention by state and agribusiness in the agricultural sector. A typical example is the fact that the design of new agrarian technologies (the green revolution packet is a good example) is nearly always based on the assumption of increasing externalization of particular agricultural tasks to industry. This results in further incorporation and commoditization and in an abrupt redistribution of produced wealth.

Commodity relationships are not neutral givens. They form an intrinsic part of the arena in which farmers find themselves face to face with state, traders, agribusiness and their advocates. In a political and ideological respect, therefore, it appears to me that theories which on a priori grounds ignore the contradictions and room for maneuver of farmers in these arenas are seriously inadequate. Then strategies such as "the seeking of maximum autonomy" (Fraslin and Simier, 1983), the "resistance paysanne" (Pernet 1982), the movement of young farmers in Italy to create maximum distance from agribusiness, or the reverse—the active pursuit of commoditization—become totally incomprehensible. Anta Pampa is in this respect an outstanding illustrative microcosmos. Commoditization proceeds here through programs such as those fashionable in international circuits of development aid. Incorporation into the capital market in turn brings with it (according to the modus operandi of the Proderm Program) incorporation into the market for "improved" genetic material. That leads to a dependency relationship with markets for fertilizers, herbicides and pesticides, to dependency on labor markets, markets for machine services, and markets for "professional knowledge." It also leads, as described earlier, to a deepening dependency on potato markets because the farmer is no longer able to choose when he will market his harvest. The time schedule is then defined by the institution that lends the credit. And added to all this is a certain reduction in soil fertility and degradation of the farmer's own genetic stock; dependency relationships, in other words, become lasting phenomena.

This whole pattern of dependency and its associated market-induced risks appear in the eyes of the farmers of Anta Pampa to be a trap; a trap to be avoided. Those who can avoid it, i.e., the *rich*, do so, despite the "measurable savings" they might achieve. In reality it is the poor who out of necessity participate in planned market incorporation. But in doing so, poverty remains their lot.

The Impact of Commoditization

The relationship between use and exchange values in the labor process and the specific interaction between commodity and non-commodity circuits appears to have a great impact on the organization and development of the labor and production process in agriculture. Different incorporation patterns are concomitant with different styles of agricultural practice. If the labor process is largely based on the "total circulation of commodities" (Marx, 1974:III,328) (i.e., if production is in essence market-dependent), then scale enlargement and relative extensification becomes dominant, not only in capitalist-organized agriculture as current dualistic theories contend, but in the sphere of family farming as well. Family farms and cooperatives, obliged by a high degree of incorporation to use an E-logic to define the farm's development, develop in the same way as capitalist agricultural enterprises. "*Tierra sin brazos*" and "*brazos sin tierra*" or, as they say in Italy, "*agricoltura di rapina,*" then become appropriate interpretations of the development of the sector which apparantly is controlled by farmers. That is to say that under the misleading guise of independence formal subsumption is achieved. Such subsumption or subordination will result from an increasing dependence on markets and from commoditization of the elements and interrelations of the labor process that goes with this.

On the basis of the case studies from Peru and Italy, the connection between commoditization and the growing dominance of scale enlargement and relative extensification,[1] can be outlined as follows:

1. The notion of benefits and costs is redefined, in the sense that with progressive incorporation, labor, labor objects, and means appear increasingly as direct costs. In the family farm this represents a structural change. Such costs, and their level of input, either separately or together become the object of continual deliberations. The "mobility" of labor, means, etc., is far higher than in historically guaranteed autonomous production, and the production factors made "mobile" are incorporated into a general strategy of cost reduction. The mobility of production factors so heavily emphasized by Friedmann (1980) in her characterization of simple commodity production is thus in no way a universal characteristic independent of time and place. This very mobility is a continually changing given that accurately reflects the degree of commoditization.

Alongside this, and again because of the high level of incorporation, these production factors and non-factor inputs, which now appear as "costs," have to be used in such a way that valorization is assured over a shorter term than before. That means that economic efficiency as defined by prevailing commodity relations becomes dominant over the striving for technical efficiency.[2] It means also that benefits which are immediately realizable become decisive and future benefits become less relevant. The "good cow" effectively leaves the stage.

2. With progressive commercialization the time span conceived and planned for becomes considerably shorter. To an increasing degree each cycle must be made to pay, must be set up and organized in such a way that it is in tune with actual price and market relations of the day. Plans and investments which might cover a number of cycles become increasingly subjected to the time limits set by loans, as well as to trends predicted in other relevant markets. This contrasts markedly with the opposite situation, where historically guaranteed autonomous production permits a substantially longer time horizon, symbolized on the one hand in the idea of "the good farm" and on the other in reproduction and acquisition over generations (that is to say, on the family as a relation of production). The father farmed on inherited land and worked so that his sons, sometimes daughters, could take over a "good" or even better farm.

3. Both of these previous developments imply that the fragility of the agrarian sector substantially increases and brings with it changing ideas concerning risk. In the past, risk referred mainly to the unpredictability of nature, but with a sharp rise in commoditization, economic risk becomes central. Market-induced risk (Huysman, 1986) emerges

and becomes general with the advance of the incorporation process. Risk-avoidance as a strategy, as I have shown, thus acquires a wholly different form.

4. If we begin now from the simplest position—namely, with a production function, as an unambiguous, curvilinear link between inputs and outputs—then a rising level of incorporation will result in a different optimum. With a change in the definition of benefits and costs and a change in the concept of risk, the slope of the price line alters and the "equalizing point" will come to lie lower on the production function. The costs, as a consequence of incorporation, will be estimated as far higher. With a readjusted definition of benefits, "future benefits" (such as higher soil fertility, improved feedstuffs, well-maintained irrigation systems, etc.) will fall outside the shortened time span which the process of incorporation introduces into the sphere of production. Economic risk avoidance, finally, implies that where prices fluctuate, and production factors and non-factor inputs appear (are interpreted) as direct costs, farmers will by and large calculate along the "bottom line."

The results include a lower input of production factors and non-factor inputs and a lower production per labor object is the result.

5. Even more fundamental is the fact that with an increasing degree of incorporation, labor undergoes a change in quality. In the framework of artisanal production (and agrarian production is artisanal to a considerable degree), quality can be defined as a specific relation between producer and labor object.

Quality is essential where labor is geared to an optimal use and development of the productive potential of labor objects. Quality appeared frequently in earlier chapters in concrete form as craftmanship, and the difference between "good" and "bad" farming was discussed on several occasions. In irrigation, for example, labor can be so organized that it is of high or poor quality. It is the same in the cow shed where quality of labor can be distinguished in numerous ways. As Moerman (1968) rightly argues: "If we are to understand a productive system and its potential for growth, we cannot regard . . . labor as a disembodied, explanatory variable." Potentially labor may carry impressive qualities. But quality demands time. Good irrigation assumes a labor input of 1 man per 40 litre/sec in sorghum cultivation. One can irrigate with less input but the work will not be well done. A well-cared-for potato field in Chacán demands 37 man/days per topo. Those who have other matters in their head (literally and figuratively) make do with 16 man days per topo. The differences can be clearly seen in the harvests. With a similar type of cow shed (*legato moderno*) an intensive farmer spends on average 25 minutes with each cow, while his colleague might spend only 10 minutes.

As already indicated, the way the production process is perceived is changed by the process of incorporation. Production factors and non-factor inputs do appear as direct costs. Labor is no exception; it is also seen increasingly as a cost. It is seen as an item like any other item for which a cost reduction becomes the norm: thus less labor will be have to be employed per labor object. In other words, labor also becomes a function of market and price relations. Thus, time as a precondition for quality becomes an absurdity because only the quantitative aspect of production remains.

The gradual elimination of quality in farm labor and the disappearance of specific use values (such as the "good cow," "own feed," "a job well done") which go with this imply a second substantial shift of the "equalizing point."

6. Changing conceptions of costs, benefits, and risks, and the gradual elimination of craftmanship as a specific relationship between producer and labor object, affect the parameters within which the farmer makes his decisions. As Pearse (1968) states, "new conditions are laid down in which peasants make their decisions." But these new conditions are only the tip of the iceberg. Not only are the external parameters of decision making changing, but the decision-making process itself is drastically altered. The process of incorporation leads, as we have shown, to an adaptation of the calculus.

We defined and researched a calculus as the structure within which a farmer specifies goals and means and their mutual relations and with which the labor process as well as the interrelations between farm and environment are regulated. A calculus is constructed and reproduced through the repeated process of observation, interpretation, understanding, and adaptation. Thus a calculus symbolizes a particular structuring of farm labor.

In Emilia Romagna we explored the I-calculus, the model that guides decisions and normatizes labor in situations which could be classified as grounded in historically guaranteed reproduction. Alongside it, a clear E-calculus could be distinguished, a model which carries a functional rationality and is characteristic of situations in which the level of incorporation is high. It was demonstrated that even when similar external parameters are assumed, this E-calculus leads to a different optimum than the I-calculus.

7. Thus, when assuming a given production function, it can be demonstrated that on account of changing parameters (benefits, costs, time perspectives, risk, quality) and changing calculi, not one but a whole gamut of "points on the production surface" (Yotopoulos, 1974)

are to be observed. The spread is not coincidental, but rather it is structurally determined by differences in the level of incorporation. However, the assumption of one given production function (although didactically useful) is not real. From the various case studies there appear to be a number of production functions (within one and the same group of farms operating under the same set of ecological and technical conditions). The nature of these functions are strongly associated with the level of incorporation.

The technical efficiency of farm labor structured as craftsmanship tends to be relatively high. Labor structured as craftsmanship results in the creation of "frontier functions" (Timmer, 1970). Craftsmanship assumes not only quality and time, but as a particular organization of labor (with its own specific knowledge, experience and norms) it also assumes both an "overview of and insight into the relevant whole" and a certain measure of "functional autonomy." The "externalization" of an increasing part of the reproduction and production process entailed in the process of incorporation, and the growing dominance of market and price relations as a regulating principle, reduces this "relevant whole" and eliminates "functional autonomy." A commoditization of *the elements used within* the labor process as well as an *external* prescription of farm tasks become fundamental characteristics which bring with them increasing entrepreneurship and a simultaneous undermining of the basis and ratio of craftsmanship. Incorporation into markets and institutionalization by market agencies hinge precisely on that point. Thus farm labor loses its role of generating renewal and progress in production. Adoption of externally developed innovations becomes the key word. It is in this way that alienation of farm labor and its formal subsumption to capital are accomplished.

Similarities and Differences

Most contributions to the commoditization debate relate either to peripheral agricultural systems or to the history of agricultural systems of the center. Chevalier's work, for instance, symbolizes the former whilst the work of Friedmann, which focuses primarily on North American grain farmers in the 1880s and the 1930s is to be seen as an eloquent example of the latter. However, the *contemporary* reality of *central* agricultural systems is seldom the object of research, let alone of comparative research that relates to both center and periphery. Empirical research of a comparative nature, is as it were, replaced by a series of assumptions which under scrutiny appear more fiction than reality. These assumptions then refer to a "fully commoditized" agrarian sector in which the commoditization process is historically "complete" and in which the mobility of production factors is understood to be unlimited. With such assumptions empirical research of commoditization in "highly developed" agricultural systems obviously becomes superfluous.

The assumption that West European and North American agriculture represents full commoditization leads in turn to a strong bias in the analysis of peripheral systems, where partial commoditization thus emerges as a central characteristic of underdevelopment, closely bound to the notion of the "intrinsic backwardness" of these systems (Bernstein, 1979). In that respect, Bernstein's position distressingly resembles the thesis of "uncapturedness" of which Hyden (1980) is an exponent: both relate the low level of commoditization of peasant agriculture in one way or another to underdevelopment and stagnation. The coincidence is not surprising. Both theoretical perspectives combine a second assumption, that development procedes in a unilinear manner from a "natural economy" to complete commoditization (for a critique of this point of view see Long, 1984; Long et al., 1986). This complete commoditization, or full market integration, would therefore be typical for agricultural systems of the center.

The way in which Friedmann relates the concepts of simple commodity production and petty commodity production is a pregnant expression of the tendency to equate development with the "historical completion" of the commoditization process. "The end point of commoditization is simple commodity production," according to Friemann, (1980:163). Petty commodity production is characterized in Friedmann's conceptualization by a not-yet-complete integration in markets, i.e., the allocation and remuneration of production factors take place partly outside the market and do not respond to the "logic of the market." "Commodity relations are limited in their ability to penetrate the cycle of production" (Friedmann, 1980:163). Simple commodity production, on the other hand, would represent "full market integration"; in other words, in simple commodity production, the logic of the market rules. Agriculture is then "controlled by definite and precise forms of capitalist regulation," according to Gibbon and Neocosmos (1985:165). Bernstein completes this view by presenting petty commodity production as a *transitional* category which relates to a few not-yet-completed realities. "The passage from 'household production' to simple commodity production is charted through full market integration" states Bernstein in summarizing Friedmann's work (1986:15). Simple commodity production, on the other hand, would be the key for understanding European, American, and "modernized" Third World agriculture (Bernstein, 1986).

This unilinear perspective points to an astonishing convergence between commoditization theory and neoclassical development economics

(Vandergeest, 1988; Long and van der Ploeg, 1988). The latter tradition takes the position that "rural development is concerned with the modernization and monetization of rural society and its transition from traditional isolation to integration with the national economy" (World Bank, 1975b:3). The view that one of the main obstacles to modernization is the incapacity or unwillingness to organize production on the basis of current market relations (Rogers, 1969) is wholly in agreement with this.

Thus far we have discussed in general the convergence between at least some of the contributions to the commoditization debate and modernization theory. Now if we look at the results of the research on Italian and Peruvian agriculture we are obliged to conclude that simple commodity production, understood as "complete" market integration, no more exists than do farmers who are wholly "homo economicus." Even in such highly developed agricultural areas as the Italian Po plain, there is no such thing as complete mobilization of production factors, or of a way of farming which reflects unequivocally and unilinearly "the logic of the market." So the fictitious "end point" of Friedmann is not reached.

If one were to use one of the above-mentioned concepts, then ironically enough it would be petty commodity production, for Italian farmers are as much petty commodity producers as their Peruvian counterparts. A partial market integration for them is no historical accident, let alone a phenomenon that they relate to assumed backwardness in the development of their farm enterprises. Those who structure their farm labor according to an I-calculus or logic consciously strive for only partial market integration. And even E-farmers achieve less than complete market integration, even though it is what they strive for, because in that striving they come up against a number of contradictions peculiar to capitalist formations: banks will never finance farms 100%; they even demand a non-mobility of various production factors.[3] A complete integration into labor and land markets is equally impossible. Prevailing market relations and the ratios between prices and costs which they entail simply prohibit that. One can even go a step further. If one examines the available data on commoditization patterns in Italian and Peruvian agriculture, one is obliged to conclude that in some respects Peruvian agriculture is more commoditized than Italian agriculture.

Table 5.1 presents the relevant data. Besides the data on Emilia Romagna on the Po plain of Italy, Anta pampa in southern Peru, and Luchadores, the cooperative in northern Peru, new data relating to Campania, an agricultural area in the south of Italy, and to dairy farming in the Netherlands are provided to fill out the picture. For

Table 5.1. Differential Incorporation in the Netherlands, Italy and Peru

Incorporation into markets for:	Netherlands dairying	Emilia Romagna dairying Plain	Emilia Romagna dairying Mountains	Campania mixed farming (S. Italy)	Peru Coop. agric. coastal plain	Peru Potato cultiv. Andes
Labor	6.6%	9.1%	0.1%	13%	100%	25%
Land	NA	28.7%	20.2%	8%	100%	21%
Short-term loans	1.9%	4.6%	1.9%		65%	27%
Medium-term loans	17.8%	11.1%	3.4%	23.2%	50%	0%
Long-term loans		2.4%	2.4%			
Machine services	20.5%	30.7%	10%	14%	70%	60%
Genetic material	13.7%	7.2%	7.6%	8%	65%	43%
Main inputs	NA	43.8%	37.8%	26.3%	85%	35%
Composite index	NA	26%	15%	NA	NA	NA

those countries where more regional data are available, i.e., in Italy and Peru, one might still conclude that central agricultural areas are more incorporated than peripheral areas. The explanation for this is obvious. Peripheral or marginal areas such as the sierra in Peru and the mountains in Italy form—because of their meager volume of production, and poor infrastructure or ecological complications—less interesting "objects" for the agencies which constitute the driving forces of incorporation. To this must be added the fact that technological models which are based on a strong externalization of the various tasks of farm labor and which both assume and encourage a high level of incorporation, are not applicable, or are at least less so, in the more marginal areas.

Crossing land frontiers, a different pattern, however, is observable. European agriculture (represented in the table by the Netherlands and Italy) is, generally speaking, less incorporated into various markets than Peruvian agriculture. If we take Peru as being indicative of peripheral agricultural systems on a world scale, one may conclude that, in general, agriculture of the periphery is more commoditized, more based on a "complete circulation of commodities," than systems of the center.

However improbable it might at first glance seem, the foregoing interpretation is supported by a number of arguments. They partly relate to the "driving forces" behind the incorporation and commoditization process; they are also partly derived from a careful analysis of various social relations of production which lie behind the data summarized in the table.

Peripheral economies are characterized as disarticulated (de Janvry, 1981). The agrarian sector is primarily of importance insofar as it supplies cheap labor and food. If this is primarily produced for the internal market and the producers are small farmers, then disarticulation applies a fortiori. Such a situation constrasts sharply with economies in which economic sectors are interlinked by more symmetrical dependency patterns through which a balanced articulation of sectors is maintained. That occurs, for example, when the agricultural sector is also an important consumer market for industrially manufactured inputs and means of production. Then, for obvious politico-economic reasons, it is important that farmers get good prices for their products in order to maintain their buying power.

Be that as it may, a large section of agriculture in disarticulated economies forms a simple "hinterland" for growth poles with which they are linked through several politico economic mechanisms (Quijano, 1977). Low farm prices and chronic inflation of costs define the structural conditions under which production must take place. Export poles in the agricultural sector are usually no exception. Instability, the long-term fall in prices, and sharply rising costs, are also found in typical export enclaves because of the international trade situation and technological dependence. This implies that it is not so much a question of a one-off "reproduction squeeze" (Bernstein, 1977), but of a *constant* drain of resources. Autonomous, historically guaranteed reproduction is being continuously eroded and increasingly substituted by market-dependent reproduction. The result is a rapid, forced, and all-embracing incorporation of one part of the agrarian sector and a rapid marginalization of the other.

Fitzgerald (1981) provides a convincing description of the disarticulated structure of the Peruvian economy and documents in detail the ever-worsening terms of trade with which the agricultural sector is confronted. Webb (1972) shows the same for Peru in the 1950s and 1960s and Billone et al. (1982) for the decade of the 1970s. The role played by Peruvian agricultural policy in this disarticulation has also been analyzed (Caballero, 1980). The general conclusion of these studies is simple: there is a *permanent and continuously deepening* "reproduction squeeze" in the Peruvian countryside.

The logical accompaniment of this is very rapid incorporation spurred on by political motives.[4] In the 1970s short-term credit allocated to agriculture in Peru rose from 13.8% of the GVP to 29.2%. Comparative figures for Italian agriculture show that at the beginning of the 1970s the total amount of short-term credit given was 12.4% of the GVP, less therefore, than in Peru. But more important, available figures show that during the 1970s the proportion of short-term credit in Italian agriculture remained constant while, as we saw, in Peruvian agriculture it more than doubled (data derived from INEA, 1977; Fabiani, 1979; Haudry, 1978; and Ministerio, 1981). Farmers respond to the permanent production squeeze by continually substituting the resources controlled and owned by themselves by resources which have to be mobilized in the various markets. The sanction for not doing so is marginalization. And because of this, the level of incorporation rises faster in disarticulated peripheral economies than in central agricultural systems which form a part of more articulated economies. That is why the level of commoditization is usually higher in agricultural systems of the Third World than in systems where some theorists assume commoditization is already "complete."

There are other driving forces discernable which together cause a more abrupt movement of and often a more penetrating form of commoditization, especially in peripheral agricultural systems. These forces are to be seen in the power arenas in which agricultural policy arises, in the influence of farmer organizations and cooperatives as countervailing powers, in the impact of culture and "farming pride" (Hofstee, 1983; Einaudi, 1975). However, glancing back at the table in which incorporation patterns in the Netherlands, Italy and Peru were compared, what appears to me most interesting is that the comparisons naturally bear a certain ambiguity. Dependence on markets for machine services in Italy, for example, is not identical and in certain respects not even comparable to a similar dependence in Peru. Behind what appear to be identical categories (which indeed bring out important quantitative differences), there lie hidden completely different social relations of production. That does not need to be a problem for the interpretation of the data because the drama becomes even clearer when differences in the social relations of production are taken into account.

Commoditization and the Social Relations of Production

Different degrees of commoditization represent differences in the social relations of production. Hence the dynamics, contradictions and problems of simple commodity production are not to be found in its "intrinsic" nature. As Marx (1974:III, 638–639) argued: "No producer, whether industrial or agricultural, when considered by himself alone, produces value or commodities. His product becomes a value and a commodity only in the context of definite social interrelations." Social relations of production are, as Poulantzas (1974) pointed out, that whole set of specific relations that constitutes the labor and production process (i.e., that gives the labor process its concrete form) and that defines the distribution of produced wealth. Given this conceptualization of the social relations of production, it is evident that levels of commoditization must be understood as an integral part of it. And this is exactly what various researchers, such as Nemchinov and others, have done.[5] Relations between farm enterprises, markets and market agencies shape the labor process to an important degree, either because they allow farmers "freedom" to exercise control over labor objects and the means of production, or because they directly condition, prescribe and sanction the organization of labor and production.

The distribution of produced wealth is closely related to the level of incorporation. Levels of incorporation not only denote the various actors and institutions related to the production process (each with a specific position in the social division of labor) but they also quantify that relation. In this way they dictate what part of farm generated wealth will fall to the banks, to landowners, to those who monopolize the supply of machine services, and to agribusiness. An average degree of incorporation in capital markets of 17.8 and 1.9% = 19.7% (see Table 5.1) means that 20–30% of the gross income of the average Dutch dairy farm flows to the bank. And then we are speaking of averages. There are farms in the Netherlands where incorporation into capital markets is substantially higher and where banks appropriate a far greater part of annual earned income.[6]

Differential degrees of commoditization form an important part of the relevant social relations of production. But social relations of production are not exhaustively defined by incorporation level alone. This point of view is especially important in *comparative* research into agricultural systems located in markedly differing politico-economic settings.

Markets are arenas, and although such arenas are reasonably uniform within a given country, they differ markedly between countries. On the basis of this observation, the data brought together in Table 5.1 should be critically reappraised.

Let us first look at the market for important non-factor inputs. In the Netherlands and Italy (at least in Emilia Romagna) these markets are characterized by stable prices—in contrast to Peru where marked price fluctuations, rising inflation, and uncertainty over the quality of the relevant inputs is part of everyday experience. In the center, farmers' cooperatives often constitute important price correcting mechanisms that are able, to some extent, to limit the impact of agribusiness. Such strong cooperatives are mostly absent on the periphery. In the center, government research stations and farmer organizations exercise constant quality control over inputs such as concentrates and fertilizer. In the Third World, where this countervailing power is lacking, it is easy to think that such products "no longer have any power"; such is often actually the case.

A similar difference operates in the market for genetic materials. In Peru it is the multinationals who control the market, and it regularly happens that cooperatives such as Luchadores are supplied with seed that has insufficient germination power. Such a thing is unthinkable in the Netherlands or Emilia Romagna, and if it should happen, then Dutch and Italian farmers are able to fall back on several mechanisms to recover their loss. In Peru, bringing a lawsuit against a multinational such as DeKalb would be a fruitless undertaking. Seed potatoes in the sierra, at least the so-called "improved varieties," degenerate within four or five years, as we saw in Chapter 3. Again, that is difficult to imagine in West European agriculture where a closely knit network for adaptive agricultural research exists. The arenas thus differ markedly.

That means also that level of incorporation—in a comparative analysis—refers to different things. If a heavy dependence on markets for non-factor inputs and genetic material is already problematic for many European farmers, it must be more so in the typical arena formed by peripheral markets. There, the countervailing powers present in European arenas are weak or sometimes even totally absent. In other words, in order to be able to judge incorporation in terms of the social relations of production, an analysis of the relevant arenas as a whole is necessary, certainly when making comparisons. The market for machine services in Italy, for example, is highly competitive. The greater part of supply stems from the small-scale farmer sector and all kinds of social links condition actual transactions. In Peru, however, machine pools are mostly controlled by former landowners. Cooperatives and small farmers have few, sometimes none, of the necessary machines at their disposal, nor do they have the means to purchase them. Relations in the market for machine services then become highly asymmetrical. It is a similar story in the capital market.

Without wanting to idealize European agricultural banks, it must be said that they are still, because of their cooperative origins, to a certain degree decentralized. Farmers usually have direct access to their administration, which is local. In addition the state has created a number of correcting mechanisms such as interest subsidies, security funds, and "safety nets," so that a bad harvest does not lead directly to bankruptcy. The relation between banks and farmers in peripheral agricultural systems is illustrated by the Colombian farmers who call themselves *patasucias* vis-à-vis the bank. "Patasucias" means muddy feet. The farmers are thus saying that they have too much mud on their feet even to step on the marble floors, let alone be able to arrange private terms with the manager. Not being able to repay loans on time frequently leads to bankruptcy. The criteria and parameters and relationships at work in such markets thus penetrate far deeper into the heart of farm operations than they do in Italy or in the Netherlands, and the need to adapt the farm is immediate and cannot be ignored. An oustanding example is to be found in land and labor markets. What is so conspicuous in cooperative agriculture of the Peruvian coast (see Table 5.1) is a complete incorporation into the land market. Friedmann's criterium for a full market integration appears here to have been achieved. During the period of expropriation and the subsequent forming of cooperatives, the state bought the land from the landowners. The state then tried to off-load the financial obligations attendent upon this onto the newly formed cooperatives. Thus arose the "agrarian debt." Only when the debt was fully repaid would the cooperatives acquire dominion and control over the land.

This agrarian debt (and the idea behind it that land is a commodity and should be bought) was so vehemently and radically opposed by farmer organizations that the government finally canceled it. However because of this the status of land has never been clearly regularized. For the government and the agrarian bank the land is still a commodity, i.e., a completely mobile production factor (which can serve as collateral for loans and can thus also be sequestered). For the cooperatives, the same land is held to be an inalienable labor object, whose control, use and access cannot be regulated by market relations. Here then we have an interesting interface: farmer organizations and the state define the land in decidedly different terms. In the end, it is mutual power relations that are crucial in determining the degree to which land does or does not appear as a commodity. The same is true for Italy. The land market there was also the object of fierce farmer struggle, both before and after the fascist period. Thanks to this struggle a tenancy law arose (which includes legal protection for tenants) that critically altered relations in the land market—to the advantage of the tenants. Social struggle is likewise an essential element for interpreting the degree of incorporation into the labor market. Formally speaking, cooperative agriculture on the Peruvian coast is fully incorporated. Those working maintain a wage-labor relation with "their" enterprise. Government agencies which in some ways view the management of cooperatives as one of their tasks (and often effectively take this role upon themselves) also directly apply "the logic of the market" to the workers. The chapter on Luchadores provided several examples of this: government services continually estimated that a part of the workforce was superfluous. They wanted to finance only half the wage bill, wished to impose rotation, etc. However, social struggle and the unwillingness to consider their own work place as a variable, as a mere derivative of changing markets, acted as a counterbalance to such tendencies. Work was considered an acquired right and defended as such, and creating employment (by means of the union's control over the cooperative enterprise) was also considered to be a value worth fighting for.

In summary, if taken at face value, the data concerning differential commoditization patterns, summarized in Table 5.1, are not strictly comparable. Because they refer to different arenas of social contradictions and different degrees of social struggle, they refer to social relations of production which are significantly different in practice. However, taking such differences into account, one is led to conclude not only that the degree of commoditization in peripheral agricultural systems is often higher than those of the center but also that the process of commoditization on the periphery subordinates farm labor to social relations of production that are far less favourable than those at the center. Hence, the consequences of the commoditization process in these typical peripheral situations are often extremely disruptive.

Commoditization and the Reproduction of the Agrarian Question

Raising the real level of commoditization creates a drastic change in the pattern of agrarian development. Progressive intensification (although sometimes blocked) gives way to increasing enlargement of scale and relative extensification. This is a general process, demonstrable in agricultural sectors of both periphery and center. It is equally a general process in that it is an aspect of the capitalist transformation of the social relations of production. Non-commodity circuits and the interests, insights and perspectives of the direct producers that they encompass are substituted by commodity relations as the guiding mechanism. In this way then, the "logic of the market" becomes indeed dominant. The labor process is subsumed to the same relations which form the rationale for capitalist-organized agriculture.

In peripheral agricultural systems this change is expressed as a deepening of agrarian underdevelopment. Although the link between a sharp increase in commoditization and the increasing dominance of scale enlargement and relative extensification is structural and therefore general, its effects differ widely.

In the first place, this is because the process of commoditization is more rapid and more far-reaching on the periphery than in the center. We also showed unmistakable differences in the social relations of production formed by particular market arenas. In addition to this, market relations, which through commoditization become dominant as social relations of production, represent an often gigantic gap between local realities and general market relations, especially in peripheral agriculture. All this means that in the Third World the effects of commoditization will be far more disruptive than in Western agriculture. In synthesis, it is not the often supposed "uncapturedness" of farmers (their being outside of commodity circuits) nor their supposed submission to the "laws of capital" which constitutes the core of the agrarian question. Central to the agrarian question, to themes of underdevelopment, rural exodus, etc., is the concrete historical process through which the interlinking of markets and farming is established and renegotiated. And as far as the Third World is concerned, one might sustain that it is above all the abrupt, massive and centrally propelled commoditization that gives rise to the "agrarian question" so omnipresent nowadays.

In the second place, we can point to technological development. Present-day agrarian technology offers to a certain, though strongly differentiated, degree the possibility of correcting at least some of the regressive effects of the commoditization process in Western agricultural systems. It is, after all, designed on behalf of these systems. In peripheral agriculture this is only possible as the exception. In Anta Pampa we were confronted with a concrete case of commoditization and simultaneous technological development in which the regressive effects of the one dominate the potentially moderating effects of the other. This makes for a second important characteristic of the "agrarian question," that is, the almost exclusive orientation of technological research and development towards interests other than, and often opposed to, those of Third World farmers. Central to this "bias" is the assumption of an increasing commoditization, an assumption made true through the very application of new technological models in "integrated rural development programs."

In the third place, there is the simple fact that peripheral economies, through their place in the international division of labor, became net importers of food, which is partly related to the fact that Europe and the United States shift their particular "agrarian question" (i.e., overproduction) to the Third World. Given a food dependency, a stagnation in agricultural growth as provoked by a rapid commoditization is far more negative than would be the case in areas with overproduction. However, there is a real possibility of self-sufficiency, with its obvious politico-economic importance. Rapid and continuous self-sustained agricultural growth which depends on raising the quantity and quality of farm labor (and with it rural employment and income-generating opportunities), appears as a first priority within this framework. However, the abrupt and wholesale commoditization of the labor process promoted by present integrated rural development programs achieves the opposite. They result in the reproduction of agrarian underdevelopment, in the slowing down of agrarian growth, in an acceleration of the rural exodus, and in the increasing subsumption of agriculture to (international) capitalist groups who control the various markets.

Hayami and Ruttan relate the direction, nature and tempo of agrarian development to relative factor prices and the degree to which primary producers and various institutions (banks, agricultural research stations, extension) manage to translate the inherent logic of relative factor prices into consistent patterns of action.

A double critique can now be formulated against such a position. To begin with, it is not so much the relative factor prices as such that determine the nature and direction of agrarian development. The commoditization process is equally decisive. It is only through commoditization that relative factor prices (and other market relations) do or do not become reality at the core of agricultural production, i.e., in the labor process. Consequently institutional reforms oriented towards a better correspondence between institutional action and the "logic of the markets" can easily become a disastrous strategy in Third World countries. A deepening of agrarian underdevelopment will be the inescapable result.

There are, however, other perspectives to be found. They may be found in forms of peasant-managed agricultural growth, which explicitly seek to protect the labor and development process from the dominance of prevailing commodity relations. Development of agriculture is often in opposition to prevailing market and price relations. This is accepted as self-evident by agrarian science and politics in the European context (de Wit, 1988) while at the same time the opposite is being advocated for Third World agriculture. A relative autonomy of agriculture vis-à-vis the prevailing market relations does not imply that farmers are to seek an illusory way out in autarkic practices. It means that the social struggle must embrace, in an explicit way, the domains of both production and circulation. The real possibility for such a perspective is highlighted in the vital practices, as we have seen, of farmers in Peru and Italy. But then, the recognition of such possibilities and perspectives implies that the "agrarian question" is to be redefined; it will have to include farmers and peasants as active actors, capable of making history, instead of defining them as passive victims or perennial losers.

Notes

1. I am fully aware of the fact that although the negative interrelations between scale and intensity was, until one and a half, maybe two decades ago, virtually a universal one (Jacob, 1971; Feder, 1973) this relation is now changing—

in a limited number of localities—into a positive one. Recently, in the northwest areas of the European Community, it has been possible to identify a segment of farm enterprises that combines scale enlargement and intensification in a positive and systematic way. This gives rise to what the French call *les grands intensifs* (Perraud, 1983). This phenomenon (described and analyzed for the Netherlands in van der Ploeg, 1987), is strongly related to scientification of the labor process in agriculture as well as to an increased division of labor between industry and agriculture. It is mostly limited to cattle breeding and dairy farming (Hairy, 1983; Fraslin and Simier, 1983). In agriculture as such it remains impossible (Crisenoy, 1983; Reboul, 1983). A similar pattern has been analyzed for US agriculture by Gregor (1982) as the "industrialization" of agriculture.

The same pattern emerged in areas where the Green Revolution proved successful. In those areas the initially negative relation between scale and intensity was replaced by a so-called U-curve.

In the agricultural areas studied in this book the negative relation between scale and intensity still forms an empirical reality: scale enlargement and relative extensification dominate as one of the important agricultural development patterns. Whether this phenomenon is to be understood as a temporary one or as a structural feature merits specific analysis. In view of the analysis represented in Chapter 3, which indirectly regards the introduction of "improved varieties" as an attempt at "scientification" of potato cultivation in the Andes, one might conclude that an attempt towards scientification of agricultural development in the Third World will often be in vain.

The same conclusion was drawn from a systematic comparison of technological developments in Dutch and Italian dairy farming (van der Ploeg, 1987). Through its interaction with politico-economic processes (and the specific hierarchization of space and the reorganization of time that it embraces), the "scientification" and hence the emergence of "grands intensifs" or "vanguard farms" will be limited to the most favoured areas ecologically, economically and institutionally. And these remain, on a world-scale, rather exceptional. The spatial distribution of "scientification" or industrialization as a development pattern in European agriculture is given in Meeus et al. 1988.

2. For an empirial study of negative interrelation between economic and technical efficiency, see Messori (1984). A theoretical exposition is to be found in Yotopoulos (1974).

3. Dutch agricultural banks quite often forbid the farmer's wife to give up her job outside the farm, so as to ensure at least some stable income. A more general illustration is the stipulation that land is not to be sold or given in tenancy without the bank's approval.

4. Bates (1981) and Mamdani (1986) give a clear analysis of this phenomenon.

5. At the local level commodity relations often imply relations between different categories (or classes) of farmers (Shanin, 1980).

6. This is especially the case in the so-called "vanguard farms" (van der Ploeg, 1987). This high dependency on external financing urges, in turn, a continuous growth, a *fuite en avant* (Hairy, 1983).

Bibliography

AID. "Small Farmer Credit." Summary Papers, *Spring Review of Small Farmer Credit*, Vol. XX, Washington, June 1973.

Akker, K.J. van den. *Van de mond der oude Middelzee, schetsen uit het oude leven op het land en uit het boerenbedrijf*, Friese Maatschappij van Landbouw (fifth edition), Leeuwarden, 1967.

Albrecht, H. *Innovationsprozesse in der Landwirtschaft, eine kritische Analyse der agrarsoziologischen "adoption" and "diffusion."* Forschung in Bezug auf Probleme der landwirtschaftlichen Beratung, Saarbrucken, 1969.

Albujar, E. Lopez. *Los Caballeros del Delito*, Juan Mejia Baca, Editorial, Lima, 1969.

Alvarez, E. *Politica Agraria y Estancamiento de la Agricultura, 1969–1977*, IEP, Lima, 1981.

Alvisi, F. "Rapporti e condizionamenti reciproci fra agricoltura e industria di trasformazione," in: *atti del XVII Convegno di Studi della Società Italiana di Economia Agraria*, Catania, 1980.

Anania, G. "Differenziazioni aziendali e modelli di classificazione: i risultati di un'applicazione di analisi fattoriale," in: *Rivista di Economía Agraria*, no. 3, 1981.

Angeli, L., and Omodei-Zorini, L. Un'applicazione dell'analisi fattoriale nell'attività di asistenza technica, in: de Benedictis, M., and Fanfani, R. (eds.), *Economia delle produzione agricola e metodi quantitativi*, Bologna, 1981.

Angladette, A. *Le riz*, Paris, 1966.

Antonello, S. "Imprenditorialità e modernizzazione in agricoltura," in: *Notiziario IPA.AT*, n.5–6, 1981.

Arensberg, C.M., and Kimball, S.T. *Family and community in Ireland*, Cambridge, 1948.

Arrús, P. *La programación linear en la cooperativa de producción Caqui*, Lima, 1974.

Ashan, A.A.M. Comments on Exploring the Gap between Potential and Actual Rice Yields: the Philippine Case, in: IRRI, *Economic Consequences of the New Rice Technology*, Los Banos, 1978.

Barigazzi, C. *L'Agricoltura Reggiana nel Settecento, le lezioni academiche di L. Codivilla, tra scienza sperimentale e "rivoluzione agronomica," 1771–1772*, Pesaro, 1980.

Barth, F. Economic Spheres in Darfur, in: R. Firth (ed.), *Themes in Economic Anthropology*, London, 1967.

Bates, R.H. *Markets and states in tropical Africa: the political basis of agricultural policies*, Berkeley, 1981.

Benavides, M.I. *Aspectos socio-economicos de la producción de papa en la unidad campesina (Valle del Mantaro)*, PUC, Lima, 1981.

Benedictis, M. de, and Cosentino, V. *Economia dell'Azienda agraria. Teoria e metodi*, Bologna, 1979.

Benedictis, M. de. "Teoria della produzione e analisi economico-agraria: riflessione critica e prospettive di sviluppo." Introduzione al Congreso di Villa Salina, Reggio Emilia, 1984.

Bennett, J. *Of time and the enterprise: North American family farm management in a context of resource marginality*, University of Minnesota Press, Minneapolis, 1981.

Benvenuti, B. "General Systems Theory and entrepreneurial autonomy in farming: towards a new feudalism or towards democratic planning?" *Sociologia Ruralis*, Vol. XV, 1/2, 47–62, 1975a.

Benvenuti, B. "Operatore agricolo e potere," in: *Rivista di Economia Agraria*, XXX, fasc. 3, 489–521, 1975b.

Benvenuti, B. "Imprenditorialità, partecipazione e cooperazione agricola, Considerazioni alla luce della situazione olandese," in: *Rivista di Economia Agraria*, XXXV, no. 1, 1980.

Benvenuti, B. "Dalla mano invisibile a quello visibile: un'analisi applicata ad alcune tendeze evolutive della agricoltura italiana," in: *La Questione Agraria* 7, 1982a: 73–116.

Benvenuti, B. "De technologies-administratieve taakomgeving (TATE) van landbouwbedrijven," in: *Marquetalia 5*, Wageningen, 1982b, 111–136.

Benvenuti, B., Bolhuis, E., and van der Ploeg, J.D. *I Problemi dell'Imprenditorialità agricola nella integrazione cooperativa*, AIPA, Bologna, 1982.

Benvenuti, B., Bussi, E., and Satta, M. *L'imprenditorialità agricola: alla ricerca di un fantasma; i risultati di una ricognizione sulle teorie in materia di imprenditorialità agricola*, Bologna, 1983.

Benvenuti, B., and van der Ploeg, J.D. "Development models of farm firms and their importance for mediterranean agriculture." Paper presented to the World Congress of Rural Sociology, Manilla, 1984.

Benvenuti, B. "On the dualism between sociology and rural sociology: some hints from the case of modernization," in: *Sociologia Ruralis*, Vol. XXV–3/4, 214–230, 1985a.

Benvenuti, B. Replica, Convegno di Bari, ottobre 1985, in: Atti del Convegno, INEA, Roma, 1985b.

Benvenuti, B., and van der Ploeg, J.D. "Modelli di sviluppo aziendale agrario e loro importanza per l'agricoltura mediterranea," in: *La Questione Agraria*, 17, 85–105, 1985.

Benvenuti, B., and Mommaas, H. *De technologies-administratieve taakomgeving van landbouwbedrijven; een onderzoeksprogramma op het terrein van de economische sociologie*, Wageningen, Landbouwhogeschool, 1985.

Benvenuti, B., Antonello, S., Sauda, E., Vidotto, G., and Gullotta, M. *Assistenza Tecnica e Stampa Agraria, due prime ricerche empiriche*, Reda, Roma, 1987.

Bernstein, H. "Notes on Capital and Peasantry," *Review of African Political Economy* No. 10, Sep.-Dec. 60–73, 1977.

Bernstein, H. "African Peasantries: a theoretical framework," in: *Journal of Peasant Studies*, 6, 421–443, 1979.

Bernstein, H. "Is there a concept of petty commodity production generic to capitalism?" Paper presented at the 13th European Congress for Rural Sociology, Braga, Portugal, 1986.

Berry, S.S. *Fathers work for their sons: accumulation, mobility and class formation in an extended Yoruba Community*, University of California Press, Berkeley, 1985.

Billone, J., Carbonetto, D., and Martinez, D. *Términos de intercambio ciudad-campo, 1970–1980: precios y excedente agrario*, Lima, 1982.

Bishop, C.E., and Toussaint, W.D. *Introduction to Agricultural Economic Analysis*, New York, 1958.

Blalock, H.M. "Theory Building and Statistical Concept of Interaction," in: *American Sociological Review*, 30, June 1965, 374–380.

Blalock, H.M. "Theory Building and Causal Inferences," Ch. 5 in Blalock and Blalock, 1968.

Blalock, H.M., and Blalock, A.B. *Methodology in Social Research*, New York, 1968.

Blalock, H.M. (ed.). *Causal Models in the Social Sciences*, Chicago, 1971.

Bloch, M. "Economie-nature ou économie-argent: un pseudo dilemme," in: *Annales d'Histoire Sociale*, I/1, 7–16, 1939.

Bolhuis, E.E. *Del analisis economico hacia la planificación de la empresa*, un estudio de la CAP Luchadores, El Ingenio, 1982.

Bolhuis, E.E., and van der Ploeg, J.D. *An empirical analysis of 24 capitalist farm firms*, CRPA, Reggio Emilia, 1982.

Bolhuis, E.E., and van der Ploeg, J.D. *Boerenarbeid en stijlen van landbouwbeoefening; een socio-economisch onderzoek naar de effecten van incorporatie en institutionalisering op agrarische ontwikkelingspatronen in Italië en Peru*, Leiden, 1985.

Boserup, E. *The conditions of agricultural growth, the economics of agrarian change under population pressure*, London, 1965.

Boudon, R. "A method of linear causal analysis: dependence analysis," in: *American Sociological Review*, 30, Juni 1965, 365–373.

Bourdieu, P. *Outline of a Theory of Practice*, Cambridge, 1982.

Box, L. "Food, Feed or Fuel? Agricultural Development Alternatives and the Case for Technological Innovation in Cassava (Manihot Esculenta Crantz) Cultivation." *Quarterly Journal of International Agriculture*, Special Issue, 34–48, 1982.

Box, L. "Cassava Cultivators and their Cultivars: preliminary results of case studies in the sierra region of the Dominican Republic," in: *Proceedings of the 6th symposium of the International Society for Tropical Root Crops*, Lima (International Potato Center), 1984.

Box, L., and Doorman, F. J. *The Adaptive Farmer: sociological contributions to adaptive research on cassava and rice cultivation in the Dominican Republic (1981–1984)*, Wageningen, Department of Rural Sociology, 1985.

Brade-Birks, S. Graham. *Modern Farming: a practical illustrated guide*, London, 1950.

Brand-Koolen, M.J.M. *Factoranalyse in het sociologisch onderzoek*, Leiden, 1972.

Braverman, H. *Labor and Monopoly Capital, the degradation of work in the 20th century*, New York, 1974.

Bray, F. *The rice economies: technology and development in Asian Societies*, Blackwell, Oxford, 1986.

Brokensha, D., Warren D.M., and Werner O. (eds.). *Indigenous knowledge systems and development*, University Press of America, Washington, 1985.

Brugnoli, A., Messori, P., Piccinini, A., Zucchi, G. *La combinazione dei fattori produkttivi nell'impresa coltivatrice*, Universitá di Bologna, Instituto di Zooeconomia, Reggio-Emilia, 1976.

Brugnoli, A. "Analisi della evoluzione strutturale ed organizzativa nel settore di produzione del formaggio parmigiano-reggiano;" *comunicazione al XVII Convegno di Studi della Società Italiana di economia agraria*, Catania, 1980.

Brusco, S. *Agricoltura Ricca e classi sociali*, Milano, 1979.

Brush, S.B., Heath, J.C., and Huamán, Z. "Dynamics of Andean Potato Agriculture," in: *Economic Botany*, 35 (1), 70–88, 1981.

Bussi, E., and Rizzi, P.L. *La struttura dell'industria del formaggio Parmigiano Reggiano, Un analisi statistica per il periodo 1956–1968*, Milano, 1974.

Caballero, J.M. *Agricultura Reforma Agraria y Pobreza Campesina*, IEP, Lima, 1980.

Caballero, J.M. *Economia Agraria de la Sierra Peruana, antes de la reforma agraria de 1969*, IEP, Lima, 1981.

Caballero, J.M., and Alvarez, E. *Aspectos Cuantitativos de la Reforma Agraria (1969–1979)*, IEP, Lima, 1980.

Cabral, A.L. "Recenseamento agrícola da Guiné—Estimativa em 1953," in: *Boletim Cultural da Guiné Portuguesa*, XI (43), 7–246, Lisboa, 1956.

Cantarelli, F., and Salghetti, A. *Evoluzione dell'azienda agraria nella pianura parmense nel decennio 1971–1980*, Parma, 1983.

Capelle, F. "L'intensification face à la reduction des couts de production," in: *Economie Rurale* (172), mars-avril, 1986.

Casaverde, J. Anta (Informe preliminar), Instituto de Estudios Peruanos, Proyecto "Reforma Agraria y Desarrollo en el Perú," Lima, ms, 1979.

Castro Pozo, H. *Del Ayllu al Cooperativismo Socialista*, Biblioteca Peruana, Ediciones PEISA, Lima, 1973.

Ccori, W. Disponibilidad de Recursos y Actividades Economicas Campesinas, Instituto UNSAAC-NUFFIC, *II Seminario de Investigacion*, Cusco, 1982.

C.E.C., Capacitación Empresarial Campesina. *Estudios Básicos DRI/SENA de 20 municipios y veredas*, Bogotá, Medellín, Cali, etc., 1976 and 1977.

CENCICAP-Anta. *Testimonio Campesino sobre las areas asociativas en las comunidades Campesinas de la micro-region de Anta*, Cusco, 1980.

Cépède, M. "The Family Farm: a primary unit of rural development," in: Weitz, 1971.

Chayanov, A.V. *The theory of peasant economy* (ed. by Thorner, D., et al.), Homewood, 1966.

Chevalier, J. *Civilization and the Stolen Gift: capital, kin and cult in Eastern Peru*, University of Toronto Press, Toronto, 1982.

Cole, J.W., and Wolf, E.R. *The Hidden Frontier: economy and ethnicity in an Alpine Valley*, New York, 1974.

Comunidad de Catacaos Plan de Trabajo, Catacaos, 1973.

Conklin, H.C. *Hanunóo Agriculture: a report on an integral system of shifting cultivation in the Philippines*, FAO, Rome, 1955.

Consorzio del Formaggio Parmigiano Reggiano. *Regolamento per la Produzione del Latte, norme per l'alimentazione*, Reggio Emilia, 1973.

Consorzio del Formaggio Parmigiano Reggiano. *Programma di Attivitd del Consorzio del Formaggio Parmigiano Reggiono per il Triennio 1980–1983*, Reggio Emilia, 1980.

Constandse, A.K. *Boer en Toekomstbeeld; enkele beschouwingen naar aanleiding van een terreinverkenning in de Noordoostpolder*, Bull. 24, Departments of Sociology, Wageningen, 1964.

Convenio para Estudios Economicos Basicos. *Aspectos Sociales y Financieros de un Programa de Reforma Agraria, para el periodo 1968–1975*, Lima, 1970a.

Convenio para Estudios Economicos Basicos. *La Reforma Agraria, un enfoque dirigido a medir su impacto en la economía provincial*, Lima, 1970b.

Cooperativa de Producción Luchadores del 2 de enero. Balance General y Anexos, Memorias de los Consejos, Ejercicio Economic 1980, Piura, 1981.

Corazza, G. La regolamentazione della qualità dei prodotti per l'industria alimentare; *comunicazione al XVII Convengno di Studi della Società Italiana di Economía Agraria*, Catania, 1980.

Cramer, G.L., and Jensen, C.W. *Agricultural Economics and Agribusiness* (2nd ed.), New York, 1982.

Crisenoy, C. de. *Pour une agriculture diversifiée*, Paris, 1983.

Crouch, B.R. "Innovation and Farm Development: a multidimensional model," in: *Sociologia Ruralis*, Vol. XII, no. 3/4, 1972, 431–449.

Dijkstra, H., and van Riemsdijk, J.F. *Uitkomsten van weidebedrijven, over 1947/48 tot en met 1950/51, ontwikkeling van de kosten en de opbrengsten over de laatste jaren*, LEI, Den Haag, 1952.

Dumont, R. *Types of Rural Economy: studies in world agriculture*, London, 1970.

Duncan, O.D. "Path Analysis: Sociological Examples," Ch. 7, in: Blalock, 1971.

Egoavil de Castillo, T. *La Cooperativa de Producción Túpac Amaru II Antapampa Ltda*, Instituto de Estudios Peruanos, Proyecto "Reforma Agraria y Desarrollo Rural en el Perú," Lima, 1978.

Eguren, F. Politica agraria y estructura agraria; in: *Estado y politica agraria*, DESCO, Lima, 1977.

Eguren, F. La tierra, su distribución y los regímenes de tenencia, Trabajo presentado al *III Seminario sobre Problemática Agraria "Aracelio Castillo,"* UN San Antonio Abad, Cusco, 1978.

Eguren. F. (ed.). *Situación actual y perspectivas del problema agraria en el Peru*, Lima, 1982.

Eguren Lopez, F. "Politicas Agrarias y Perspectivas." Paper presented to Congress: "Situación Actual y Perspectivas del Problema Agrario," DESCO, Lima, December 1981.

Einaudi, L. *Scritti*, Milano, 1975.

Eisenstadt, S.N. *The political systems of empires*, London, 1963.

Eizner, N. *Les paradoxes de l'agriculture française; essai d'analyse à partir des Etats Généraux de Développement Agricole*, avril 1982–février 1983, Harmattan, Paris, 1985.

Equiplan. *Diagnósto de la microregión de Anta*, Cusco, 1979.

Fabiani, G. *L'Agricoltura in Italia tra sviluppo e crisi (1945–1977)*, Bologna, 1979.

Fals Borda, O. *Campesinos de los Andes, Estudio sociologico de Saucio*, Bogota, 1961.

Fals Borda, O. "Formation and deformation of cooperative policy," in: *Latin America, Bulletin 7*, ILO, Geneva, 1970.

Fanfani, R. "L'applicazione della Programazioni Lineaire sulla Realtà Cooperativistia." Unpublished working document, Bologna, N.D.

Feder, E. *Gewalt und Ausbeutung*, Lateinamerikas Landwirtschaft, Hamburg, 1973.

Fernandez, A.R. *El maiz en San Pedro de Casta*, Serie Estudios no. 1, Direccion de analisis y Estudios, Ministerio de Alimentacion, Lima, 1977.

Figueroa, A.A. *El empleo Rural en el Peru*, Informe preparado para la Organización Internacional del Trabajo, Lima, 1975.

Figueroa, A. *La Economia Campesina de la Sierra del Peru*, PUC, Lima, 1982.

Fioravanti, E. *Latifundio y Sindicalismo Agrario en el Perú*, IEP, Lima, 1974.

Firth, R. "Capital, Saving and Credit in Peasant Societies: A Viewpoint from Economic Anthropology," in: Firth, R., and Yamey, B.S., eds., *Capital, Saving and Credit in Peasant Societies*, Chicago, 1964.

Fitzgerald, E.V.K. *La Economía Politica del Perú*, IEP, Lima, 1981.

Franco, E., et al. *Evaluación agro-economica de ensayos coducidos en campos de agricultores en el Valle del Mantaro*, Peru, campana 1978/79, CIP, Lima, 1980.

Franco, E., and Horton, D. *Producción y utilización de la papa en el valle del mantaro*, Peru, CIP, Lima, 1981.

Franco, E., Moreno, C., and Alarcon, J. *Producción y Utilizazión de la papa en el region del Cusco*, CIP, mayo 1981.

Fraslin, J.H., and Simier, J.P. *Diversité des systèmes d'exploitations laitières du pays virois*, Vol. I, Paris, 1983.

Friedmann, H. "Household production and the national economy: Concepts for the analysis of agrarian formations," in: *Journal of Peasant Studies*, 7, 158–184, 1980.

Frouws, J., and van der Ploeg, J.D. *Over de landbouwvoorlichting, materiaal voor een kritiek op de voorlichtingskunde en de agrariese sociologie*, Wageningen, 1973.

Galeski, B. *Basic Concepts of Rural Sociology*, Manchester, 1972.

Galizzi, G. Sistema agro-alimentare e linee di politica agraria; *atti del XVII Convegno di Studi, Società Italiana di Economia Agraria*, Catania, 1980.

Galjart, B.F. "The Future of Rural Sociology," in: *Sociologia Ruralis,* Vol. XIII, no. 3/4, 1973, 254–263.

Galjart, B. *Seeking the good deeds that lead to obscurity,* University of Leiden, The Netherlands, 1981.

Galletti, R., Baldwin, K.S., and Dina, I.O. *The Nigerian Cocoa Farmer,* Oxford, 1956.

Garoglio, P., and Mosso, A. "Ricerca di fattori di aggregazione in un gruppo di 23 aziende produttrice di latte tecnologicamente avanzate," in: *Rivista di Economia Agraria,* XLI, no. 2, 173–220, 1986.

Geertz, C. *Agricultural Involution: the process of ecological change in Indonesia,* Berkeley, 1963.

Gibbon, P., and Neocosmos, M. Some problems in the political economy of "African Socialism," in: Bernstein, H., and B.K. Campbell (eds.), *Contradictions of Accumulation in Africa: Studies in Economy and State,* Sage Publications, Beverly Hills, 1985.

Giddens, A. *Central Problems in Social Theory: action, structure and contradiction in social analysis,* London, 1979.

Giddens, A. "Class, social theory and modern sociology," in: *American Journal of Sociology,* 81, 4, 703–729.

Giddens, A. *A contemporary critique of historical materialism,* Vol. 1, Power, property and the state, London, 1981.

Goldberger, A.S. "Continuities, on Boudon's method of linear causal analysis," in: *American Sociological Review,* 97–101, 1970.

Gorgoni, M. "Una analisi della strutture dell'agricoltura italiana," *Rivista di Economia Agraria,* 1973, no. 6.

Gorgoni, M. "Sviluppo economico, progresso technologico e dualismo nell' agricoltura italiana," in: *Rivista di economia agraria,* 2/77, Rome, 1977.

Gouldner, A. *The coming crisis of Western Sociology,* London, 1970, 213–216.

Gregor, H.F. *Industrialization of US Agriculture: an interpretive atlas,* Westview Press, Boulder, 1982.

Grigg, D.B. *The agricultural systems of the world: an evolutionary approach,* Cambridge, 1974.

Grillenzoni, M., and Toderi, G. *La risicoltura italiana nella prospottiva comunitaria,* IRVAM, Rome, 1974.

Grillenzoni, M., and Cipriani, L. *Produttivitd ed economie di scala, azienderisicole e risicole-zootechniche del Vercellese e del Pavese (triennio 1970–72),* Bologna, 1975.

Gudeman, S. *The demise of a rural economy, from subsistence to capitalism in a Latin American village,* London, 1978.

Guillèn, J. *El desarrollo agricola en el departamente de Cuzco:* contribuci6n al seminario PRODERM, agosto 1982, Cuzco, 1982.

Hairy, D. *Endettement des exploitations et intensification de la production laitiere,* vol. 4 van de serie: La production laitiere dans l'ouest (enquete INRA-CCAOF), Paris, 1983.

Hardeman, J. *Selectieve innovatie door kleine boeren in Mexico,* Contribution to Sociale Geografie en Planologie, nr. 8, Vrije Universiteit Amsterdam, 1984.

Harman, H.H. *Modern Factor Analysis,* Chicago, 1976.

Haudry de Soucy, R. *El Crédito Agropecuario en el Perú, 1966–1976,* BS Thésis, PUC, Lima, 1978.

Haudry de Soucy, R. Situación del programa de credito Proderm al 31/XII78 y propuestas te acción, Cuzco, 1984.

Hayami, Y., and Ruttan, V. *Agricultural Development: An International Perspective,* Baltimore, 1971.

Hayami, Y., Ruttan, V., and Southworth, F. (eds.). *Agricultural Growth in Japan, Taiwan, Korea and the Philippines,* Honolulu, 1979.

Hazell, P.B.R., et al. *The Importance of Risk in Agricultural Planning Models,* World Bank Staff Working Paper, no. 3, 307, Nov. 1978.

Heady, E.O. *Economics of Agricultural Production and Resource Use,* London, 1952.

Heady, E.O., and Jensen, H.R. *Farm Management Economics,* Englewood Cliffs, 1954.

Heemst, H.D.J. van, et al. *Modelling of agricultural production: weather, soils and crops* (international post graduate training course), Wageningen, Genève, 1983.

Herrera, E. de. *Agricultura General, que trata de la labranza del campo y sus particularidades, crianza de animales y propriedades de las plantas.* Madrid, 1513. (Re-edited by Terròn, E., Servicio de Publicaciones del Ministerio de Agricultura, Madrid, 1984).

Hibon, A. *Transfert de Technologie et Agriculture Paysanne en Zone Andine: Le Cas de la Culture du Mais dans les systems de Production du Cusco (Perou),* Tome I et II (These de docteur-ingenieur), Toulouse, 1981.

Hinken, J. *Ziele und Zielbildung bei Unternehmern im Gartenbau,* Ein Beitrag zur betriebswirtschaftlichen Zielforschung, Hannover, 1974.

Hofstee, E.W. *Groningen van grasland naar bouwland, 1750–1938, een agrarisch-economische ontwikkeling als probleem van sociale verandering,* Wageningen, 1985.

Hofstee, E.W. "Grondlegger van de agrarische sociologie in Nederland," in: *Spil,* 1982–1983/29–30: 3–14.

Horton, D., Tardieu, F., Benavides, M., et al. *Tecnologia de la producción de papa en en valle del Mantaro,* Peru, CIP, Lima, 1980.

Hosier, A.J., and Hosier, F.H. *Hosier's Farming System,* London, 1951.

Huysman, A. *Choice and uncertainty in a semi-subsistence economy: a study of decision-making in a Philippine village,* PUDOC, Wageningen, 1986.

Hyden, G. *Beyond Ujamaa in Tanzania,* London, 1980.

INEA. *Risultati Economici di Aziende Agrarie,* Roma, 1962.

INEA. *Le Aziende Agrarie italiane, dati strutturali ed economici,* Roma, 1969.

INEA. *Risultati Economici, 1975, 1976, 1977,* Roma.

Ishikawa, S. *Essays on technology, employment and institutions in economic development: Comparative Asian experience,* Tokyo, 1981.

Jacoby, E. *Man and Land: the fundamental issue in development,* London, 1971.

Janvry, A. de. *The Agrarian Question and Reformism in Latin America,* Baltimore, 1981.

Kervin, B. *La dinamica de la pequena producción campesina;* contribución al seminario PRODERM, agosto 1982, Cuzco, 1982.

Koning, N. "Agrarische gezinsbedrijven en industrieel kapitalisme," in: *Tijdschrift voor Politieke Economie,* 6/1, 35–66, Sept. 1982.

Lacroix, A. *Transformations du procès de travail agricole, incidenced del'industrialisation sur les conditions de travail paysannes,* Grenoble, 1981.

Latour, B. "Give me a laboratory and I will raise the world," in: Knorr-Cetina, K.D., and Mulkay, M. (eds.), *Science observed: perspectives on the social study of science,* Sage Publications, London, 1983.

Leeuwis, C. *Marginalization misunderstood: different farm development strategies in the West of Ireland,* Wageningse Sociologische Studies, 26, Landbouwuniversiteit, Wageningen, 1988.

Lenin, V.I. The Agrarian Question and the "critics of Marx," in: *Collected Works,* V, Moskou, 1961.

Le Roy, X. *L'introduction des cultures de rapport dans l'agriculture vivriere Senoufo,* Abidjan, 1979.

Lipton, M. "The Theory of the Optimising Peasant," in: *The Journal of Development Studies,* 4 (3), 327–351, 1968.

Long, N. *An Introduction to the Sociology of Rural Development,* Tavistock Publications and West View Press, London, 1977.

Long, N., and Roberts, B. (eds.). *Peasant Cooperation and Capitalist Expansion in Central Peru,* University of Texas Press, Austin, 1978.

Long, N. Multiple enterprise in the central highlands of Peru, in: Greenfield, S.M., et al. (eds.), *Entrepreneurs in cultural context,* University of New Mexico Press, Albuquerque, 1979.

Long, N. *Creating space for change: a perspective on the sociology of development,* Wageningen, 1984.

Long, N., and Roberts, B. *Miners, peasants and entrepreneurs: regional development in the central highlands of Peru,* Cambridge University Press, Cambridge, 1984.

Long, N., van der Ploeg, J.D., Curtin, C., and Box, L. *The commoditization debate: labour process, strategy and social networks.* Papers of the Departments of Sociology 17, Agricultural University, Wageningen, 1986.

Long, N., and van der Ploeg, J.D. "New challenges in the sociology of rural development, a rejoinder to Peter Vandergeest," in: *Sociologia Ruralis,* Vol. XXVIII, 30–41, 1988.

Luna Vargas, A. "Ensenanzas de la actual etapa de la lucha de clases en el campo Piurano," in: *Critica Marxista Leninista,* no. 6, Lima, 1973. This article was published later in a slightly changed form, in *Journal of Peasant Studies,* Vol. 2, no. 1, 1974.

Madueno, A.M. *Avio Individual,* PRODERM, Cusco, 1980.

Mamdani, M. "The agrarian question and the struggle for democracy in Africa," in: *Bulletin of the Third World Forum,* no. 6, 1986.

Marasi, V., and Salghetti, A. "Alcuni aspetti economici e finanziari della gestione di cooperative di conduzione terreni," in: *Rivista di Economia Agraria,* XXXV, no. 1, 1980.

Marx, K. *Capital,* Vol. 1 and 3, Progress Publishers, Moscow, 1974.
Maso, B. *Rood en zwart, bedrijfsstrategieën en kennismodellen in de Nederlandse melkveehouderij.* Papers of the Departments of Sociology, 18, The Agricultural University, Wageningen, 1986.
Matos Mar, J., and Mejia, J.M. *La reforma Agraria en el Perú,* IEP, Lima, 1980.
Mayer, E., and Zamalloa, C. "Reciprocidad en las relaciones de producción," in: *Reciprocidad e intercambio en los Andes peruanos,* Peru Problema 12, IEP, Lima, 1974.
Mayer, E. *Uso de la tierra en los Andes, ecología y agricultura en el valle del mantaro del Peru con referencia especial a la papa,* CIP, Lima, 1981.
McKinnon, R.L. *Money and Capital in Economic Development,* Washington, 1973.
Medici, G. *Aziende tipiche nei comprensori di Reggio Emilia e Parma,* Bologna, 1934.
Meeus, J., van der Ploeg, J.D., and Weyermans, M. *Changing agricultural landscapes in Europe,* Rotterdam, 1988.
Mellor, J.W. in: *International Encyclopedia of the Social Sciences,* Vol. 1, London, 1968.
Mellor, J.W. Comment on the papers for the *AID Spring Review of Small Farm Credit,* Washington, 1973.
Mendras, H. *La fin des Paysans,* Paris, 1967.
Mendras, H. *The Vanishing Peasant: innovation and change in French agriculture,* Cambridge, 1970.
Messori, F. *Il costo di produzione del latte per Parmigiano-Reggiano nella impresa coltivatrice,* Università di Bologna, Instituto di Zooeconomia, 1981.
Messori, F. "La valutazione dell'efficienza attraverso indici sintetici," in: *Rivista di Economia Agraria,* no. 4, 707–726, Dec. 1984.
Minderhoud, G. *Inleiding tot de Landhuishoudkunde,* Haarlem, 1948.
Ministerio de Agricultura (MAA). Memoria Descriptiva de la Cooperativa agraria de producción "Buenos Aires," Piura, 1974.
Ministerio de Agricultura y Alimentación (MAA). "Estructura y Costo Real de Producción agricola por estratos de productores nucleados en el sistema de producción agropecuario (cultivos, campana '76-'77); subsistema 2: costos de producción," *Boletin Estadistica,* n. 6, OSEI/MAA, Lima, 1978.
Ministerio de Agricultura (MAA). *Boletín Estadístico del Sector Agrario,* Lima, 1981.
Moerman, M. *Agricultural Change and Peasant Choice in a Thai Village,* Berkeley, 1968.
Muggen, G. "Human Factors and Farm Management: a review of the literature," *World Agricicultural Economics and Rural Sociology,* Abstracts, Vol. 11, no. 2, June, 1969.
Newby, H. "The rural sociology of advanced capitalist societies," in: Newby, H. (ed.), *International Perspectives in Rural Sociology,* Chichester, 1978a.
Newby, H., Bell, C., Rose, D., and Saunders, P. *Property, Paternalism and Power: class and control in rural England,* London, 1978b.

Nielson, J. "Managerial Requirements of Farm Firms," in: *Implications of structural change in the economy of the commercial farm firm,* Chicago, 1965.

Noordwijk, M. van. Bodemvruchtbaarheid en duurzame landbouw in de tropen, deel 2, ecologisch onderzoek. Paper given in the symposium "Ecologie en ontwikkelingssamenwerking," Amsterdam, 1985.

Nove, A. *The Economics of Feasible Socialism,* London, 1983.

Oasa, E.K. *The International Rice Research Institute and the Green Revolution: a case study on the politics of agricultural research,* University of Hawaii, Honolulu, 1981.

Ochoa, R.G.J. "Empleo en las cooperativas costenas: entre las haciendas y la empresa autogestionaria," *Cuadernos de Investigación,* no. 3, CESIAL, Lima, 1980.

O.E.C.D. *Impact des entreprises multinationales sur les potentiels scientifiques et techniques nationaux. Industries alimentaires,* Paris, 1979.

Ortiz, S.R. de *Uncertainties in Peasant Farming: a Colombian case,* New York, 1973.

Pearse, A. "Metropolis and Peasant: the expansion of the urban industrial complex and the changing rural structure." Paper for the Latin American Seminar, Royal Institute of International Affairs, London, 30 Oct. 1968.

Pearse, A. *The Latin American Peasant,* Library of Peasant Studies, no. 1, London, 1976.

Pearse, A. "Technology and Peasant Production: Reflections on a Global Study," in: *Development and Change,* 8, 125–159, 1977.

Peltre-Würtz, J., and Steck, B. *Influence d'une societé de developpement sur le milieu paysan, coton et culture attelée dans la regiӧn de la Bagoué,* 1979.

Pernet, F. *Resistances Paysannes,* Grenoble, 1982.

Perraud, D. *Intensification et systèmes de production,* Grenoble, 1983.

Ploeg, J.D. van der. *De Gestolen Toekomst, landhervorming, boerenstrijd en imperialisme in Peru,* Wageningen, 1977.

Ploeg, J.D. van der. "Small Farmer Credit in Colombia," in: *Approach,* 1978, 6, 21–26 (Wageningen, International Agricultural Center).

Ploeg, J.D. van der. "Landbouwontwikkeling en verborgen klassenstrijd: Guiné-Bissau," in: *Marquetalia* 2, Wageningen, 1981.

Ploeg, J.D. van der. "Regelmaat en breuk: voor het bestaansrecht van het absurde," in: *Een Andere in een Ander,* Liber Amicorum voor R.A.J. van Lier, Leiden, 1982.

Ploeg, J.D. van der. "De multinational als ordenend principe in Mozambique," in: *Marquetalia,* 6, Wageningen, 1983.

Ploeg, J.D. van der, and Bolhuis, E. "Scelte techniche e incorporamento delle aziende zootechniche nelle strutture esterne: una indagine nella realtà emiliana," *quaderno di studio,* Parma, Sept. 1983.

Ploeg, J.D. van der. "Patterns of farming logic, structuration of labour and impact of externalization: changing dairy farming in northern Italy," in: *Sociologia Ruralis* XXV, no. 1, 5–25, 1985.

Ploeg, J.D. van der. *La Ristrutturazione del lavoro agricolo; gli effectti dell'incorporament e dell'istituzionalizzazione sullo sviluppo dell'azienda agri-*

cola; con una presentazione di Guiseppe Barbero e con una postilla di Bruno Benvenuti, *La Reda,* Roma, 1986.

Ploeg, J.D. van der. *De verwetenschappelijking van de landbouwbeoefening,* Papers of the departments of sociology, 21, Agricultural University, Wageningen, 1987.

Polanyi, K. *The Great Transformation,* New York, 1957.

Poulantzas, N. *Les classes sociales dans le capitalisme aujourdhui,* Seuil, Paris, 1974.

Poulantzas, N. *Klassen in het huidige kapitalisme, de internationalisatie van de kapitalistische verhoudingen en de nationale staat,* Nijmegen, 1976.

Prakken, R. "Erfelijkheid en veredeling," in: *Honderd jaar Mendel,* Symposium Biologische Raad/KNAW, PUDOC, Wageningen, 147–167, 1965.

Quijano, A. "Die Agrarreform in Peru," in: Feder, E., 1973.

Quijano, A. *Imperialismo y "Marginalidad" en America Latin,* Mosca Azul Editores, Lima, 1977.

Rabbinge, R. "Een eeuw landbouwkundige ontwikkeling in vogelvlucht; selectieve ontwikkeling: noodzaak! Maar ook mogelijk?" *Spil,* febr./maart 1979, 6, 148–151.

Rambaud, P. "Organisation du travail agraire et identités alternatives," in: *Cahiers Internationaux de Sociologie,* Vol. LXXV, 305–320, 1983.

Reboul, C. "L'Adaptation de l'agriculture de la RDA aux matériels à grand rendement. Facilités mécaniques, difficultés agronomiques," in: *Economie Rurale,* 156, 27–33, juillet-aout 1983.

Reynolds, L.G. "The spread of Economic Growth to the Third World: 1850–1980," in: *Journal of Economic Literature,* Vol. XXI, Sept. 1983.

Riemann, F. *Ackerbau und Viehaltung im Vorindustriellen Deutschland,* Beihefte zum Jahrbuch der Albertus Universität zu Königsberg, Kitzingen-Main, 1953.

Rocca, S., Bachrach, M., and Servat, J. *Participatory Processes and Action of the Rural Poor in Anta,* Peru, WEP Research, working paper 12, Lima, 1980.

Rocca, S., Bejar, H., and Diaz Veronica. The transition towards participation and self-management in Peru, 1970–1981, Institute of Social Studies, Den Haag, 1982a.

Rocca, S. "The Economic Performance of the Cooperatives" (Chapter 4 of an unpublished study on sugar cooperatives), Institute of Social Studies, Den Haag, 1982b.

Rogers, E.M. "The peasantry as a subculture," in: Wharton, 1969.

Ruttan, V. *Induced Technical and Institutional Change and the Future of Agriculture,* Sao Paolo, 1973.

Ruttan, V.W. "Induced innovation and agricultural development," in: *Food Policy,* August 1977, 196–216.

Sabogal Wiesse, J. *El maiz en Chacán,* UNA, Lima, 1966.

Sahlins, M. *Stone Age Economics: sociology of primitive exchange,* London, 1974.

Salaverry Llosa, J.A. "El crédito agrario en el Perú: situacion actual y perspectivas." Paper given to Congress: "Situacion Actual y Perspectivas del Problema Agrario," DESCO, Lima, 1981.

Samaniego, C. "Peasant movements at the turn of the century, and the rise of the independent farmer," in: Long and Roberts, 45–71, 1978.

Sauda, E., and Antonello, S. *Regional distribution of TATE-patterns and their correlation with types of agriculture,* Frascati 1983.

Schultz, Th.W. *Transforming traditional agriculture,* New Haven, 1964.

Scobie, G.M., and Posada, T.R. *The Impact of High Yielding Rice Varieties in Latin America, with special emphasis on Colombia,* CIAT, Cali, 1977.

SEDES. *Region de Korhogo, etude de developpement socio-economique;* rapport agricole, Abidjan, 1965.

Shanin, T. Measuring Peasant Capitalism: the operationalization of concepts of political economy: Russia's 1920's, India's 1970's, in: Hobsbawn, E.J., et al. (eds.), *Peasants in history, essays in honour of Daniel Thorner,* Oxford, 1980.

Sinamos, ORAMS I. Exposicoón de Motivos, Central de Cooperativas, Alto Piura, Chulucanas, 1975.

Skar, H.O. *The Warm Valley People: duality and land reform among the Quechua Indians of Highland Peru,* Gotenborg, 1981.

Slicher van Bath, B.H. *Boerenvrijheid,* Inaugural Lecture, Groningen, 1948.

Slicher van Bath, B.H. *De agrarische geschiedenis van West-Europa 500–1850,* Utrecht, 1960.

Spaan, H. *De Stalle Sociali, een nieuw soort cooperatie in de melkveehouderij in Italië,* Wageningen, 1982.

Spaggiari, P.L. (a cura di). *Insegnamenti di Agricoltura Parmigiana del XVIII secolo,* Parma, 1964.

Spahr van der Hoek, J.J., and Postma, O. *Geschiedenis van de Friese Landbouw,* Vol. 1, Leeuwarden, 1952.

Spijkers, P. *Rice peasants and rice research in Colombia,* Wageningen, 1983.

Thomas, H., and Logan, C. *Mondragon, An Economic Analysis,* Allen and Unwin, London, 1982.

Thurstone, L.L. *Multiple-factor analysis,* New York, 1950.

Timmer, C.P. "On measuring technical efficiency, in: Food Research Institute Studies in Agricultural Economics," *Trade and Development,* Vol. IX, no. 2, 99–171, 1970.

Tupayachi, E.B. *Economia Campesina y Mercados del Trabajo: Caso del Valle Sagrado de los Incas,* UN San Antonio Abad de Cusco, Cusco, 1982.

Turbati, E. *Risultati economici di Aziende Agrarie, 1965–1969,* INEA, Roma, 1971.

Ullrich, O. *Technik und Herschaft, Von Handwerk zur verdinglichten Blockstruktur industrieller Produktion,* Frankfurt, 1979.

Unilever Italiano. *Nota sul settore ortofrutticolo,* Milaan, 1978.

Vandergeest, P. "Commercialization and commoditization: a dialogue between perspectives," in: *Sociologia Ruralis,* Vol. XXVIII-1, 7–29, 1988.

Vela Suarez, C. "Piura, la más grave sequia en su historia," in: *Agro Andino,* 34–35, agosto 1982.

Villasanta, M. *El Campesino y su difereciacion en las Provincias Altas,* Instituto UNSAAC-NUFFIC, II Seminario de investigación, Cusco, 1982.

Vincent, J. "Agrarian Society as Organized Flow: processes of development, past and present," in: *Peasant Studies* VI, no. 2, 56–65, 1977.

Walters, A.A. "Production and Cost Functions: an econometric survey," in: *Econometrica,* Jan-April 1963.

Warman, A. *Y venimos a contradecir, Los campesinos de Morelos y el Estado Nacional,* Mexico, 1976.

Watson, A.M. *Agricultural Innovation in the Early Islamic World: the diffusion of crops and farming techniques,* 700–1100, Cambridge, 1983.

Webb, R. *The Distribution of Income in Peru,* London, 1972.

Wharton, C.R. (ed.). *Subsistence agriculture and economic development,* Chicago, 1969.

Williams, S., and Miller, J.A. *Credit Systems for Small Scale Farmers: Case Histories from Mexico,* Austin, Texas, 1973.

Wit, C.T. de, and Heemst, H.D.J. van. Aspects of agricultural resources, in: Koetsier, W. (ed.), *Chemical engineering in a changing world,* Amsterdam, 125–144, 1976.

Wit, C.T. de. "Oude wijn in nieuwe zakken," in: *Landbouwkundig Tijdschrift,* 93, 10, 257–262, 1981.

Wit, C.T. de. "Landbouw en milieu in de Europese Gemeenschap," in: *Spil,* 63–64/65–66, 17–25, 1988.

World Bank (IBRD). *Agricultural Credit,* Sector Policy Paper, Washington, D.C., 1975a.

World Bank. "Rural Development," in: *The Assault on World Poverty,* London and Baltimore, 1975b.

Yotopoulos, P.A."Rationality, Efficiency and Organizational Behavior Through the Production Function, Darkly," in: *Food Research Institute Studies,* Vol. XIII, 3, 263–274, 1974.

Yotopoulos, P.A."The population problem and the development solution," in: *Food Research Institute Studies,* XVI, 1, 1–119, 1977a.

Zachariasse, L.C. Publications 5, 6 en 7 of the Department of Agrarian Business Economics, Agricultural University, Wageningen, 1972a, 1972b, 1974a.

Zachariasse, L.C. *Boer en Bedrijfsresultaat,* Wageningen, 1974b (publ. 8).

Zachariasse, L.C., et al. *Boer en Bedrijfsresultaat na 8 jaar ontwikkeling,* LEI publ. 3.86, juli 1979, Den Haag.

Zanden, J. L. van. *De economische ontwikkeling van de Nederlandse landbouw in de negentiende eeuw, 1800–1914,* Wageningen, 1985.

About the Book and Author

Focusing on the complex and often contradictory relationships between agricultural production and markets, *Labor, Markets, and Agricultural Production* examines the micro-macro linkages between farm production, farm labor issues, and the degree of autonomy or dependency vis-à-vis markets. By comparing the case of farmers in Peru, generally regarded as peripheral agricultural producers, with that of European farmers able to easily access the centralized markets of the EEC, Dr. van der Ploeg is able to draw general conclusions about the ongoing process of commoditization of agriculture and the roles farmers play in agrarian development.

Jan Douwe van der Ploeg is professor in the Department of Rural Sociology and Development, Landbouwuniversiteit Wageningen, The Netherlands.

Index

Adaptation, double, 187
Agrarian Bank (Peru), 171, 172, 180, 214–15, 218
"Agrarian debt" (Peru), 280
"Agrarian question," 143–44
  commoditization and, 281–83
Agribusiness, 112, 267
Agricultural systems
  developmental level of, 2–3, 7–8, 10–11
  peripheral vs. central, 275–77, 280–81
  See also Cooperative(s); Dairy farming; Potato cultivation
Agriculture as "intrinsically backward," 32
Alto Piura cooperatives. *See Luchadores del Dos de Enero*
Alvaro Castillo cooperative, 222, 253–55
American model of intensification, 6
Anta Pampa, 267–68, 282. *See also* Potato production in Peruvian highlands
Art de la localité, ix, 185, 187
Artificial insemination of cows, 88
Autonomy
  functional, 272
  as prerequisite for intensification, 252–55
  strife and, 255, 255(fig.)
Ayni system, 160–62, 163, 164, 202(n1)
  capital mobilization through, 173–74
  wage labor (*jornaleros*) vs., 194–95
Ayuda, 166–67

Banana cultivation, 216
Banco Interamericano de Desarollo (BID), 256(n3)
Banks, Western European vs. Third World, 279–80. *See also* Credit and credit programs
Barbecho, 193
Belaunde, 221
Bell'azienda, la (the beautiful farm) concept, 59–60, 66, 91
Benefits
  defined, 145(n7)
  incorporation and redefinition of, 269
  See also Cost/benefit relations
BID. *See* Banco Interamericano de Desarollo
Bracciantile cooperative, 138–40
Breeding, 81, 82(fig.)
  externalization of, 20
  plant, 188–93, 192(fig.)
  scientification of, 188–93, 192(fig.)

Calculi, 39–40
  to account for heterogeneity, 72
  cattle density and, 133
  defined, 271
  employed by Chacán community, 166–69, 166(fig.), 171
  incorporation and, 263
  intensity of animal production and, 133
  See also E-calculus; Farming logic; I-calculus
Calving, 86, 106
Campania, Italy, 274–75, 275(table)
Capital
  E-calculus and market incorporation for, 117
  mobilization of, by Peruvian farmers, 163–74
  working, 78
Capital input per hectare, 127, 128(table), 129(table)
Capitalism
  contradictions of, 274
  in dairy farming in Emilia Romagna, 138–44, 142(table), 143(fig.)
  heterogeneity as uneven development of, viii
  social relations of production and, 281

See also Commoditization; Extensification; Incorporation; Market dependency; Scale enlargement
Capital markets, 22, 199–202, 200(fig.), 268
Carrasco cooperative, 222, 253–55
Caseifici (cheese) factories, 147(n19)
Cattle
density of, 72, 133
feeding of, 81, 82(fig.)
I- vs. E-farmers' views on improving, 84–89, 89(table)
See also Cow(s)
Ceja de la selva, 152
Centro International de Papa (CIP), 190
Chacán community, 154
ayni system in, 160–62, 163, 164, 173–74, 194–95, 202(n1)
calculus employed by, 166–69, 166(fig.), 171
capital needs of, 163–74
commoditization in, 168–69, 172–73
compania relations in, 157–58, 164, 173–74, 202(n1)
credit market in, 170–74, 177–83, 181(table)
falling potato yields, 176–77
fertilizer use in, 177–81, 178(fig.)
heterogeneity of, 168
incorporation in, 162–63
labor organization in, 159–63
land distribution in, 157–59, 157(fig.)
migration from, 159–60
mobilization of factors of production and nonfactor inputs, 174, 175(table)
oxen, livestock, and dung in, 163
poor farmers in, 155
reciprocal relationships in, 157–58, 160–61, 162(fig.)
secondary occupations in, 155, 164, 202(n1)
social strata in, 161–62, 202(n2)
See also Potato production in Peruvian highlands
Chacritas, 184
Chayanov, A. V., 29
Cheese, Parmesan, 37–38
Cheese factories, 109–11, 147(n19)
Chemicals, application of, 195
Chevalier, J., 263–67
CIP. *See* Centro International de Papa
"Circuits of value exchange," 202(n1)
Cities, intensification around, 15
Cognitive schema. *See* Calculi
Colombian farmers, 279–80
Combatants of the Second of January. *See* *Luchadores del Dos de Enero*
Commercialization, 154, 171
despachamamamización and, 156
farming time frame and, 269
Commitment, worker, 249–50. *See also* *Voluntad* (with a will) concept
Commoditization, viii, ix, 24–25, 94, 124, 259–84
acceleration of, 191
ayni system and, 162
in Chacán community, 168–69, 172–73
credit mechanisms and, 172–73
cross-cultural comparison of, 272–77, 275(table)
differential, 25
as differential process, 262–68
E-calculus and, 116–17
exchange requirement for, 263–64
extensification and, 154, 269–72
farming strategies and, ix
heterogeneity and, 263
impact of, 268–72, 282
incorporation level and, 22
neoclassical development economics and, 273–74
in peripheral vs. central agricultural systems, 275–77, 280–81
in Peru (general), 154, 273–77, 275(table)
among *los pobres*, 164–65
reproduction of agrarian question and, 281–83
risk perceptions and, 269–70
scale enlargement and, 269–72
scientification and, 190–91
of seed potato reproduction, 188–90
social relations of production and, 277–81
due to state and agribusiness intervention, 267, 280–81
structuring of farm labor process and, 81
TATE and, 119, 147(n23)
technological development and, 282
unilinear evolutionary view of, 272–75

See also Commercialization; Incorporation
Commodity, subsistence, 263, 266
Commodity production, 14–15
simple and petty, 273, 274
Commodity relations
as dominant social relation of production, 120
styles of farming and, 126
Community domain, 28, 30
Compania relations, 157–58, 164, 173–74, 202(n1)
Concentrates, 23, 265
Concepts, folk. *See* Calculi
Contract farming, 108
Control
externalization of, 111
over production process, 264
work and, 244–45
Cooperative(s), x, 278
autonomy in, 253–55, 253(fig.)
boundaries of, 251
Bracciantile, 138–40
in Chacán pampas, 176
dairies, 140
in Emilia Romagna, 138–44, 142(table), 143(fig.)
employment pressures in, 251
heterogeneity in, 8
incorporation of, 268, 280
machine shortages in, 279
manpower levels for, 212–13
Mondragon, 206
Morropon, 222, 253–55
production in, 139–40, 222, 228–29
profitable, 226
progressive intensification and, 249–51
in Ravenna, 250
social strife in, 253–55, 253(fig.)
sugar (Peru), 229
See also Luchadores del Dos de Enero
Cost(s)
defined, 145(n7)
heterogeneity/diversity due to, 5
incorporation and redefinition of, 269
labor as, 271
Cost/benefit relations, 28
E-calculus and, 69–71, 70(table), 80
income determined from, 69
Cost reduction, 101
I- vs. E-options and, 43–44
Luchadores plan for, 246–48
Cow(s)
artificial insemination of, 88
calving, 86
labor time per, 83–84, 84(fig.), 106–7
use value of, 264
Craftsmanship, 27, 39, 56–57, 78–95, 120, 127, 270–71
in capitalist-organized farms, 140–42
cattle improvement and, 84–89, 89(table)
collective, 230–32
defined, 78–80
E-calculus and, 90–91, 119–20
economic and social dimensions of, 95–96
entrepreneurship and, 80–81, 95, 126, 149(n26)
fertilizer application and, 198–99, 199(fig.)
frontier function and, 92–95, 93(fig.), 272
goal factors' influence on, 90
heterogeneity/diversity due to, 6
I- and E-options related to, 83, 83(table)
I-calculus and, 120
incorporation and, 199–201, 200(fig.)
indices of, 82–83, 83(table)
intensity of cultivation and, 197–99, 198(table)
labor time per cow and, 83–84, 84(fig.)
milk yield and, 83
nature of, 83
in potato production, 197–99, 198(table), 199(fig.), 202
production expectations and, 89–90
soil fertility and, 197–99, 198(table)
TATE and, 115–16, 116(table)
technical efficiency and, 272
See also Cura
Credit and credit programs
buying fertilizer and, 177–80
in Chacán community, 170–74, 170(table), 177–83, 181(table)
commoditization and, 172–73
crop rotation plans and, 181–83
extensification and, 180, 201, 238
future-dependent reproduction and, 170–71
internationalization behind, 173

scientification and, 191
See also Loans
Cropping pattern, 254–55
Cura, 56–57, 58, 60, 61, 64, 67, 68, 96, 120, 127. *See also* Craftsmanship
Cuzco, Department of, 152–53, 153(fig.)

Dairy farming
cooperative, 140
farm labor in, 18
heterogeneity in, 9–10, 9(table)
in the Netherlands, 274–75, 275(table), 278
relatively autonomous, historically guaranteed reproduction of, 13–14
Dairy farming in Emilia Romagna, Italy
BOLKAP data set, 38, 39, 140, 142, 150(n33)
capitalist and cooperative, 138–44, 142(table), 143(fig.)
cheese-making factories and, 109–11
COOP data set, 38, 39
development patterns in, 10–11, 11(table), 135–38
ERSA data set, 38–39, 45–47, 48, 49(table), 77, 83–84, 89–90, 94, 100, 101, 102, 103, 113, 117, 124, 142, 144(n1)
farming logic in, 39–40, 52–54
heterogeneity in, 39–40, 134–37, 135(fig.)
I- and E-options in, 40–54, 45(fig.)
incorporation in, 22–24, 23(table)
Parma data set, 10–11, 11(table), 38, 39, 45, 55, 78, 81, 93, 100, 101, 103, 142
plains vs. mountain, 124–26, 131–34
production process in, 81, 82(fig.)
styles of farming, 126–31
See also Craftsmanship; Entrepreneurship; Technical administrative task environment (TATE)
Debt, agrarian (Peru), 280
Decapitalization, 214–15, 222–23, 243–45, 253
Decision making
incorporation and, 271
TATE and, 111–12, 123–24
See also Calculi
Despachamamamización, 155–56
Development patterns, 2, 3, 7–8, 12. *See also* Cooperative(s); Dairy farming; Extensification; Intensification; Potato cultivation; Scale enlargement
Differential commoditization, 25
Diversity. *See* Heterogeneity; Styles of farming
Domains of farming, coordination of, 28–33
Domination, 111
Drought of 1982 (Alto Piura, Peru), 208, 218–19, 221

E-calculus, 67–73, 69(fig.), 74, 111, 113, 116–20, 124, 251, 271
in capitalist-organized farms, 140–43, 142(table)
capital market incorporation and, 117
cattle density and, 133
commoditization and, 116–17
cost/benefit ratio and, 69–71, 70(table), 80
craftsmanship and, 90–91, 119–20
expropriation and tailorization of labor and, 124
extensivity and, 131, 131(fig.)
feed input per cow and, 75
I-calculus and, 68, 123
incorporation and, 76–78, 116–19, 118(fig.), 120, 121(table), 131
institutionalization and, 116–19, 118(fig.), 120, 121(table), 131
intensity of animal production and, 133
long-term loans and, 77–78
market dependency and, 67
scale and, 68, 69, 71–73, 80, 131, 131(fig.)
spread of, 134–35
system interweaving required by, 123
See also E-farmers; Entrepreneurship; E-option
Ecological specialization, 189
Ecology, heterogeneity/diversity due to, 5
Economic efficiency, technical efficiency vs., 269
"Economic holding" concept, 212–13
Economic relations, 28, 31
I-calculus and, 66–67
Economics
"application of," vs. spending, 239–42
neoclassical agricultural, 3–7, 273–74
"Economizing on labor," 216
E-farmers, ix, 54

on cattle improvement, 85–88
farm performance and, 106, 106(table)
production functions of, 93–94, 93(fig.)
See also E-calculus; Entrepreneurship; E-option
Efficiency, technical. *See* Technical efficiency
E-logic. *See* E-calculus
Emilia Romagna, Italy, 9, 37, 271. *See also* Dairy farming in Emilia Romagna, Italy
Employment, 228–29, 250–51. *See also* *Luchadores del Dos de Enero;* Wage labor
Ente Regionale di Svilluppo Agricola (ERSA), 38–39
Entrepreneurship, 39, 95–105, 127, 202(n1)
attributes of, 98–99
conscious activity of, 96–98
craftsmanship and, 80–81, 95, 126, 149(n26)
current theories on, 96–100
defined, 80
E-option and, 105
external innovations and, 146(n17)
factor analysis of, 103–5, 104(table), 105(table)
heterogeneity and, viii, 260–62
income use and expectations, 103–4
incorporation and, 80
institutionalization and, 102
investment profile, 103
market emphasis of, 100, 101
meaning of, 97–98
price orientation of, 100–101
profit maximization and, 97–98
risk taking and, 102–3
as role, 100
as social activity, 100–103
technical and economic dimensions of, 95–96
technology and, 102
See also Technical administrative task environment (TATE)
Environment, institutional, 5–6. *See also* Technical administrative task environment (TATE)
E-option, 40–54
cost reduction and, 43–44, 48
craftsmanship and, 83, 83(table)
entrepreneurship and, 105
factor analysis of, 45–48, 49(table)
hectarage expansion and, 42–43, 43(table)
incorporation/institutionalization-goals relationship and, 48–52, 51(fig.)
in mountain dairy farming, 125–26
relative weighting of planning elements and, 42–44, 42(table)
variable cluster analysis of, 50
ERSA. *See* Ente Regionale di Svilluppo Agricola
European Community, heterogeneity in farming practice/organization in, 9
Exchange, 262, 263–64, 266. *See also* Commoditization; Incorporation
Exchangeability concept, 263–65, 266
Exchange calculations, 263–64, 267
Exchange values, and use values, 268
Extensification, 2, 3, 12, 268
commoditization and, 154, 269–72
credit and, 180, 201, 238
E-calculus and, 131, 131(fig.)
fertilizer efficacy and, 181
incorporation and, 130
negative connotations of, 3–4
of potato production, 180
produzione concept and, 53–54
progressive intensification and, 281
scale enlargement with, 10, 73
thin and fat cows phenomenon and, 73–76
See also E-calculus; E-option
Externalization, 18–21, 19(fig.)
of breeding, 20
defined, 19
new relations arising from, 20–21
of resource allocation and control, 111
of seed potato reproduction, 188
TATE and, 107–8
See also Incorporation

Factor prices, relative, 7, 283
Fallow period, 176–77
Family domain, 28–31
Family farms, 39, 141, 142–44, 142(table), 143(fig.), 268
"Farming freedom" concept, 265–66
Farming logic, 31
in dairy farms in Emilia Romagna, 39–40, 52–54

heterogeneity and, viii
See also Calculi; E-calculus; I-calculus; Strategies, farming
Farm labor. *See* Labor
Farm Management Economics (Heady & Jensen), 71–72
Fat and thin cows phenomenon, 73–76
Fat and thin years, cycle of, 16–17
Feed
input per cow, 73–76
self-produced, 264–65
Fertility, soil
craftsmanship and, 197–99, 198(table)
fertilizing and, 177–78
production of, 174–83, 197–99, 198(table)
reproduction of, 19–20, 185
rotation system and, 176–77
Fertilizers, 177–78
Chacán community's use of, 177–81, 178(fig.)
chemical, 191
craftsmanship and application of, 198–99, 199(fig.)
efficacy of, and extensification, 181
Fodder
market for, 23
production of, 81, 82(fig.)
Folk concepts. *See* Calculi
Folk taxonomies, 184–85
Food industry, 112–13
Franco cooperative, 222, 253–55
Friedmann, H., 272, 273
Friesland, 15–16
Frontier function, 92–95, 93(fig.), 272
"Functional autonomy," 272
Future-dependent reproduction, credit use and, 170–71

Genetic materials, market for, 23, 279
Golpear, 193

Haciendas, 174, 207, 208–10
Hay tedding, 26–27, 28
Hectarage expansion, 42–43, 43(table), 72
Herbicides, 238
Heterogeneity, viii
agricultural system's level of development and, 2–3, 7–8, 10–11
of Chacán community, 168
commoditization and, 263
in cooperatives, 8
in dairy farming, 9–10, 9(table), 37, 39–40, 134–37, 135(fig.)
entrepreneurship and, viii, 260–62
in European Community, 9
farm logic and, viii
incorporation and, 263
intensification-extensification framework and, 3–4
multiple calculi to account for, 72
neoclassical agricultural economics and, 5–6
in potato production, 8–9, 184, 197–202
production per labor force and, 6–7
relations of production and, 259–60
as residual phenomenon, 5–7
"stages" of agrarian development and, 260, 261(fig.)
as uneven development of capitalism, viii
Homogeneity, defined, 7
Huancabamba river, 221

I-calculus, 54–67, 55(fig.), 56(fig.), 111, 124, 232, 271
la bell'azienda (beautiful farm) concept and, 59–60, 66
cattle density and, 133
craftsmanship and, 120
cura and, 56–57, 60, 61, 64, 120
E-calculus and, 68, 123
economic relations and, 66–67
feed input per cow and, 75
historical continuity and, 134
impegno and, 58–59, 62–64
incorporation and, 78, 120–24, 121(table), 122(fig.)
institutionalization and, 120–24, 121(table), 122(fig.)
intensity of animal production and, 133
knowledge and, 57
long-term loans and, 77–78
passione and, 57
potential benefits in, 65–66
price fluctuation and, 64
production function, 62–65, 64(fig.)
produzione and, 55–56, 61, 62, 120
rejection of market dependency in, 65
relatively autonomous, historically guaranteed reproduction and, 67

"self-evidence" of, 60
self-sufficiency (*autosuficienza*) and, 59–60, 61, 66
statistical definition of, 120
structure of, 59(fig.)
substantive rationality of, 60–61
time horizon of, 91
See also Craftsmanship; I-farmers; Intensification; I-option
"Ideal plant type," 190–91
I-farmers, ix
on incorporation, 77
long-term perspective of, 66
on market dependency, 61–62
production functions of, 93–94, 93(fig.)
I-logic. *See* I-calculus
Impegno, 58–59, 62–64, 68, 77, 127
Income
determination of, 69
entrepreneurs' use and expectations, 103–4
produzione and, 55–56
scale and, 55, 56, 69
Incorporation, viii, 123, 124, 263
into capital markets, 199–202, 200(fig.), 268
in Chacán community, 162–63
of cooperatives, 268, 280
craftsmanship and, 199–201, 200(fig.)
cross-cultural comparison of, 272–77, 275(table)
decision making and, 271
defined, 50
distribution of produced wealth and, 278
E-calculus and, 76–78, 116–19, 118(fig.), 120, 121(table), 131
entrepreneurship and, 80
external prescription of investment decisions and, 119
of family farms, 268
I-calculus and, 78, 120–24, 121(table), 122(fig.)
I-farmers on, 77
I- vs. E-options and, 48–52, 51(fig.)
into labor market, 199–201, 200(fig.)
labor quality and, 270–71
markets as structuring principles and, 262–63
nature of, 25
production functions and, 92
production optimum and, 61–62, 270
produzione and, 78, 79(table)
redefinition of costs and benefits due to, 269
scale and extensivity and, 130
social struggle and, 280
due to state and agribusiness intervention, 267
styles of farming and, 130–31, 132(fig.), 268
TATE and, 119, 147(n23)
time horizon for production and, 90–92
See also Externalization; Institutionalization
Incorporation, degree of, 21–26, 21(fig.), 40
comparative analysis of, 278–80
in dairy farms (Emilia Romagna), 22–24, 23(table)
defined, 21–22
feed input per cow and, 74–75, 75(fig.)
heterogeneity and, 263
lowering, 230
market dependency and, 24, 76
relatively autonomous, historically guaranteed reproduction and, 76
technical efficiency and, 94–95
See also Commoditization
Industrial labor, 27
INEA. *See* Istituto Nazionale di Economia Agraria
Inheritance, 157
Innovations, external, 133–34
entrepreneurs' dependence on, 146(n17)
heterogeneity/diversity due to, 5
Insemination, artificial, of cows, 88
Institutionalization, viii, 94, 102
defined, 50
E-calculus and, 116–19, 118(fig.), 120, 121(table), 131
I-calculus and, 120–24, 121(table), 122(fig.)
of investment decisions, 114
I- vs. E-options and, 48–52, 51(fig.)
structuring of farm labor process and, 81
styles of farming and, 130–31, 132(fig.)
TATE and, 109–11
Institutional relations, 28, 31
Intensification, 1–2, 3, 12, 120, 123, 124, 127, 128(table), 129(table)

American model of, 6
around cities, 15
based on technological development, 134
cura and, 56–57
employment and, 250–51
Japanese model of, 6
in *Luchadores del Dos de Enero*, 205, 212–13, 215–16, 224, 226–27, 228–46
manpower levels and, 212–13
mountain dairy farmers and, 124–26
neoclassical agricultural economics and, 4
peasant struggles and, 205, 212–13, 232–33
production process under, 81
produzione concept and, 52–53
secondary occupations to further, 156–57
social struggle and autonomy as prerequisites for, 252–55
See also I-calculus; I-option; Progressive intensification
Intensification scientifique, 134
Intensitätsinsel, 15
Intensive agriculture, 1–2
animal production, 133
craftsmanship and, 197–99, 198(table)
indicators of, 254–55
positive connotations of, 3–4
potato cultivation, 197–99, 198(table), 201, 202
rice cultivation, 236–38, 237(fig.)
scale and, 283(n1)
size and, 137–38
Investment decisions
entrepreneurship and, 103
external prescription of, 119
institutionalization of, 114
TATE and, 114–15, 115(table)
I-option, 40–54
collectively carried, 252
cost reduction and, 43–44, 48
craftsmanship and, 83, 83(table)
factor analysis of, 45–48, 49(table)
hectarage expansion and, 42–43, 43(table)
incorporation/institutionalization-goals relation and, 48–52, 51(fig.)
relative weighting of planning elements and, 42–44, 42(table)
variable cluster analysis of, 50
Iowa-Mission of the USA, 212
Iron, 15
Istituto Nazionale di Economia Agraria (INEA), 39, 136
Italian agriculture, commoditization patterns of (general), 273–77, 275(table). *See also* Dairy farming in Emilia Romagna, Italy
Italy, markets in, 278–80

Japanese model of intensification, 6
Jehovah's Witnesses, 194
Jornaleros, 194–95
Junta Interventora, 207, 216–17, 219, 225

Labor
coordination of, 18–19
as cost, 271
in dairy farms, 18
defined, 12, 30–31
E-calculus and, 124
economizing on, 216
externalization and, 20–21
industrial vs. farm, 27
new "optima" within process of, 32–33
organization of, in Chacán community, 159–63
quality of, and incorporation, 270–71
reproduction and, 13, 24, 32, 35(n17)
scientification of, 283(n1)
social relations of production and, 278
structuring of, 81
wage labor, 194–95
See also Craftsmanship; Entrepreneurship
Labor input, 127, 128(table), 129(table)
Labor market, 280
incorporation into, 199–201, 200(fig.)
Lampa, 195
Land, rented, 91
Land market, 280
Land reforms, Peruvian, 212–14
Law of diminishing returns, 248–49
Law of increasing returns, 248
Libro genealogico, 114
Liebig-function, 248–49
Linear programming, 250
Loans, 91
long-term, 77–78

"modernization" conditions attached to, 125
See also Capital markets; Credit and credit programs
Luchadores del Dos de Enero (Combatants of the Second of January), 253–55
bank dealings of, 214–15, 218, 221, 222, 244–45, 256(n3)
conflicting norms in, 239–46
cultivated areas in hectares over time, 219–22, 219(fig.)
decapitalization of, 214–15, 222–23, 243–45, 253
drought of 1982 and, 208, 218–19, 221
employment development of, 208–19, 209(fig.)
engineer-administrator and, 227–28
establishment as cooperative (1973), 207, 210–14
government interventions in, 280–81
government land reforms and economic policies and, 212–14
hacienda period (up to 1968) of, 207, 208–10
intensification in, 205, 212–13, 215–16, 224, 226–27, 228–46
Junta Interventora and, 207, 216–17, 219, 225
meaning of "social strife" in, 230
mechanization issue, 245–46
membership problems in, 217–18, 256(n4)
parceleros period (1968–1973), 207, 210
power relations between farmers and state apparatus, 224–25
production in, 222–25, 223(fig.), 225(fig.)
profit and loss analysis, 225–28, 226(fig.)
reestablishment as cooperative (1977–), 207, 217–18
rice cultivation in, 234–38
Rospigliosi family and, 206, 208, 209, 210
self-administration of, 207, 214–16
strike of 1978, 217
temporary workers and small farmers (*chacreros*) of, 218–19
the "twenty-one weeks" and, 214–15
unionism in, 209–10, 211, 216, 218, 246–51, 252, 255
uniqueness of, 206–7
water problems of, 219–21, 224
working day in, 243
See also Progressive intensification

Machine services, market for, 279
Maizales, 176
Mantaro Valley, 189
Manure, labor reduction and spreading of, 92
Marginalization, 277
Market
as outlet in relatively autonomous, historically guaranteed reproduction, 14–15, 18
as structuring principle, 262–63
universal, 173
Market dependency
E-calculus and, 67
feed input per cow and, 73–74
I-farmers on, 61–62
incorporation and, 24
rejection of, under I-calculus, 65
risk factor and, 65
Market-dependent reproduction, 17–26, 17(fig.), 276
credit use and, 170–71
externalization and, 20
incorporation and, 76
production factors in, 17
Marketing, 121–23
Market relations
agrarian development and, 283
calving and, 106
local vs. general, 281–82
See also Incorporation; Institutionalization
Marriage, 157
Mechanization, 81, 82(fig.), 139, 245–46
Medios, los, 161, 162(fig.), 170–74, 183
Mercantilización, 171
Mezzadria system, 149(n30)
Milk yield, craftsmanship and, 83
Minderhoud, G., 3–4
Modernization theory, 272–74
Mondragon cooperative, 206
Morropon cooperative, 222, 253–55
Muggen, G., 98–99

Neoclassical agricultural economics, 3–7, 273–74
Netherlands
dairy farming in, 274–75, 275(table), 278
markets in, 278–80
Northeast Polder, 259, 260
New Husbandry tradition, 188
Non-factor inputs
market for, 278–79
mobilization of, 174, 175(table)

Objects of labor, reproduction of, 32
Occupations, secondary, 155, 164, 202(n1)
Optimum/optima, production
in incorporation framework, 61–62, 270
labor process and, 32–33
neoclassical agricultural economics and, 4–5

Pacha mama, 166, 167
Pampa lands, 174–76
Papa de regalo, 184, 190
Papa mejorada, 190–91
Parceleros, 207, 210
Parmesan cheese, 37–38
Part-time farming, 72
Passione, 57
Patasucias, 279–80
Peasant-managed agricultural development. *See* Cooperative(s)
Peasant struggles, 152, 205, 212–13, 232–33. *See also Luchadores del Dos de Enero;* Struggle, social
Peasant unions, 205, 209–10, 211, 216, 218, 246–51, 252, 255
Peripheral economies, 272–77
disarticulated structure of, 276–77
as net importers of food, 282–83
Peru
commoditization patterns in (general), 154, 273–77, 275(table)
farmers and enterprises in, 154–59
gross value of production (GVP) for products from, 152–54, 153(fig.)
income levels in, 151–52
land reform in (DL 17716), 212–14
markets in, 279
potential for intensification in, 152
productivity in, 151–52
sugar cooperatives, 229
See also Chacán community; *Luchadores del Dos de Enero*
Petty commodity production, 273, 274
Phenotypical conditions (potato), 186–88
for "modern"/"improved" seed potatoes, 190–93
Plant breeding, 188–93, 192(fig.)
Pobres, los, 161, 162(fig.), 164–65, 182, 183
Potato cultivation
production-reproduction unity in, 13
seed potatoes, 183–93, 186(fig.), 192(fig.), 197, 279
Potato production in Peruvian highlands
craftsmanship in, 197–99, 198(table), 199(fig.), 202
credit financing of, 173
extensification of, 180
fallow period for, 176–77
harvest estimates and per hectare yields, 196, 197–99, 198(table), 199(fig.)
heterogeneity in, 8–9, 184, 197–202
intensity of cultivation and, 197–99, 198(table), 201, 202
process of, 193–96
production of soil fertility and, 197–99
secondary occupations and, 155, 156(fig.)
seed potato commoditization and, 188–90
seed potato reproduction, 183–93, 186(fig.), 197
See also Chacán community
Potato varieties, "modern," 188, 190–91
Power relations, 111
land as commodity and, 280
between Peruvian farmers and state, 224–25
Prescriptions, external, 111, 119, 123, 272
Price(s)
agrarian development and, 283
calving and, 106
entrepreneurs' orientation toward, 100–101
factor, 283
heterogeneity/diversity due to, 5
I-calculus and, 64
strife over, 101
Proderm Program, 171, 174, 179, 190, 201
Production, 121–23

animal, 133
control over, 264
in cooperatives, 139–40, 222, 228–29
craftsmanship and expectations for, 89–90
in dairy farms, 81, 82(fig.)
domain of, 28, 29, 30–31
factors of, 17, 174, 175(table), 269
family domain and, 29–30
of fodder, 81, 82(fig.)
heterogeneity and, 6–7
in *Luchadores del Dos de Enero*, 222–25, 223(fig.), 225(fig.)
of soil fertility, 174–83, 197–99, 198(table)
time horizons for, 90–92
unity of reproduction and, 13, 14
See also Optimum/optima, production; Social relations of production
Production functions, 270
of I- and E-farmers, 93–94, 93(fig.)
validity of, 145(n6)
Produzione, 52–54, 61, 62, 67, 68, 70, 95, 96, 120, 127
in capitalist-organized farms, 142
cura and, 56–57
income and, 55–56
incorporation and, 78, 79(table)
Profit maximization, 97–98, 260
Profit maximizer enterprise, 229–30, 230(fig.)
"Progress," farmers' striving for, 31–32
Progressive intensification, 228–46
collective craftsmanship and, 230–32
cooperative structure and, 249–51
extensification and, 281
hectare yields and, 233–38
law of diminishing returns and, 248–49
Luchadores plan for, 246–51
production goals and, 229–30, 230(fig.)
scale enlargement and, 281
"spending" vs. "application of economics" argument and, 239–42
voluntad vs. skiving argument and, 242–46
water management and, 234–35
worker commitment and, 249–50
work-struggle relationship and, 232–33
Protestant groups, 194

Quality of labor, 270–71. *See also* Craftsmanship

Rationality
forms of, 12
substantive, 60–61
Ravenna cooperative, 250
Reciprocal relationships, 157–58, 160–61, 162(fig.)
Regolamento per la Produzione del Latte (Consorzio), 109, 110(fig.)
Regressive substitution, 238
Relatively autonomous, historically guaranteed reproduction, 13–17, 14(fig.), 180, 276
la bell'azienda concept and, 91
I-calculus and, 67
incorporation and, 76
market as outlet in, 14–15, 18
los ricos and, 166, 169–70
Rented land, 91
Reproduction, 12–26
community domain and, 30
development patterns and, 12
domain of, 28, 29, 30–31
externalization of, 18–21
of labor, 13, 24, 32, 35(n17)
of seed potatoes, 183–93, 186(fig.), 197
of soil fertility, 19–20, 185
unity of production and, 13, 14
See also Market-dependent reproduction; Relatively autonomous, historically guaranteed reproduction
Resource allocation, externalization of, 111
Rice cultivation, 234–38, 237(fig.)
Ricos, los, 161, 162(fig.), 165–70, 181–82, 183, 202(n1)
relatively autonomous, historically guaranteed reproduction scheme of, 166, 169–70
Risk
commoditization and incorporation and, 269–70
entrepreneurship and, 102–3
heterogeneity/diversity due to, 5
market dependency and, 65
Rospigliosi family, 206, 208, 209, 210
Rotation system
credit and, 181–83
soil fertility and, 176–77
Roughage, 265
Rural development, integrated, 201

Savings, measurable, 266–67

Savings quota, 255(n2)
Scale, 127, 128(table), 129(table)
  E-calculus and, 68, 69, 71–73, 80, 131, 131(fig.)
  income and, 55, 56, 69
  incorporation and, 130
  intensity and, 283(n1)
  size vs., 137–38
  social relations of production and, 134–38
Scale enlargement, 268
  cattle density and, 133
  commoditization and, 269–72
  E-calculus and, 131, 131(fig.)
  forms of, 72–73
  neoclassical model of, 6–7
  progressive intensification and, 281
  with relative extensification, 10, 73
  thin and fat cows phenomenon and, 73–76
  See also E-calculus; E-option
Schema, cognitive. *See* Calculi
Scientification
  commoditization and, 190–91
  credit and, 191
  of labor process, 283(n1)
  of seed potato breeding, 188–93, 192(fig.)
Seed potatoes
  "improved varieties" of, 279
  reproduction of, 183–93, 186(fig.), 192(fig.), 197
Self-management, 251. *See also* *Luchadores del Dos de Enero*
Self-provisioning, 24
Self-sufficiency, 282–83
  feed input per cow and, 73–74
  I-calculus and, 59–60, 61, 66
Sembrio, 193
Semilleristas, 189–90
Simple commodity production, 273, 274
Size of farm, 136–38
Skiving, 242–46
Small farmers, 218–19, 279
Social relations of production
  capitalism and, 281
  commodity relations as, 120, 277–81
  defined, 277–78
  heterogeneity and, 259–60
  labor and, 278
  scale and, 134–38
  TATE as, 120
  See also Incorporation; Institutionalization
Social scientists, "backwardness" of, 257(n7)
Soil fertility. *See* Fertility, soil
Soil use. *See* Styles of farming
Specialization, 72–73
  ecological, 189
"Spending" vs. "application of economics" argument, 239–42
Standardization, 113
Strategies, farming, ix. *See also* Farming logic
Struggle, social
  autonomy and, 255, 255(fig.)
  defined, 252
  incorporation and, 280
  as prerequisite for intensification, 252–55
  work and, 232–33
  See also Luchadores del Dos de Enero; Peasant struggles
Styles of farming
  defined, 11–12
  factor analysis of, 127–30, 128(table), 129(table)
  farm labor distinguished from, 12
  hectare yields for different, 2–3, 2(table)
  incorporation and, 130–31, 132(fig.), 268
  institutionalization and, 130–31, 132(fig.)
  as social constructions, 126–31
  use values and, 265
  See also E-calculus; Extensification; I-calculus; Intensification
"Subsistence commodities," 263, 266
Substitution, regressive, 238
Sugar cooperatives, 229
Surplus, 14, 96

Task(s)
  coordination of, 32
  external prescription of, 111, 119, 123, 272
  TATE and unity of, 107–8
TATE. *See* Technical administrative task environment
Taxonomies, folk, 184–85
Technical administrative task environment (TATE), 30, 106–16, 133

craftsmanship and, 115–16, 116(table)
as dominant social relation of production, 120
external forces–internal responses dynamic and, 111–12
externalization and, 107–8
farmer's decision-making and maneuvering spaces and, 111–12, 123–24
incorporation/commoditization and, 119, 147(n23)
influence of, 50–52
information flow from, 112, 113–14
institutionalization and, 109–11
investment decisions and, 114–15, 115(table)
of mountain dairy farmers, 124
styles of farming and, 126
unity of tasks and, 107–8
Technical efficiency, 1, 33(n1, n2), 93, 145(n8), 230–31
craftsmanship and, 272
economic efficiency vs., 269
Technical management, social coordination of, 28
Technological change, induced, 7
Technology
commoditization and, 282
entrepreneurship and, 95–96, 102
intensification based on, 134
Temporales, 176
Temporary workers, 218–19
Thin and fat cows phenomenon, 73–76
Thin and fat years, cycle of, 16–17
Time perspective, 65–66
commercialization and, 269
of I-farmers, 66, 91
incorporation and, 90–92

"Uncapturedness" thesis, 273, 282
Unilever, 112
Unions. *See* Peasant unions
Upgrading, 186, 186(fig.)
Use values, 66, 263, 264–65
exchange values and, 268
styles of farming and, 265

Valle de Convencion y Lares, 152
Value as differential process, 262–68
Value exchange, circuits of, 202(n1)
Variability. *See* Heterogeneity; Styles of farming
Voluntad (with a will) concept, 242–46, 249, 250

Wage labor, 194–95
Water management, 234–35
Wealth, distribution of, 277–78
Weeding, 195
Work
control and, 244–45
struggle and, 232–33
voluntad vs. skiving, 242–46
See also Labor
Working capital, 78

Yields, variation in, 10–11, 33(n5)

Zachariasse, L. C., 96–97
Z-score, 146(n12)

For Product Safety Concerns and Information please contact our EU representative GPSR@taylorandfrancis.com Taylor & Francis Verlag GmbH, Kaufingerstraße 24, 80331 München, Germany

Batch number: 10725887

Printed by Printforce, the Netherlands